Springer Complexity

Springer Complexity is an interdisciplinary program publishing the best research and academic-level teaching on both fundamental and applied aspects of complex systems – cutting across all traditional disciplines of the natural and life sciences, engineering, economics, medicine, neuroscience, social and computer science.

Complex Systems are systems that comprise many interacting parts with the ability to generate a new quality of macroscopic collective behavior the manifestations of which are the spontaneous formation of distinctive temporal, spatial or functional structures. Models of such systems can be successfully mapped onto quite diverse "real-life" situations like the climate, the coherent emission of light from lasers, chemical reaction-diffusion systems, biological cellular networks, the dynamics of stock markets and of the internet, earthquake statistics and prediction, freeway traffic, the human brain, or the formation of opinions in social systems, to name just some of the popular applications.

Although their scope and methodologies overlap somewhat, one can distinguish the following main concepts and tools: self-organization, nonlinear dynamics, synergetics, turbulence, dynamical systems, catastrophes, instabilities, stochastic processes, chaos, graphs and networks, cellular automata, adaptive systems, genetic algorithms and computational intelligence.

The three major book publication platforms of the Springer Complexity program are the monograph series "Understanding Complex Systems" focusing on the various applications of complexity, the "Springer Series in Synergetics", which is devoted to the quantitative theoretical and methodological foundations, and the "SpringerBriefs in Complexity" which are concise and topical working reports, case-studies, surveys, essays and lecture notes of relevance to the field. In addition to the books in these two core series, the program also incorporates individual titles ranging from textbooks to major reference works.

Understanding Complex Systems

Founding Editor: S. Kelso

Future scientific and technological developments in many fields will necessarily depend upon coming to grips with complex systems. Such systems are complex in both their composition – typically many different kinds of components interacting simultaneously and nonlinearly with each other and their environments on multiple levels – and in the rich diversity of behavior of which they are capable.

The Springer Series in Understanding Complex Systems series (UCS) promotes new strategies and paradigms for understanding and realizing applications of complex systems research in a wide variety of fields and endeavors. UCS is explicitly transdisciplinary. It has three main goals: First, to elaborate the concepts, methods and tools of complex systems at all levels of description and in all scientific fields, especially newly emerging areas within the life, social, behavioral, economic, neuro- and cognitive sciences (and derivatives thereof); second, to encourage novel applications of these ideas in various fields of engineering and computation such as robotics, nano-technology and informatics; third, to provide a single forum within which commonalities and differences in the workings of complex systems may be discerned, hence leading to deeper insight and understanding.

UCS will publish monographs, lecture notes and selected edited contributions aimed at communicating new findings to a large multidisciplinary audience.

For further volumes:
http://www.springer.com/series/5394

SB: To my parents, Saraswati and Raghunath Banerjee
LR: To my parents, Annita and Elio Rondoni

Preface

This is the third volume of the collection of essays entitled *Applications of Chaos and Nonlinear Dynamics in Engineering and Science*, which has been designed to introduce nonspecialists to the applications of the modern theory of dynamical systems and complexity. Indeed, we believe that the exchange of ideas which takes place when different disciplines make use of a common language, such as that provided by the modern mathematical techniques of dynamical systems and complexity, is beneficial to both the experts of the different disciplines and the experts of the language itself. This requires that such a language be presented in a digested form, which takes it immediately closer to the expected applications. We have thus decided to pursue this purpose by means of essays written in the form of tutorials. Volume 1 consists of ten chapters organized in five parts, each concerning one field of present-day engineering:

I. Nonlinearity and Computer Simulations
II. Chaos and Nonlinear Dynamics in Electrical Engineering
III. Chaos and Nonlinear Dynamics in Building Mechanism and Fluid Dynamics
IV. Chaos in Robotics
V. Chaos and Nonlinear Dynamics in Communication

Volume 2 also contains ten chapters, subdivided into the following five sections:

I. Nonlinearity in Control Systems and Geo Engineering
II. Nonlinear Dynamics and Chaos in Electronics
III. Nonlinear Dynamics in Stochastic Networks
IV. Nonlinear Dynamics in Transport and Mechanical Engineering
V. Chaos Theory in Communication and Cryptography

The ten chapters of this volume cover a wider spectrum of disciplines, organized in five sections, as follows:

I. Fluctuation Relations and Chaotic Dynamics in Physics
II. Monsoon Chaos and Wind Turbine System

III. Fractal and Its Applications in Epileptic Seizure
IV. Chaos Synchronization A: Communications and Symbolic Analysis
V. Chaos Synchronization B: Systems and Circuits

The first contribution to this volume, by Colangeli and Rondoni, addresses a problem of current interest, in the development of response theories suitable in biophysical and nanotechnological contexts. The second contribution, by Bakka and Karimi, uses a bond graph methodology as a graphical approach for modeling wind turbine generating systems. The third paper, by P. Carl, concerns monsoon dynamics and some hypothesis on the potential evolution of the climate. The fourth paper, by Uthayakumar, tries to understand electroencephalograms of epilectic siezures from the point of view of a fractal analysis of nonlinear time series. The fifth paper, by Jalan, Atay, and Jost, gives one method to detect global synchrony in coupled systems. The sixth contribution, by A.A. Dmitriev, A.S. Dmitriev, Andreyev, Efremova, Antoniades, Miliou and Anagnostopoulos, is devoted to chaotic synchronous response for the purpose of cryptography. The seventh paper, by Hegazi and Matouk, illustrates synchronization of a chaotic system and the behavior of a modified Van der Pol-Duffing circuit. The eighth contribution, by Middya, Basak, Ray, and Roychowdhury, considers experimental verifications of synchronization within a network and among different networks of Rössler Oscillators.

Chapter 9, by Stavrinides and Anagnostopoulos, is devoted to experimental tests of the robustness and stability of synchronization and shows how chaotic-synchronized communication systems may desynchronize. The tenth contribution, by Theesar, Banarjee, and Balasubramaniam, completes this book with a study of synchronization of delayed chaotic systems, providing analytical conditions for the occurrence of projective synchronization of time-varying delayed chaotic systems.

As for the articles published in volumes 1 and 2, we hope that this third collection of papers will be useful to professionals as well as to undergraduate and graduate students of applied sciences.

We will conclude the series with volume 4, which will be published in 2013.

We wish to express our gratitude to the staffs of Springer Verlag, for their invaluable help and support throughout this work. In particular, we would like to thank Dr. Christian Caron (editor, Springer Physics), Gabriele Hakuba (editorial assistant), Benjamin Feuchter, and the other members associated with this project.

Finally we would like to thank Prof. M. Mitra, who is not actively associated in this present volume but her inspiration has always been there throughout the project.

Torino
December 2012

S. Banerjee
L. Rondoni

Contents

Part V Chaos Synchronization: Systems and Circuits

Part I
Fluctuation Relations and Chaotic Dynamics

Chapter 1
Fluctuation Relations and Nonequilibrium Response for Chaotic Dissipative Dynamics

Matteo Colangeli and Lamberto Rondoni

Abstract In a recent paper [Colangeli and Rondoni, Physica D 241:681, 2011] it was argued that the Fluctuation Relation for the phase space contraction rate Λ could suitably be extended to non-reversible dissipative systems. We review here those arguments, by discussing the properties of a simple irreversible nonequilibrium baker model. We also consider the problem of the extension of the Fluctuation-Dissipation Theorem to dissipative deterministic dynamical systems, which enjoy a nonvanishing average phase space contraction rate. As noted by Ruelle, the statistical features of the perturbation and, in particular, of the relaxation, cannot be understood solely in terms of the unperturbed dynamics on the attractor. Nevertheless, we show that the singular character of the steady state does not constitute a serious limitation in the case of systems with many degrees of freedom. The reason is that one typically deals with projected dynamics, and these are associated with regular probability distributions in the corresponding lower dimensional spaces.

1.1 Introduction

One of the central aims of nonequilibrium statistical physics is to find a unifying principle in the description of nonequilibrium phenomena [1]. Nonequilibrium fluctuations are expected to play a major role in this endeavour, since they are ubiquitous, are observable in small as well as in large systems, and a theory about them is gradually unfolding, cf. [2–7] for recent reviews. A number of works have been devoted to the derivation and test of Fluctuation Relations (FRs), of different nature [8–16]. It is commonly believed that, although nonequilibrium phenomena concern a broad spectrum of seemingly unrelated problems, such as hydrodynamics and

M. Colangeli (✉) · L. Rondoni
Dipartimento di Matematica, Politecnico di Torino, Corso Duca degli Abruzzi 24, 10129 Torino, Italy
e-mail: matteo.colangeli@polito.it; lamberto.rondoni@polito.it

S. Banerjee and L. Rondoni (eds.), *Applications of Chaos and Nonlinear Dynamics in Science and Engineering 3*, Understanding Complex Systems,
DOI 10.1007/978-3-642-34017-8_1,

turbulence, biology, atmospheric physics, granular matter, nanotechnology, gravitational waves detection, etc. [6, 17–19], the theory underlying FRs rests on deeper grounds, common to the different fields of application. This view is supported by the finding that deterministic dynamics and stochastic processes of appropriate form obey apparently analogous FRs [6,7,13,14], and by the fact that tests of these FRs on systems which do not satisfy all the requirements of the corresponding proofs typically confirm their validity. Various works have been devoted to identify the minimal mathematical ingredients as well as the physical mechanisms underlying the validity of FRs [7, 15, 20, 21]. This way, the different nature of some of these, apparently identical but different, FRs has been clarified to a good extent [4,7,14,15]. However, analytically tractable examples are needed to clearly delimit the range of validity of FRs, and to further clarify their meaning. In this work, the assumptions of time reversal invariance, required by certain derivations of FRs for deterministic dynamical systems, are investigated by means of simple models that are amenable to detailed mathematical analysis. In particular, we will introduce, in Sect. 1.2, a deterministic map in which one may distinguish *relevant* from *irrelevant* variables, as long as the observable of interest is the phase space contraction rate, denoted by Λ. Next, we will add a source of irreversibility in the dynamics, which concerns only the irrelevant variable but also affects the structure of the steady state probability distribution.

We will discuss, hence, in Sects. 1.3 and 1.4, the implications on the condition of detailed balance in the projected space and on the validity of the FR for the quantity Λ, hereafter called Λ-FR.

A further crucial aspect of the emerging nonequilibrium theories concerns the application of the Fluctuation-Dissipation Theorem (FDT) in beyond-equilibrium systems. The celebrated FDT was originally developed in the context of Hamiltonian dynamical systems, slightly perturbed from equilibrium, and it was later extended to stochastic systems obeying a Langevin Equation [6,22]. The importance of the FDT rests on the fact that it sheds light on the crucial relation between the response $R_V(t)$ of a system to an external perturbation and a time correlation function computed at *equilibrium* [23]. In other words, having perturbed a given Hamiltonian H_0 with an external field h_e, to obtain the perturbed Hamiltonian $H_0 - h_e V$, where V is an observable conjugated with h_e, the FDT allows us to compute nonequilibrium quantities, such as the transport coefficients [24–26], solely in terms of the unperturbed equilibrium state. On the other hand, a generic dynamical system is not Hamiltonian: for phenomenological practical purposes, one typically deals with dissipative dynamics, as in the important case of viscous hydrodynamics [6]. The classical approach, which considers infinitesimal perturbations out of the stationary state of interest, still applies to these cases, if the steady state is represented by a regular probability density, as appropriate in the presence of noise, cf. [6,27]. Nevertheless, the invariant measure of a chaotic dissipative system, μ say, is typically singular with respect to the Lebesgue measure and is usually supported on a fractal attractor. This is not just a mathematical curiosity, it is a potential source of difficulties for the applicability of the FDT in dissipative systems. Indeed, the standard FDT ensures that the statistical features of a perturbation are related to the statistical properties of the unperturbed system, but that cannot be the case in general, in dissipative systems. The reason is that, given an initial state $\underline{x}(0)$ on the attractor and a generic perturbation $\delta\underline{x}(0)$, the perturbed initial state $\underline{x}^{(p)}(0) = \underline{x}(0)+\delta\underline{x}(0)$ and its time evolution may lie outside the support of μ, hence their statistical properties cannot be expressed by μ, which attributes vanishing probability to such states. From the mathematical point of view, this fact is rather transparent. On the other hand, it should not be a concern in statistical mechanics, except in pathological cases. Indeed, a series of numerical investigations of chaotic dissipative systems shows that the standard FDT holds under rather general conditions, mainly if the invariant measure is absolutely continuous with respect to Lebesgue, cf. [6] for a review. Moreover, although dissipative systems have singular invariant measures, any small amount of noise produces smooth invariant measures, which allow generalized FDTs to be expressed solely in terms of the unperturbed states, analogously to the standard equilibrium case. Apart from technical aspects, the intuitive reason for which the FDT in systems with noise can be expressed only in terms of the invariant measure, is that $\underline{x}^p(0)$ remains within the support of this measure.

Thus, in Sects. 1.5 and 1.6, we will shed light on the extension of the FDT to dissipative dynamical systems. We will discuss the case of chaotic deterministic maps and show that FDT continues to hold as long as one works in some suitable projected space of the full phase space. Indeed, marginals of singular phase space measures, on spaces of sufficiently lower dimension than the phase space, are usually regular [28, 29].

1.2 Global Conservativity vs. Local Dissipativity

Let us introduce the dynamical system $(\mathscr{U}, M, \mu)$, with phase space $\mathscr{U} := \mathbb{T}^2 := \mathbb{R}^2/\mathbb{Z}^2$ and mapping $M : \mathscr{U} \to \mathscr{U}$ defined by:

$$\begin{pmatrix} x_{n+1} \\ y_{n+1} \end{pmatrix} = M \cdot \begin{pmatrix} x_n \\ y_n \end{pmatrix} = \begin{cases} \begin{pmatrix} \frac{1}{2\ell}x_n + \frac{1}{2} \\ \left(\frac{1}{2}-q\right) y_n + \left(\frac{1}{2}+q\right) \end{pmatrix} & \text{for } 0 \le x_n < \ell \\ \begin{pmatrix} \frac{x_n}{1-2\ell} - \frac{\ell}{1-2\ell} \\ (1-2\ell-q)y_n + (2\ell+q) \end{pmatrix} & \text{for } \ell \le x_n < \frac{1}{2} \\ \begin{pmatrix} 2x_n - \frac{1}{2} \\ \left(\frac{1}{2}+q\right) y_n \end{pmatrix} & \text{for } \frac{1}{2} \le x_n < \frac{3}{4} \\ \begin{pmatrix} 2x_n - \frac{3}{2} \\ (2\ell+q)y_n \end{pmatrix} & \text{for } \frac{3}{4} \le x_n \le 1 \end{cases} \tag{1.1}$$

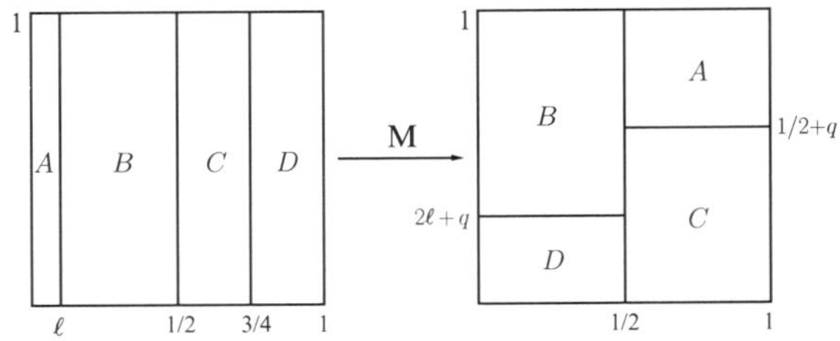

Fig. 1.1 The map M in (1.1) for general values of the parameters ℓ and q

for $q \in [0, \frac{1}{2}]$ and $\ell \in (0, \frac{1}{4}]$, cf. Fig. 1.1, and with natural measure μ. The Jacobian determinant of this map takes the values:

$$J_M(\underline{x}) = \begin{cases} J_A = \dfrac{1}{4\ell} - \dfrac{q}{2\ell} & \text{for } 0 \le x < \ell \\[2ex] J_B = 1 - \dfrac{q}{1-2\ell} & \text{for } \ell \le x < \dfrac{1}{2} \\[2ex] J_C = 1 + 2q & \text{for } \dfrac{1}{2} \le x < \dfrac{3}{4} \\[2ex] J_D = 4\ell + 2q & \text{for } \dfrac{3}{4} \le x \le 1 \end{cases} \tag{1.2}$$

in the four different regions of $\mathcal{U}$. This model generalizes the one introduced in [30], as it features the two parameters ℓ and q, which can be tuned to produce different forms of "equilibrium", i.e. of natural measures μ, which are called non-dissipative steady states because are characterized by vanishing phase space contraction rates. In our case, it proves convenient to determine the projection of the invariant probability density on the x-coordinate of this map. This can be accomplished by integrating over the y-direction the Perron-Frobenius equation [31] for the measures defined on the square. This, indeed, yields the evolution equation for the probability measures defined on the x axis which are evolved by the map of the interval $[0, 1]$ obtained by projecting M on the x axis.

The calculation of the marginal invariant probability measure can the be performed by introducing a Markov partition of the unit square, consisting of two regions: $[0, 1/2)$ and $[1/2, 1]$, respectively furnished with the invariant densities $\rho_l(x)$ and $\rho_r(x)$. Then, the transfer operator T associated with the projected dynamics, via the projected Perron-Frobenius equation, can be written as:

$$\begin{pmatrix} \rho_l(x_{n+1}) \\ \rho_r(x_{n+1}) \end{pmatrix} = T \cdot \begin{pmatrix} \rho_l(x_n) \\ \rho_r(x_n) \end{pmatrix} \tag{1.3}$$

where T is defined by:

$$T = \begin{pmatrix} 1-2\ell & 1/2 \\ 2\ell & 1/2 \end{pmatrix} \tag{1.4}$$

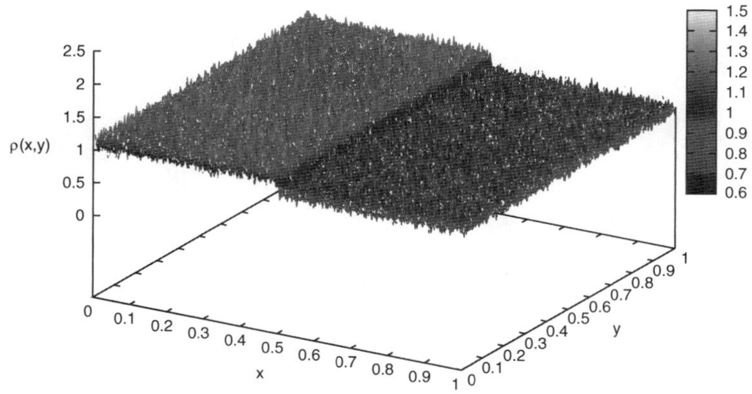

Fig. 1.2 Result of a numerical simulation for the invariant density of the equilibrium reversible map derived from (1.1) with $q = 0$ and $\ell = 0.15$, obtained by evolving a set of $2 \cdot 10^7$ initial conditions randomly (and uniformly) chosen on the square $[0,1] \times [0,1]$. This shows an invariant density $\rho(x,y)$ which is uniform along the y-coordinate and piecewise constant along the x-coordinate, attaining the values $\rho_l(x)$ and $\rho_r(x)$ for, respectively $x \in [0,1/2)$ and $x \in [1/2,1]$, given in (1.5)

The matrix T satisfies the Perron-Frobenius Theorem, hence its largest eigenvalue, $\lambda = 1$, is separated from a spectral gap from its other eigenvalue. Then, the calculation proceeds by evaluating the eigenvectors of the transfer operator corresponding to the dominant eigenvalue. The result of this procedure shows that the invariant probability density of the map (1.1), projected onto the x-axis, depends on the value of ℓ, but not on q, because q only affects the dynamics along the vertical direction. The corresponding projected density, cf. Fig. 1.2, is given by the piecewise constant function:

$$\rho(x) = \begin{cases} \rho_l(x) = \dfrac{2}{1+4\ell} & \text{for } 0 \le x < \dfrac{1}{2} \\ \\ \rho_r(x) = \dfrac{8\ell}{1+4\ell} & \text{for } \dfrac{1}{2} \le x \le 1 \end{cases}, \tag{1.5}$$

The marginal probability density ρ suffices to compute the statistical properties of phase functions such as the phase space contraction rate $\Lambda(x,y) = -\log J(x,y)$, because of the special form of the Jacobian determinants (1.2), which depend on the x-coordinate only. In this case, one has:

$$\langle \Lambda \rangle = -\int_{\mathcal{U}} \log J(x,y) \mu(dx \times dy) = -\int_0^1 \log J(x) \rho(x) dx$$

The quantity $\langle \Lambda \rangle$ is represented in Fig. 1.3 as a function of the parameters (ℓ, q). The figure shows that "equilibrium", i.e. by definition the condition in which $\langle \Lambda \rangle$ vanishes, holds only for $q = 0$, independently of the value of ℓ. The dynamical

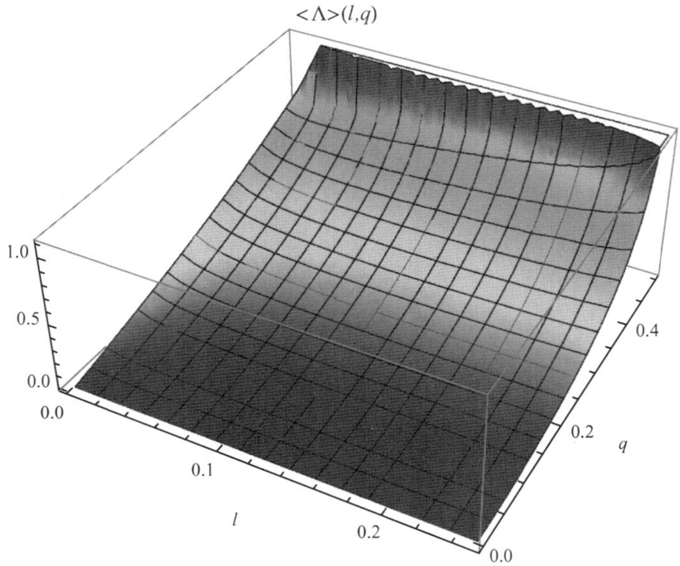

Fig. 1.3 The average phase space contraction rate $\langle \Lambda \rangle$, as a function of the parameters ℓ and q

system of [30] can be seen as a special case of our map, in which $q = \frac{1}{2} - 2\ell$ and, correspondingly, the only possible equilibrium state of that map is given by the further choice $\ell = \frac{1}{4}$.

In this section, we focus on the case $q = 0$ and begin by considering the average phase space contraction rate which, in this case, vanishes $\forall \ell \in (0, \frac{1}{4}]$ and can be written as:

$$\langle \Lambda \rangle = -\ln J_A \rho_l \ell - \ln J_D \frac{\rho_r}{4} = 0 \tag{1.6}$$

The choice $\ell = \frac{1}{4}$, in particular, ensures that the mapping is locally conservative, i.e. that $\Lambda(x, y) = 0$ uniformly on $\mathcal{U}$, as all Jacobians (1.2) are unitary. In particular, $\ell = \frac{1}{4}$ leads to a uniform invariant measure μ, which we call "microcanonical" by analogy with the statistical mechanics of an isolated particle system with given total energy. The projection of μ is then uniform along the x-axis.

For $\ell \neq \frac{1}{4}$ and $q = 0$, the invariant density along the x-axis remains smooth, except at one point of discontinuity, $x = 1/2$ and, in spite of the fluctuations of the phase space volumes, the invariant measure is still uniform along the stable manifold, i.e. the vertical direction, as illustrated by a numerical simulation reported in Fig. 1.2. Thus, for $\ell \neq \frac{1}{4}$, we obtain a form of *fluctuating equilibrium*, which we call "canonical" by analogy with the statistical mechanics of a particle system in equilibrium with a thermostat at a given temperature [17].

Let us also observe that our equilibrium dynamics ($q = 0$) are time reversal invariant, according to the standard dynamical systems notion of reversibility [32], because there exists an involution $G : \mathcal{U} \to \mathcal{U}$ such that

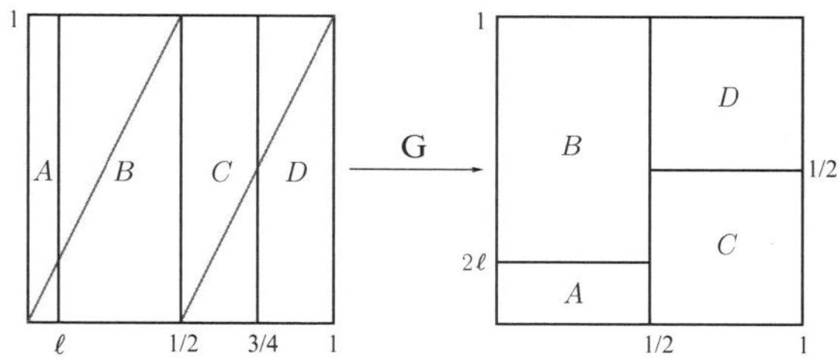

Fig. 1.4 Involution G defined in (1.8). *Blue lines*: the two diagonals, along which the map G reflects the two halves of the phase space

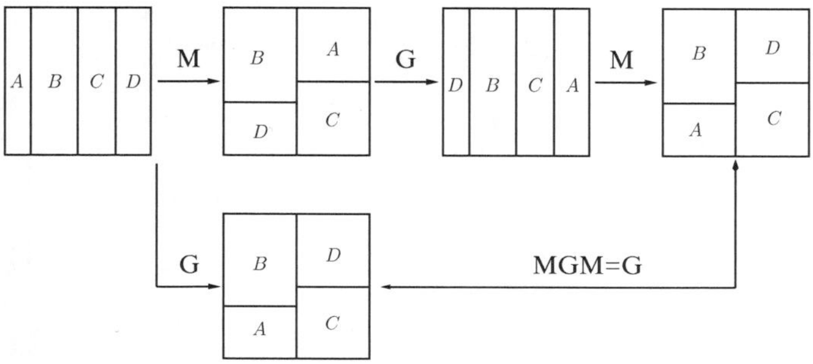

Fig. 1.5 Check of reversibility for the map (1.1)

$$MGM = G \tag{1.7}$$

which attains the form:

$$\begin{pmatrix} x_G \\ y_G \end{pmatrix} = G \cdot \begin{pmatrix} x \\ y \end{pmatrix} = \begin{cases} \begin{pmatrix} 2x \\ \frac{1}{2}y \end{pmatrix} & \text{for } 0 \le x < \frac{1}{2} \\ \begin{pmatrix} 2x - 1 \\ \frac{1}{2}(y+1) \end{pmatrix} & \text{for } \frac{1}{2} \le x \le 1 \end{cases} \tag{1.8}$$

The mapping G in (1.8) reflects the half squares $[0, 1/2)$ and $[1/2, 1]$ along the respective diagonals, drawn from their lower left to their upper right corners, cf. Fig. 1.4.

Hence, according to the definition (1.7), the baker model in (1.1) with $q = 0$ is T-symmetric for all values of the parameter ℓ, as shown in Fig. 1.5.

Consider, now, a trajectory of n time steps, $\{\underline{x}_0, \underline{x}_1, \ldots, \underline{x}_n\}$, along which the average phase space contraction rate is given by:

$$\overline{\Lambda}_n(\underline{x}_0) = -\frac{1}{n} \sum_{k=0}^{n-1} \ln J_M(M^k \underline{x}_0) \tag{1.9}$$

The average phase space contraction rate over the time reversed path is given by:

$$\begin{aligned}\overline{\Lambda}_n(GM^n(\underline{x}_0)) &= -\frac{1}{n}\sum_{k=0}^{n-1}\ln J_M(M^kGM^n\underline{x}_0)\\ &= -\frac{1}{n}\sum_{k=0}^{n-1}\ln J_M(GMM^{(n-1)-k}\underline{x}_0)\end{aligned}\tag{1.10}$$

Then, the relation $J_M(\underline{x}) = J_M^{-1}(GM\underline{x})$, yields the known result [30]:

$$\overline{\Lambda}_n(GM^n\underline{x}_0) = -\overline{\Lambda}_n(\underline{x}_0).$$

Since the Jacobians (1.2) depend only on the x-coordinate and are piecewise constant, the expressions (1.9) and (1.10) take the simple form:

$$\overline{\Lambda}_n(i_0) = -\frac{1}{n}\sum_{k=0}^{n-1}\ln J_M(i_k)\tag{1.11}$$

$$\overline{\Lambda}_n(Qi_{n-1}) = -\frac{1}{n}\sum_{k=0}^{n-1}\ln J_M(Qi_{(n-1)-k}) = -\overline{\Lambda}_n(i_0)\tag{1.12}$$

with i_k the region containing the point $M^k\underline{x}_0$, out of the four regions $\{A, B, C, D\}$ and $Q = GM$, where

$$QA = D, \quad QD = A, \quad QB = B, \quad QC = C.\tag{1.13}$$

Thus, the computation of $\overline{\Lambda}_n$ for the forward (respectively, time reversed) path, can be conveniently performed by keeping track only of the *coarse-grained* sequences of visited regions:

$$\{i_k\} = (i_0, i_1, \ldots, i_{n-1})\tag{1.14}$$

$$\{Qi_{(n-1)-k}\} = (Qi_{n-1}, Qi_{n-2}, \ldots, Qi_0),\tag{1.15}$$

rather than relying on the more detailed knowledge of the sequence of points $\{M^k\underline{x}\}$ and $\{GM^{n-k}\underline{x}\}$ in the phase space. The last regions in each sequence, i_n and Qi_{-1}, need not be taken into account, as they are unessential in the evaluation of the $\overline{\Lambda}_n$, see [30] for details. A considerable amount of information, regarding the microscopic trajectory in the phase space, is lost by passing from the phase space deterministic dynamics to the effectively stochastic process arising from the projection of the dynamics onto the x-axis. Such a process is described by a Markov jump process, which yields sequences such as those of (1.14) and (1.15). Nevertheless, this loss of information is irrelevant to compute the phase space contraction rate of sets of phase space trajectories. In particular, we may disregard variations of the dynamics internal to the single regions, as long as the resulting internal, or "hidden", dynamics preserve phase space volumes and do not affect (1.11) and (1.12). These observations are relevant for the stochastic descriptions of physical phenomena, which are thought to be based on reduced (projected) dynamics of phase space deterministic dynamics. Indeed, the projected Perron-Frobenius equation (1.3) is a time-integrated Master equation for the probability densities ρ_l and ρ_r over the coarser, projected, state space defined by the chosen Markov partition. These considerations are reminiscent of the fact that thermodynamics, to a large extent, does not depend on the details of the microscopic dynamics, hence is consistent with many different phase space evolutions. This is consequence of the fact that thermodynamics describes the object of interest by a few observable quantities, i.e. in a space of reduced dimensionality, which can be seen as a projection of the whole phase space. Thus, in order to compute the quantities of interest, one may conveniently choose the detailed microscopic dynamics which most easily represent the phenomenon under investigation. For instance, in our idealized setting, the equilibrium dynamics and the existence of some symmetries relating forward and reversed paths in a subspace of the phase space, may be investigated by means of the map M with $q = 0$, which is reversible via the involution G depicted in Fig. 1.4, and by means of the projections of M on the relevant directions. The class of dynamics which are equivalent from a given, restricted or projected, standpoint could include maps which are not even time reversal invariant. Indeed, one may consider microscopic dynamics obtained from map (1.1) introducing an irreversible transformation N which does not contract nor expand phase space volumes. This can be simply done by e.g. letting N flip the y-coordinates of the phase space points of a vertical strip of width ϵ in the region B:

$$\begin{pmatrix} x_{n+1} \\ y_{n+1} \end{pmatrix} = N \begin{pmatrix} x_n \\ y_n \end{pmatrix} = \begin{cases} \begin{pmatrix} x_n \\ 1 - y_n \end{pmatrix}, & \text{for} x \in [\tilde{x}, \tilde{x} + \epsilon] \text{ and} y \in \left[0, \frac{1}{2}\right) \\ \begin{pmatrix} x_n \\ y_n \end{pmatrix} & \text{for} x \in [\tilde{x}, \tilde{x} + \epsilon] \text{ and} y \in \left[\frac{1}{2}, 1\right] \end{cases} \tag{1.16}$$

cf. Fig. 1.6 for a graphical representation.

As pointed out in [30], the composed map $K = NM$ is irreversible because it does not admit an inverse. Moreover, the irreversible mechanism of the dynamics described in (1.16) gives rise to an invariant measure which is fractal along the vertical direction, cf. Fig. 1.7, and which is strongly at variance with its reversible equilibrium counterpart shown in Fig. 1.2. Nevertheless, it is clearly seen that (1.11) and (1.12) still hold true for the map K, despite the irreversible feature of the equations of motion. In fact, although the irreversible dynamics K no longer admits an involution (hence, (1.9) and (1.10) can no longer be fulfilled), (1.11) and (1.12) remain unaltered, because N maps a point $\underline{x} \in i$ into a point $\underline{x}' \in i$, and it neither contracts nor expand phase space areas. Thus, if we replace the dynamics M with the dynamics K and, accordingly, we take $Q = GK$, the sequences (1.14) and

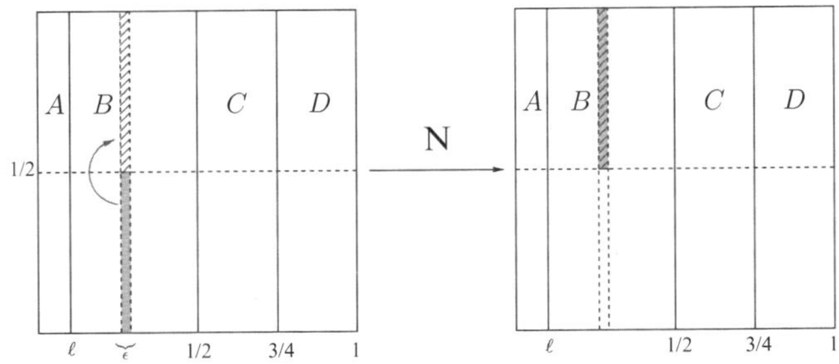

Fig. 1.6 The map N defined in (1.16), which spoils the reversibility of the model

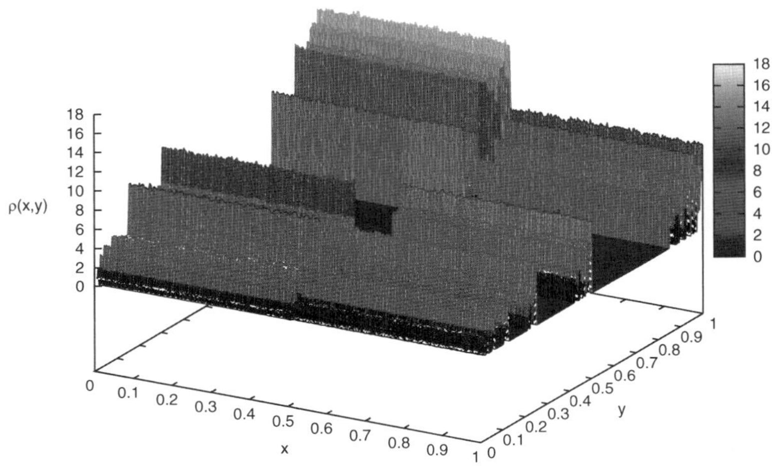

Fig. 1.7 Result of a numerical simulation for the invariant density of the equilibrium irreversible map derived from (1.1) and (1.16) by setting $q = 0$, $\ell = 0.15$, $\tilde{x} = \ell$ and $\epsilon = \frac{1}{2} - \ell$, obtained by evolving a set of $2 \cdot 10^7$ initial conditions randomly (and uniformly) chosen on the square $[0, 1] \times [0, 1]$. The density $\rho(x, y)$ is not smooth along the y-coordinate, which is a signature of the strongly irreversible dynamics given by the map N in (1.16)

(1.15) remain unaltered, since they are invariant under the action of an irreversible perturbation of the y-coordinate.

Now, let $\omega(i_0, n-1) \subset i_0$ and $\omega(Qi_{n-1}, n-1) \subset Qi_{n-1}$ denote, respectively, the sets of points corresponding to the forward (1.14) and to the time reversed (1.15) sequences. Trivially, the sets of points corresponding to these symbolic sequences have invariant measures $\mu(\omega(i_0, n))$ and $\mu(\omega(Qi_{n-1}, n))$, as in the reversible case. Therefore, in spite of the irreversible modification N, we may say that the dynamics enjoy a form of reversibility which is weaker than the standard reversibility in phase space, but which cannot be distinguished from that if observed from the stochastic (reduced) viewpoint of the projections on the horizontal direction.

As a matter of fact, time reversibility is contemplated in stochastic dynamics and amounts to the requirement that a sequence of events have positive probability if its reverse does [33]. Because our projected dynamics are not affected by the action of N, on the level of the stochastic description, the phase space reversible dynamics of M are equally *stochastically reversible* as the phase space irreversible dynamics of K. Furthermore, the fact that N reshuffles phase space points within vertical strips of the square implies that the redistributions of mass due to the phase space contraction rate of M and to the rearrangement of phase space volumes produced by N are indistinguishable on the projected horizontal space and can be quantified by the same observable Λ. In particular, (1.5) holds for both M and K, hence the statistics of all projected observables is the same. Only in the case that one is interested on observables which explicitly concern the y coordinate would the two dynamics be distinguishable, but as long as one focuses on quantities which do not depend on y, or which result from a projection on the x axis, M and K lead to the same conclusions: the corresponding reduced stochastic evolution is exactly the same, as in the case of *felt dynamics* introduced in [34].

Therefore, we may state that the map K with $q = 0$ represents a kind of *equilibrium* but *irreversible* dynamics. Although this might appear contradictory, it is simply explained by the observation that the equilibrium behavior concerns the level of the horizontal projection, which is stochastically reversible, while the irreversibility concerns the phase space. This situation differs from that of [32], in which time reversible nonequilibrium systems are recognized to be common—see, e.g. the standard models of nonequilibrium molecular dynamics—while equilibrium irreversible systems are thought to be rare, as far as phase space is concerned.

Our study concerns, instead, the bridge between deterministic and stochastic-like descriptions. In particular, we are going to show that the x-projection of K, being an *equilibrium* model, satisfies the principle of detailed balance (DB), from the point of view of the stochastic dynamics, in spite of its irreversibility in the phase space. Note that DB is often referred to as the *principle of microscopic reversibility* [35], which is, then, to be understood as a notion of reversibility in the reduced space, not in the full phase space. This naturally connects with the distinction between *relevant* and *irrelevant* coordinates which underlies the statistical mechanics reduction of deterministic descriptions in the phase space to stochastic descriptions, which typically concern the one-particle space [36, 37]. Clearly, our dynamical system is too simple to allow a physically meaningful distinction between relevant and irrelevant variables. Therefore, we merely focus on one of them (the x-coordinate) and regard the other as irrelevant (the y-coordinate), without implying that one subspace of our subspace is endowed with any special meaning.

1.3 From Phase Space to Detailed Balance

In this section we introduce the notion of *detailed balance in the phase space* (PSDB) which constitutes a strong concept of equilibrium dynamics. We will show that the standard notion of DB descends from PSDB through a projection of the phase space dynamics onto a suitable subspace. It will also be shown that the

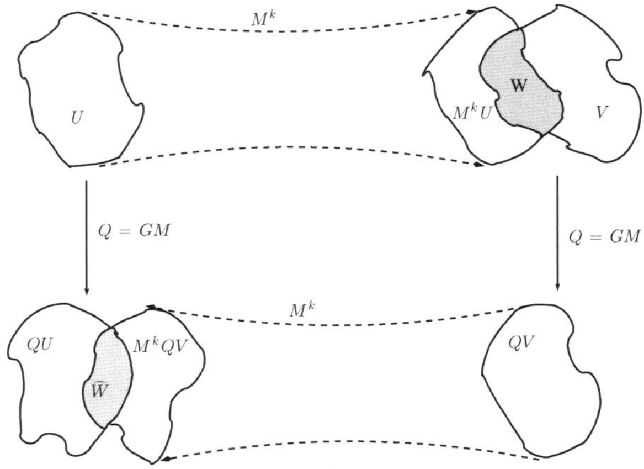

Fig. 1.8 Set of points, in phase, belonging to the sets U and V in the forward (*upper path*) and in the time reversed (*lower path*) trajectories

DB condition is insensitive to the reversibility properties of the full phase space equilibrium dynamics. Let

$$\underline{x}_{n+1} = M\underline{x}_n, \quad \underline{x} \in \mathscr{U}$$

be the microscopic dynamics, where M is time reversal invariant, with involution G. For sake of simplicity, we deal with discrete time, $t \in \mathbb{Z}$, but flows S^t, $t \in \mathbb{R}$ could be treated similarly.

Consider two sets in phase space, $U, V \subset \mathscr{U}$. Let $W = M^kU \cap V = \{\underline{x} \in V : M^{-k}\underline{x} \in U\}$ be the set of final points of trajectory segments starting in U, which fall in V after k iterations of M, and let $M^{-k}W$ be the corresponding set of initial conditions. Take an invariant measure μ for M, so that $\mu(W) = \mu(M^{-k}W)$, and let $\Lambda_k(\underline{x})$ be the phase space contraction along the trajectory starting in $\underline{x} \in M^{-k}W$. Its time reverse trajectory, which gives rise to the opposite phase space contraction, cf. (1.10) and (1.12), starts in $\underline{\hat{x}} = GMM^k\underline{x} \in GMW$. Call

$$\widehat{W} = M^kGMW = M^kGM\left(M^kU \cap V\right) = GMU \cap M^kGMV$$

the set of final conditions of all trajectories which are time reverses of those ending in W, cf. Fig. 1.8, where the second equality comes from the definition of W and the third equality follows from the property $M^kG = GM^{-k}$ of the involution G. The measure of this set is given by:

$$\begin{aligned}
\mu(\widehat{W}) &= \mu(M^kGMV \cap GMU) \\
&= \mu(M^kM^{-1}GV \cap M^{-1}GU) = \mu(M^{k-1}(GV \cap M^{-k}GU)) \\
&= \mu(M^{k-1}(GV \cap GM^kU)) = \mu(GV \cap GM^kU) \\
&= \mu(G(V \cap M^kU)) = \mu(GW)
\end{aligned} \tag{1.17}$$

We define the *phase space detailed balance* (PSDB) as the condition for which the probability of having opposite phase space contractions are equal:

$$\mu(M^{-k}W) = \mu(M^{-k}\widehat{W}) \quad \text{i.e.} \quad \mu(W) = \mu(\widehat{W}) \tag{1.18}$$

Because of (1.17), this condition may also be written as

$$\mu(W) = \mu(GW) \tag{1.19}$$

which is to say that PSDB requires the M-invariant measure μ to be also G-invariant, since W may be any subset of $\mathscr{U}$. Calling a function Φ odd with respect to time reversal if $\Phi(G\underline{x}) = -\Phi(\underline{x})$ for all $\underline{x} \in \mathscr{U}$, and defining equilibrium the situation in which the mean value of all such odd observables vanishes, we obtain that (1.19) implies equilibrium. Indeed, take any Φ which is odd with respect to the time reversal and choose $W = \{\underline{x} \in \mathscr{U} : \Phi(\underline{x}) > 0\}$, so that $GW = \{\underline{x} \in \mathscr{U} : \Phi(\underline{x}) < 0\}$ and $\Phi(\underline{x}) = 0$ for all $\underline{x} \in \mathscr{U} \setminus (W \cup GW)$. Then the following holds:

$$\begin{aligned}
\langle \Phi \rangle &= \int_W \Phi(\underline{x}) d\mu(\underline{x}) + \int_{GW} \Phi(\underline{x}) d\mu(\underline{x}) \\
&= \int_W \Phi(\underline{x}) d\mu(\underline{x}) + \int_W \Phi(G\underline{y}) J_G(\underline{y}) d\mu(\underline{y}) \\
&= \int_W \Phi(\underline{x}) d\mu(\underline{x}) - \int_W \Phi(\underline{y}) d\mu(\underline{y}) = 0
\end{aligned} \tag{1.20}$$

where $J_G = 1$. In particular, the PSDB implies that the average of the phase space contraction rate $\Lambda(\underline{x}) = \log J_F(\underline{x})^{-1}$ vanishes, as required for equilibrium in Sect. 1.3.

Is there any relation between PSDB and the standard DB, which implies equilibrium on the level of the projected dynamics (the caricature of the one particle or μ-space)? We notice, in passing, that some authors distinguish DB dynamics from detailed balance steady state, which proves to be a convenient tool in the analysis of stochastic processes [22]. In this case, a given evolution law is called DB dynamics if its steady state is a DB state. To derive standard DB, let us eliminate the irrelevant coordinates, which only contribute to noise, by projecting (1.19) on the subspace of *relevant* coordinates, which we call μ-space. This can be done for sets of the form $W = W^{(1)} \times \mathscr{C}_{n-1}$ and $\widehat{W} = \widehat{W}^{(1)} \times \widehat{\mathscr{C}}_{n-1}$, where $W^{(1)}$ and $\widehat{W}^{(1)}$ denote the projections of the sets W and $\widehat{W}$ onto the μ-space and $\mathscr{C}_{n-1}, \widetilde{\mathscr{C}}_{n-1}$ span the remaining, noisy, space, cf. the hyper-cylinders illustrated in Fig. 1.9, where $W^{(1)}$ and $\widehat{W}^{(1)}$ denote, respectively, the event on the μ-space. Then, if we denote by $\mu^{(1)}$ the measure in the μ-space induced from the invariant measure on the phase space, we have that:

$$\mu^{(1)}\left(W^{(1)}\right) = \int_{\mathscr{C}_{n-1}} \mu(d\underline{x}), \quad \mu^{(1)}\left(\widehat{W}^{(1)}\right) = \int_{\widehat{\mathscr{C}}_{n-1}} \mu(d\underline{x}) \tag{1.21}$$

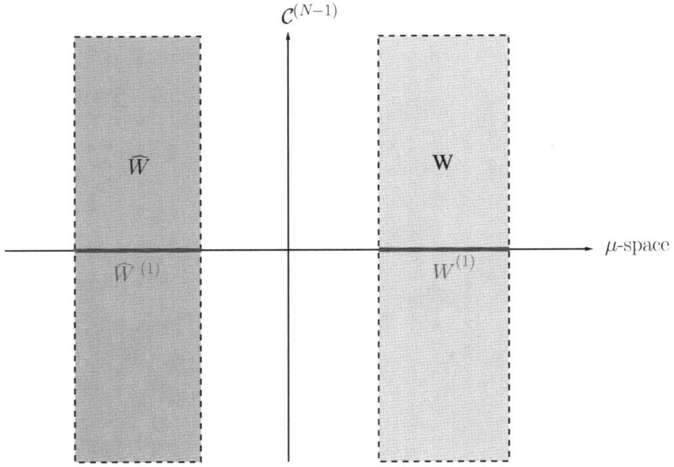

Fig. 1.9 The sets $W^{(1)}$ and $GW^{(1)}$, after reduction from phase space to the μ-space

Detailed balance holds if

$$\mu^{(1)}\left(W^{(1)}\right) = \mu^{(1)}\left(\widehat{W}^{(1)}\right), \quad \text{or} \quad \mu^{(1)}\left(W^{(1)}\right) = \mu^{(1)}\left(GW^{(1)}\right) \tag{1.22}$$

Thus, PSDB implies the standard DB, since DB amounts to the condition of PSBD restricted to sets W and $\widehat{W}$ of the form here introduced. Moreover, the projection procedure typically smoothes out singularities, hence the induced invariant measure $\mu^{(1)}$ usually is regular and has an invariant density $\rho^{(1)}$. For instance, the map (1.1) with $h = \frac{1}{2} - 2\ell$ which, for arbitrary ℓ, is dissipative but still equipped with a projected invariant density which is smooth along the unstable manifold, [30]. In the derivation of the stochastic description as a projection of some deterministic phase space dynamics, the DB condition (1.22) is usually assumed to be the consequence of the time reversal invariance of the microscopic dynamics, once equilibrium is reached. In our investigation this amounts to require the existence of the involution G, defined in the phase space. We are now going to see that this requirement may be relaxed in simple cases, such as those of the irreversible dynamics K discussed above. Using the same notation, (1.19) can then be written as:

$$\mu(Mi \cap j) = \mu(MGMj \cap GMi) \tag{1.23}$$

where $\mu(Mi \cap j)$ is the conditional probability of being in the region j one time step after having been in region i, with $i, j \in \{A, B, C, D\}$, our finite state space. The quantity $\mu(Mi \cap j)$ may be rewritten as $\mu(Mi \cap j) = p(j|i) = p_{ij}$, with notation reminiscent of stochastic descriptions. The p_{ij}'s then constitute the elements of the transition matrix:

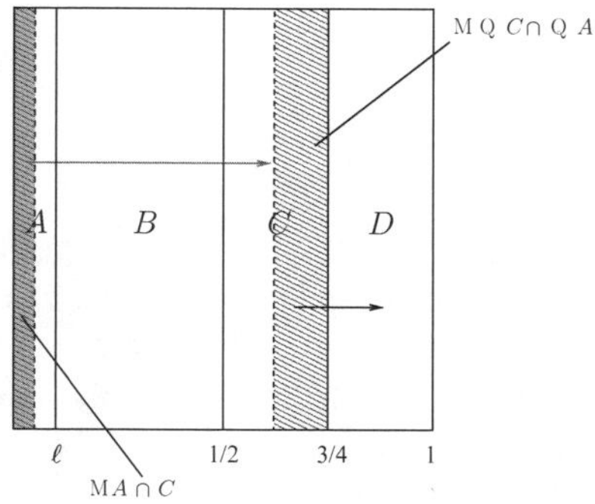

Fig. 1.10 Sets undergoing the forward path ($A \rightarrow C$, *blue arrow*) and the time reversed path ($QC \rightarrow QA$ *black arrow*) for the map (1.1) with $q = 0$

$$P = \begin{pmatrix} 0 & 0 & \frac{1}{2} & \frac{1}{2} \\ 2\ell & 1-2\ell & 0 & 0 \\ 0 & 0 & \frac{1}{2} & \frac{1}{2} \\ 2\ell & 1-2\ell & 0 & 0 \end{pmatrix} \tag{1.24}$$

which defines a *stochastic* process for the dynamics of the state in $\{A, B, C, D\}$. Application of the Perron-Frobenius theorem to the matrix P reveals the existence of one left eigenvector of P, associated with the eigenvalue $\lambda = 1$ (i.e. the geometric multiplicity of such eigenvalue is 1), which, thus, implies the existence of a unique (coarse-grained) steady state measure $(\mu_A, \mu_B, \mu_C, \mu_D)$, where:

$$\mu_i = \begin{cases} \dfrac{2\ell}{1+4\ell}, & \text{if} \quad i = A, C, D; \\ \\ \dfrac{1-2\ell}{1+4\ell}, & \text{if} \quad i = B. \end{cases} \tag{1.25}$$

Equation (1.25) highlights the fact that the measure in the full phase space is uniform along the stable manifold and piecewise constant along the unstable one, as previously illustrated in Fig. 1.2. This comes from the fact that, since the regions $\{A, B, C, D\}$ are determined only by the x-coordinate, the stochastic transition matrix P does not depend on the dynamics along the stable manifold, and, hence, $p_{ij} = (J_j^u)^{-1}$. As a result, the invariant measure in (1.25) depends only on x and is constant on y.

Consider, for instance, the one-step transition $A \rightarrow C$, whose probability is the measure of $W = MA \cap C$, which equals that of the set of the corresponding initial conditions $M^{-1}W = A \cap M^{-1}C$. The probability of the reverse transition $QC = C \rightarrow QA = D$, where we have recalled the relations (1.13), is the measure of the set $\widehat{W} = MQC \cap QA$, cf. Fig. 1.10. Since the sets W and $\widehat{W}$ span the whole range $[0, 1]$ in the vertical direction, as in Fig. 1.9, the measures $\mu(W)$ and $\mu(\widehat{W})$ can be calculated just in terms of the transition probabilities (1.24) and of the projected invariant measures (1.25). The result is

$$\mu(MA \cap C) = \mu_A p_{AC} = \frac{\ell}{1+4\ell} = \mu_C p_{CD} = \mu(MC \cap D) \tag{1.26}$$

for any ℓ. Hence, DB holds as expected, because we are dealing with an equilibrium case, although the underlying dynamics is irreversible. It is now interesting check what happens when the microscopic dynamics is pulled out of equilibrium. For instance, consider $K = NM$, with N as in (1.16) and M the time reversible mapping (1.1), with $q = \frac{1}{2} - 2\ell$. For this map M, we may consider an involution G which is consistent with the following equalities (cf. (36) in [30]):

$$QA = A, \quad QB = C, \quad QC = B, \quad QD = D \tag{1.27}$$

hence which differs from the G in (1.8). Then, the time reverse of the transition $A \to C$ is given by $QC \to QA$ i.e. $B \to A$, and we get:

$$\mu(MA \cap C) = \mu_A p_{AC} = \frac{\ell}{1+4\ell} \tag{1.28}$$

$$\mu(MB \cap A) = \mu_B p_{BA} = \frac{(1-2\ell)^2}{1+4\ell} \tag{1.29}$$

which shows that (1.28) and (1.29) do not coincide, and that PSDB and DB are violated for $q \neq 0$, i.e. outside the equilibrium defined via the chosen G. In this case, only $\ell = \frac{1}{4}$ leads to the equilibrium state which is, in addition, microcanonical. In fact, the map with $q = \frac{1}{2} - 2\ell$ attains equilibrium for $q = 0$, which gives $\ell = \frac{1}{4}$, corresponding, as discussed in [30], to a microcanonical equilibrium distribution.

1.4 The Fluctuation Relation and Nonequilibrium Response

The fluctuation relation for Λ, the Λ-FR, originally proposed by Evans et al. [8], and developed by Gallavotti and Cohen [9, 10], concerns the statistics of the mean phase space contraction rate $\overline{\Lambda}_n$, over the steady state ensemble of phase space trajectory segments of a large number of steps, n. Equivalently, it concerns the statistics of $\overline{\Lambda}_n$, computed over segments of a unique steady state phase space trajectory, broken in segments $\{x_1, \ldots, x_n\}$ of length n.

The dynamics are called dissipative if

$$\langle \Lambda \rangle = \int_{\mathcal{U}} \Lambda(\underline{x})\, \mu(\mathrm{d}\underline{x}) > 0$$

where $\langle \Lambda \rangle$ is the steady state mean of Λ, i.e. it is computed with respect to the natural measure μ on $\mathcal{U}$. It is convenient to introduce the dimensionless phase space contraction rate $e_n = \overline{\Lambda}_n/\langle\Lambda\rangle$ because its range, $[-\Lambda_{\max}/\langle\Lambda\rangle, \Lambda_{\max}/\langle\Lambda\rangle]$, does not change with n, while the values taken by e_n tend to more and more densely fill it as n grows. Further, we denote by $\pi_n(B_{p,\delta})$ the probability that e_n, computed over a segment of n-steps of a typical trajectory, falls in the interval $B_{p,\delta} = (p-\delta, p+\delta)$, for some fixed $\delta > 0$. In other words, one may write

$$\pi_n(B_{p,\delta}) = \mu(\omega_{\Lambda,n}), \quad \text{where} \quad \omega_{\Lambda,n} = \{\underline{x} \in \mathcal{U} : e_n \in B_{p,\delta}\} \tag{1.30}$$

For growing n, π_n peaks around the mean value $\langle e_n\rangle = 1$, but fluctuations about this mean may occur with positive probability at any finite n. In particular, under certain conditions, [9, 10, 14, 15, 38], π_n obeys a large deviation principle with a given rate functional ζ, in the sense that the limit

$$\lim_{n\to\infty} \pi_n(B_{p,\delta}) = e^{-n[\zeta(p)+\epsilon_\delta]} \tag{1.31}$$

exists, with $\epsilon_\delta \leq \delta$. In particular, if the support of the invariant measure is the whole phase space $\mathcal{U}$, time reversibility guarantees that the support of π_n is symmetric around 0, and one can consider the ratio

$$\frac{\pi_n(B_{p,\delta})}{\pi_n(B_{-p,\delta})}.$$

In our case, this ratio equals the ratio of the measures of a pair of sets conjugated by time reversal, as in (1.11) and (1.12). Then, the validity of the Λ-FR means that there exists $p^* > 0$ such that

$$p - \delta \leq \lim_{n\to\infty} \frac{1}{n\langle\Lambda\rangle} \log \frac{\mu(\{x : e_n(x) \in B_{p,\delta}\})}{\mu(\{x : e_n(x) \in B_{-p,\delta}\})} \leq p + \delta \tag{1.32}$$

if $|p| < p^*$ and $\delta > 0$.

So far, the proofs of this and other FR's appeared in the literature, notably those for the fluctuations of the Dissipation Function Ω, [4, 15, 39], rely on the existence of an involution representing time reversal in phase space, while they rely on the principle of microscopic reversibility in the state space of stochastic processes. So, whatever the context, the relevant notion of time reversibility has always been used. Therefore, if time reversibility is broken, but is broken as in the case of the map K, which enjoys a weaker form of reversibility requiring only the existence of the pairs of conjugate trajectories (1.14) and (1.15), the Λ-FR should remain valid. Indeed, the existence of this weaker reversibility is consistent with the principle of microscopic reversibility, adopted in the stochastic approach by e.g. Lebowitz and Spohn [13].

Let us then investigate the validity of the Λ-FR for the deterministic model $K = NM$ with N given by (1.6) and M by (1.1), which may or may not lead to equilibrium, depending on the value of the parameter q. As discussed in Sect. 1.2,

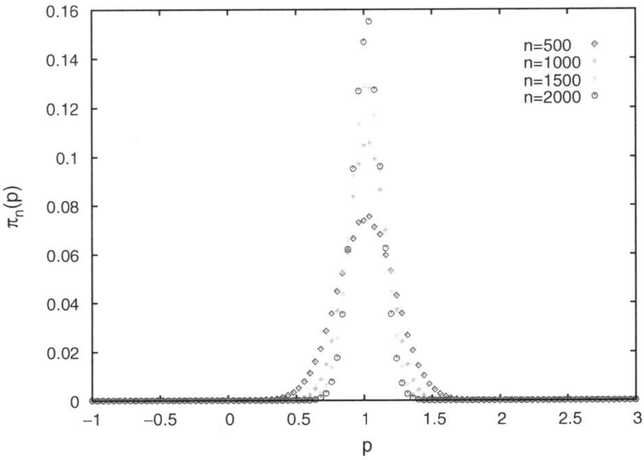

Fig. 1.11 Probability measure $\pi_n(B_{p,\delta})$

K is irreversible, M is reversible and the mapping $Q = GK$, appearing in the definition of the time reversed path (1.15), is properly defined in both cases. Take $q = \frac{1}{2} - 2\ell$, with $\ell \neq \frac{1}{4}$, consistently with [30]. Then, the Λ-FR may be written as:

$$p - \delta \leq \lim_{n\to\infty} \frac{1}{n\langle\Lambda\rangle} \log \frac{\mu(\omega(i_0, n-1))}{\mu(\omega(Qi_{n-1}, n-1))} \leq p + \delta \tag{1.33}$$

with $|p| < p^*$, for some $p^* > 0$ and any $\delta > 0$. To prove (1.33) for the map $K = NM$, one must compute the invariant probability measure π_n and let n grow without bounds.

This is guaranteed by the proof of the validity of the Λ-FR given for M in [30], which only relies on the invariant measure in the projected space.

We illustrate this result by means of numerical simulations, which we have performed for different values of n, with $\ell = 0.15$, cf. Fig. 1.11. The numerical simulations, shown in Fig. 1.12, show how the large deviation rate functional ζ is generated, and that it is smooth and strictly convex in the whole range of observed fluctuations, as required by the theory of the Λ-FR. In Fig. 1.13 we also plotted the expression in the center of (1.33), which is consistent with the validity of the Λ-FR. Once the Λ-FR is proven to hold for the map K, one may be tempted to assess the validity of the Green-Kubo formulas as well as of the Onsager reciprocal relations, by following e.g. the strategies of [40–43], in the limit of small external drivings, as summarized in [4, 6]. To this end, we consider Λ as the entropy production rate, although this identification must be done *cum grano salis*, as explained in e.g. [5,40,44–46], and we briefly summarize the argument, for sake of completeness.

The main steps are the following [6,41]:

- Assume that the system is subjected to k fields $F = (F_1, F_2, .., F_k)$, that Λ vanishes when all drivings vanish and that

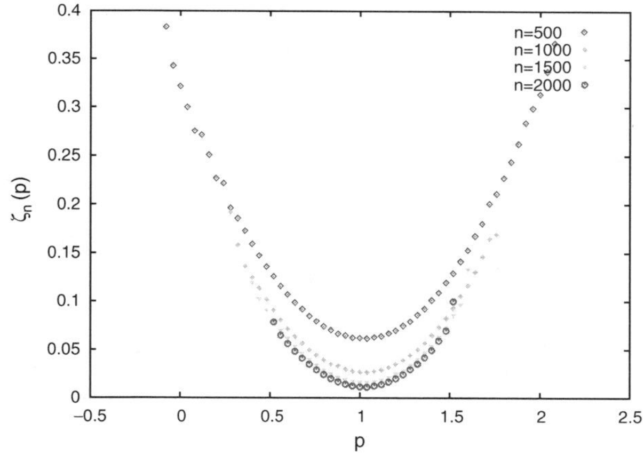

Fig. 1.12 Rate functional ζ_n associated with $\pi_n$for different values of n. As expected from theory, the curves ζ_n are expected to move downwards for growing n, so that, in the $n \to \infty$ limit, $\zeta(p)$ intersects the horizontal axis only in $p = 1$

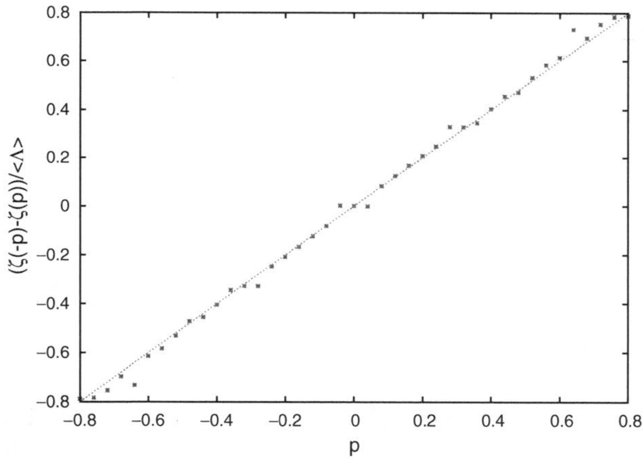

Fig. 1.13 Check of the Λ-FR, (1.33) for the irreversible dynamics dictated by the map $K = NM$. *Blue points*: results of the numerical simulations with $n = 2 \cdot 10^2$

$$\Lambda(\underline{x}) = \sum_{\ell=1}^{k} F_\ell J_\ell^0(\underline{x}) + O(F^2) \tag{1.34}$$

which defines the currents J_r^0, which are proportional to the forces F_r.

- The decay of the Λ autocorrelation function required for the Λ-FR to hold, leads to the following expansion for the rate function ζ:

$$\zeta(p) = \frac{\langle \Lambda \rangle^2}{2C_2}(p-1)^2 + O((p-1)^3 F^3) \tag{1.35}$$

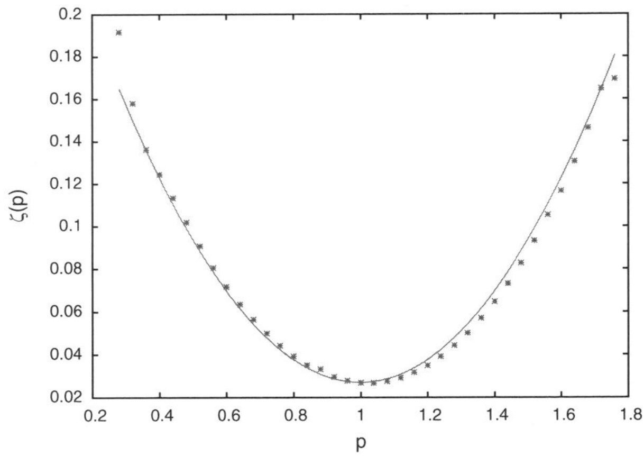

Fig. 1.14 Rate functional $\zeta(p)$ associated to $\pi_n(B_{p,\delta})$. *Blue points*: results of the numerical simulation with $n = 10^3$. *Red line*: fitting of numerical data with the parabola $a(x-1)^2 + b$, with parameters $a = 2.66 \cdot 10^{-1}$ and $b = 2.68 \cdot 10^{-2}$

where C_2 is related to the time autocorrelation of Λ. In other words, the rate functional is quadratic for small deviations from the mean $p = 1$, in accord with the Central Limit Theorem.

- Introduce the nonlinear currents as $J_\ell(\underline{x}) = \partial_{F_\ell}\Lambda(\underline{x})$, and the transport coefficients $L_{\ell r} = \partial_{F_r}\langle J_\ell\rangle|_{F=0}$. Then, one obtains:

$$\langle \Lambda \rangle = \frac{1}{2}\sum_{\ell,r=1}^{k}(\partial_{F_r}\langle J_\ell\rangle + \partial_{F_\ell}\langle J_r\rangle)|_{F=0}F_\ell F_r = \frac{1}{2}\sum_{\ell,r=1}^{k}(L_{\ell r} + L_{r\ell})F_\ell F_r \quad (1.36)$$

to second order in the forces.

- Equation (1.33) implies $\langle \Lambda \rangle = \frac{C_2}{2}$. Thus, equating the latter expression with (1.36) and by considering $(L_{\ell r} + L_{r\ell})/2$ with $\ell = r$, one obtains the Green-Kubo relations.

In our case, the rate functional is clearly quadratic, as shown by our simulations of the dynamics of K. In particular, the red quadratic curve in Fig. 1.14 reproduces nicely the behavior of the numerical data for the rate functional ζ_n, corresponding to trajectory segments of $n = 200$ steps. The necessity for a parameter $b \neq 0$ in the parabola is due the finiteness of n: indeed $b \to 0$ when $n \to \infty$. However, in spite of the validity of the Λ-FR for the irreversible map K, which entails that the irreversible map behaves to some extent equivalently to the reversible map M, the argument of [47] leading to the Green-Kubo relations cannot be reproduced here. In fact, it relies on the differentiability of the SRB measure as well as on the reversibility of the microscopic dynamics, which are both violated in the case of K. Alternatively, one may think of deriving linear response from the Λ-FR through the approach of [40] (SRE, hereafter), which does not explicitly require the differentiability of the invariant measure and the reversibility of the dynamics. In particular, SRE deals with a Nosé-Hoover thermostatted N-particle system and obtains:

$$n\sigma^2_{\overline{\Lambda}_n}(F_e) = \left(\frac{F_e V}{2K_0}\right)^2 \left[\frac{2L(F_e)k_B T}{V} + O\left(\frac{F_e^2}{nN}\right)\right] \tag{1.37}$$

for the variance $\sigma^2_{\overline{\Lambda}_n}$ of $\overline{\Lambda}_n$, provided Λ can be identified with the dissipation function Ω. Here, where K_0 is the target kinetic energy for the thermostatted particles, which corresponds to the inverse temperature β, N is the number of particles, V is the volume and F_e denotes the external force (i.e. the bias) acting on the system. The quantity $L(F_e)$ is defined by

$$L(F_e) = \beta V \int_0^\infty dt \langle (\Psi(t) - \langle \Psi \rangle)(\Psi(0) - \langle \Psi \rangle) \rangle \tag{1.38}$$

and $L(0) = \lim_{F_e \to 0} L(F_e)$ is the linear transport coefficient. The derivation of the Green-Kubo formulae is completed by comparing (1.37) with the relation $\langle \Lambda \rangle = \langle \Psi \rangle F_e = \frac{n}{2}\sigma^2_{\overline{\Lambda}_n}(F_e)$, which is implied by the FR, which yields:

$$L(0) = \lim_{F_e \to 0} \frac{\langle \Psi \rangle}{F_e} = \beta V \int_0^\infty dt \langle \Psi(t) \Psi(0) \rangle \tag{1.39}$$

In our framework of simple dynamical systems, the "current" Ψ could be defined as:

$$\Psi(\underline{x}) = \begin{cases} 0 & \text{for } \underline{x} \in A, D \\ 1 & \text{for } \underline{x} \in B \\ -1 & \text{for } \underline{x} \in C \end{cases} , \tag{1.40}$$

which implies an average current $\langle \Psi \rangle = (1 - 4\ell)/(1 + 4\ell)$, cf. (41) in [30], where the role of the external force F_e was played by the *bias* $b = 2 - 1/(1 - 2\ell)$. Nevertheless, following SRE may be problematic, as N is required to be large enough in order to derive (1.39) from (1.37), something which cannot be granted in low-dimensional systems as the map under consideration. Indeed, our numerical simulations reveal that an interesting scenario arises in the computation of the quantity $L(F_e)$, which may be conveniently approximated by:

$$L(F_e) \simeq \frac{1}{N_{ens}} \sum_{k=0}^{(N_{iter}-1)} \sum_{j=1}^{N_{ens}} \left[\Psi(\underline{x}_k^{(j)})\Psi(\underline{x}_0^{(j)}) - \langle \Psi \rangle^2\right] \tag{1.41}$$

where the upper limit of the integral in (1.38) is replaced by N_{iter} and the correlations are computed over an ensemble of fully decorrelated initial conditions $\{\underline{x}_0^{(j)}\}$, with $j = 1, \ldots, N_{ens}$, picked at random so that they occur in the ensemble

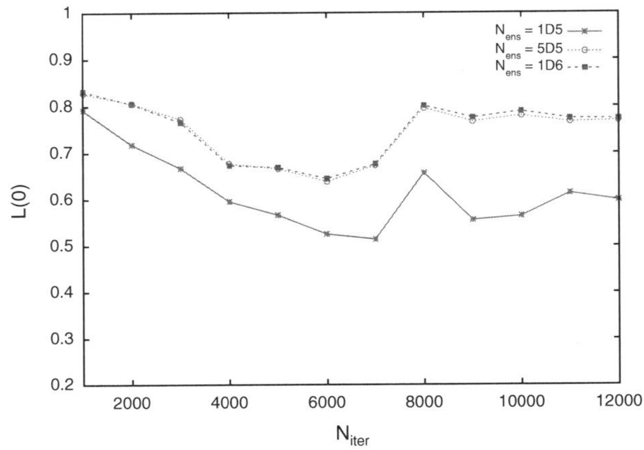

Fig. 1.15 Numerical computation of $L(0)$ from (1.41), for the irreversible map $K = MN$ derived from (1.16) and (1.1), with $q = \frac{1}{2} - 2\ell$, for different values of N_{ens}

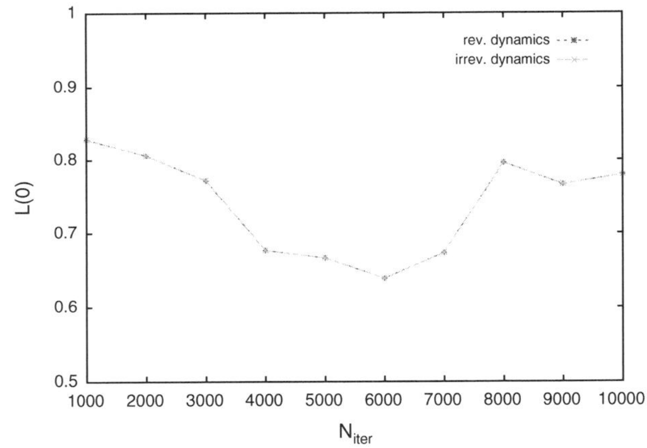

Fig. 1.16 Computation of $L(0)$ for the irreversible, $K = MN$, and the corresponding reversible map, $K = M$, with $N_{ens} = 5 \cdot 10^5$. The irreversibility does not affect the convergence in (1.41)

with the frequency corresponding to the natural invariant measure of the dynamical system. Then, the coefficient $L(0)$ may be computed from (1.41) by considering the limit of vanishing bias, i.e. by taking $\langle \Psi \rangle = 0$ and a microcanonical equilibrium ensemble of initial conditions $\{\underline{x}_0^{(j)}\}$. The values N_{ens} and N_{iter} must be chosen with care, in order to guarantee the convergence of the sums in (1.41), cf. Fig. 1.15.

Our simulations show that the irreversible character of the dynamics, which affects the "irrelevant" variable y, has no influence on the convergence of the Green-Kubo, formula (1.41), cf. Fig. 1.16.

To study the linear response for the irreversible map, that is the existence of the limit $L(0) = \lim_{F_e \to 0} L(F_e)$, we varied the value of bias within four different

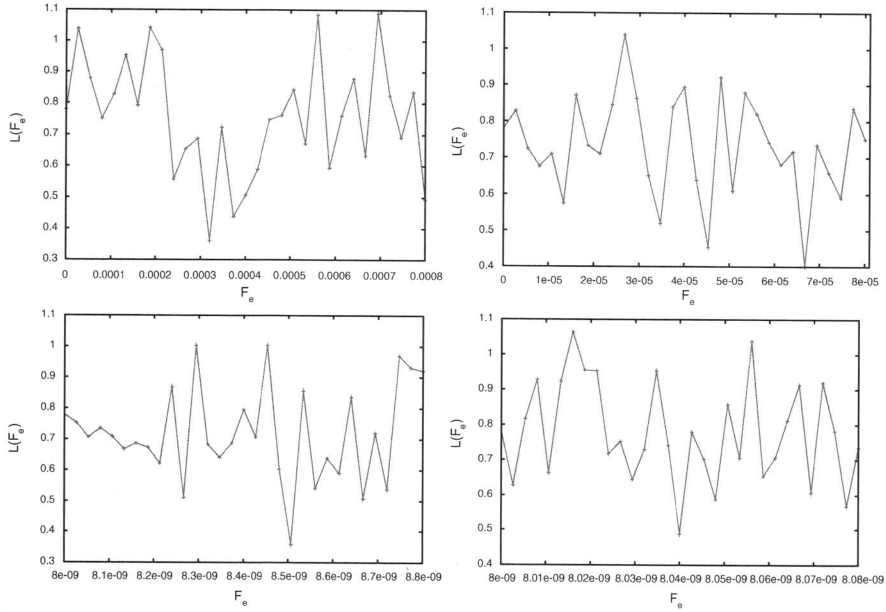

Fig. 1.17 Numerical computation of the coefficient $L(F_e)$ as a function of the bias F_e at different orders of magnitude of the bias. The coefficient $L(F_e)$ presents a very irregular structure, which may prevent the existence of the limit $L(0) = \lim_{F_e \to 0} L(F_e)$

windows of magnitude, cf. Fig. 1.17, and obtained that, in spite of the validity of the Λ-FR, no linear response can be claimed for our low-dimensional system, unless this is verified at exceedingly small bias. This fact cannot be blamed on the irreversible nature of the evolution, since we have observed that the response of the irreversible map coincides with the response of the corresponding T-symmetric one, cf. Fig. 1.18. It is more related to the irregularity typical of transport phenomena in low dimension [5]. Therefore, the dynamics of the map $K = MN$ proves that its irreversible component, the map N, affects neither the validity of the Λ-FR nor the response of the system to an external bias.

Per se, the fact that neither the approach of [47] nor that of SRE are applicable does not imply that no linear response can be established. However, this a clear observation for our model, which does not enjoy many properties of the systems of [47] and many others of the systems of [40]. More importantly, as pointed out in e.g. [17, 43, 48], the physical linear response relies on the occurrence of *local equilibrium*, in the sense that real space may be thought of as a "collection" of cells, each of which contains a statistically significant number of interacting particles. Clearly, our two-dimensional dynamical system may mimic only a few features of a real N-particle system, and the local equilibrium property is out of question.

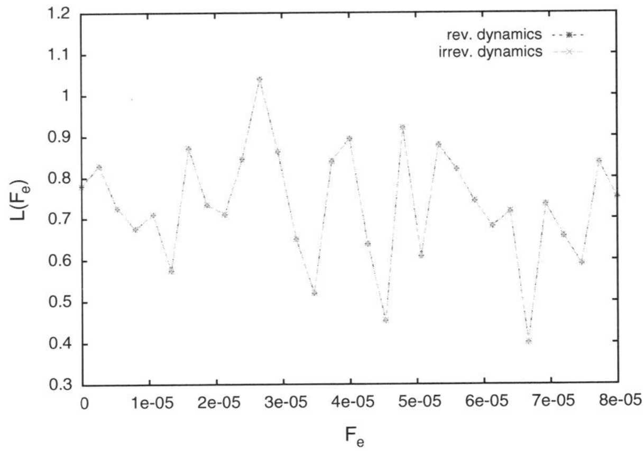

Fig. 1.18 Transport coefficient $L(F_e)$ for the irreversible, $K = MN$, and the corresponding reversible map, $K = M$, with $N_{ens} = 5 \cdot 10^5$. The irreversible component of the dynamics, N, leaves the response of the system unaffected

1.5 Some Results on FDT in Chaotic Dissipative Systems

We now turn our attention on the FDT, which, since its early developments, due mainly to the works of Onsager and Kubo [37, 49–51], represents a cornerstone along the construction of a theory of nonequilibrium phenomena [52]. To this aim, let us briefly review Ruelle's approach to linear response in deterministic dissipative dynamical systems [53]. Let $(\mathscr{M}, S^t, \mu)$ be a dynamical system, with $\mathscr{M}$ its compact phase space, $S^t : \mathscr{M} \to \mathscr{M}$ a one parameter group of diffeomorphisms and μ the invariant natural measure. Following Ruelle [53], who considers axiom A systems, one may show that the effect of a perturbation $\delta F(t) = \delta F_{\parallel}(t) + \delta F_{\perp}(t)$ on the response of a generic (smooth enough) observable A attains the form:

$$\delta \overline{A}(t) = \int_0^t R_{\parallel}^{(A)}(t-\tau)\delta F_{\parallel}(\tau)d\tau + \int_0^t R_{\perp}^{(A)}(t-\tau)\delta F_{\perp}(\tau)d\tau \tag{1.42}$$

where the subscript $_{\parallel}$ refers to the dynamics on the unstable tangent bundle (along the attractor), while $_{\perp}$ refers to the transversal directions, cf. left panel of Fig. 1.19. Ruelle's central remark is that $R_{\parallel}^{(A)}$ may be expressed in terms of a correlation function evaluated with respect to the unperturbed dynamics, while $R_{\perp}^{(A)}$ depends on the dynamics along the stable manifold, hence it may not be determined by μ, and should be quite difficult to compute numerically [6].

To illustrate these facts, the authors of [54] study a 2-dimensional model, which consists of a chaotic rotator on a plane and, for such a system, succeed to numerically estimate the $R_{\perp}^{(A)}$ term in (1.42). Nevertheless, in the next section, we argue that $R_{\perp}^{(A)}$ may spoil the generalized FDT only if the perturbation is carefully

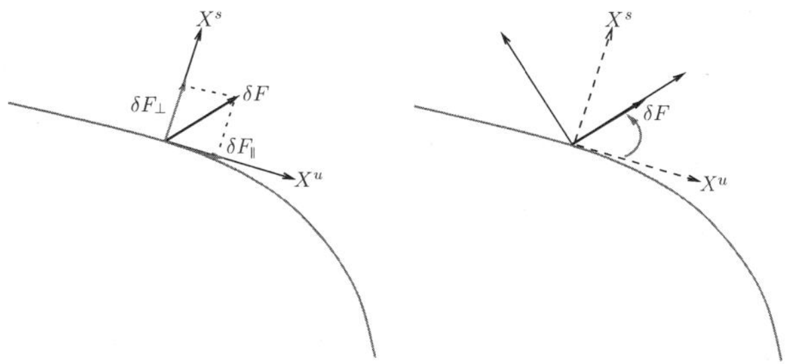

Fig. 1.19 *Left panel*: In Ruelle's approach, the perturbation is expressed as the sum of one component parallel to the unstable manifold and one parallel to the stable manifold. *Right panel*: In the present work, the reference frame is rotated so that the direction of the perturbation coincides with one of the basis vectors

oriented with respect to the stable and unstable manifolds. This is only possible in peculiar situations, such as those of [54], in which the invariant measure is the product of a radial and angular component and, furthermore, the perturbation lies on the radial direction, leaving the angular dynamics unaffected.

A different approach to the FDT has been proposed in [27], which concerns deterministic dynamics perturbed by stochastic contributions. Here, the invariant measure μ can be assumed to have density ρ: $\mathrm{d}\mu(\underline{x}) = \rho(\underline{x})d\underline{x}$. Then, if the initial conditions are modified by an *impulsive* perturbation $\underline{x}_0 \to \underline{x}_0 + \delta\underline{x}_0$, the invariant density $\rho(\underline{x}_0)$ is replaced by a perturbed initial density $\rho_0(\underline{x}_0; \delta\underline{x}_0) = \rho(\underline{x}_0 - \delta\underline{x}_0)$, where the subscript 0 denotes the initial state, right after the perturbation. This state is not stationary and evolves in time, producing time dependent densities $\rho_t(\underline{x}_0; \delta\underline{x}_0)$, which are assumed to eventually relax back to $\rho(\underline{x}_0)$. Thus, given the transition probability $W(\underline{x}_0, 0 \to \underline{x}, t)$ determined by the dynamics, the response of coordinate x_i is expressed by:

$$\overline{\delta x_i(t)} = \int\int x_i \left[\rho(\underline{x}_0 - \delta\underline{x}_0) - \rho(\underline{x}_0)\right] W(\underline{x}_0, 0 \to \underline{x}, t) d\underline{x}_0 d\underline{x} \qquad (1.43)$$

and one may introduce the response function R_{ij} as [27]:

$$R_{ij}(t) = \frac{\overline{\delta x_i(t)}}{\delta x_j(0)} = -\left\langle x_i(t) \left. \frac{\partial \log\rho}{\partial x_j}\right|_{t=0} \right\rangle \qquad (1.44)$$

which is a correlation function computed with respect to the unperturbed state. It is worth to note that it makes no difference in the derivation of (1.44) whether the steady state is an equilibrium state or not; it suffices that ρ be differentiable.

Let us consider again (1.43) and, for sake of simplicity, assume that all components of $\delta\underline{x}(0)$ vanish, except the i-th component. Then, the response of x_i may also be written as:

$$\overline{\delta x_i}(t) = \int x_i \left\{ \int [\rho(\underline{x}_0 - \delta \underline{x}_0) - \rho(\underline{x}_0)] W(\underline{x}_0, 0 \to \underline{x}, t) d\underline{x}_0 \prod_{j \neq i} dx_j \right\} dx_i$$
$$\equiv \int x_i B_i(x_i, \delta \underline{x}_0, t) dx_i \quad (1.45)$$

where $B_i(x_i, \delta \underline{x}_0, t)$, defined by the term within curly brackets, may also be written as:

$$B_i(x_i, \delta \underline{x}_0, t) = \widetilde{\rho_t}(x_i; \delta \underline{x}_0) - \widetilde{\rho}(x_i) \quad (1.46)$$

where $\widetilde{\rho}(x_i)$ and $\widetilde{\rho_t}$ are the marginal probability distributions defined by:

$$\widetilde{\rho}(x_i) = \int \rho(\underline{x}) \prod_{j \neq i} dx_j, \qquad \widetilde{\rho_t}(x_i; \delta \underline{x}_0) = \int \rho_t(\underline{x}; \delta \underline{x}_0) \prod_{j \neq i} dx_j.$$

As projected singular measures are expected to be smooth, especially if the dimension of the projected space is sensibly smaller than that of the original space, one may adopt the same procedure also for dissipative deterministic dynamical systems, as also shown in [55]. Indeed, the response function $B_i(x_i, \delta \underline{x}_0, t)$ in (1.46) is also expected to be smooth, and to make the response of x_i computable from the invariant measure only. In the next section we investigate this possibility.

1.6 Coarse Graining Analysis

In terms of phase space probability measures, the response formula (1.43) reads:

$$\overline{\delta x_i}(t) = \int x_i \, \mathrm{d}\mu_t(\underline{x}; \delta \underline{x}_0) - \int x_i \, \mathrm{d}\mu(\underline{x}) \quad (1.47)$$

where $\mathrm{d}\mu_t(\underline{x}; \delta \underline{x}_0)$ is the time evolving perturbed measure whose initial state is given by

$$\mathrm{d}\mu_0(\underline{x}_0; \delta \underline{x}_0) = \rho_0(\underline{x}_0; \delta \underline{x}_0) \, \mathrm{d}\underline{x}_0 = \rho(\underline{x}_0 - \delta \underline{x}_0) \, \mathrm{d}\underline{x}_0.$$

Because dissipative dynamical systems do not have an invariant probability density, it is convenient to introduce a coarse graining in phase space, to approximate the singular invariant measure μ by means of piecewise constant distributions.

Let us consider a d-dimensional phase space $\mathscr{M}$, with an ϵ-partition made of a finite set of d-dimensional hypercubes $\Lambda_k(\epsilon)$ of side ϵ and centers $\underline{x}_k$. Introduce the ϵ-coarse graining of μ and of μ_t defined by the probabilities $P_k(\epsilon)$ and $P_{t,k}(\epsilon; \delta \underline{x}_0)$ of the hypercubes $\Lambda_k(\epsilon)$:

$$P_k(\epsilon) = \int_{\Lambda_k(\epsilon)} \mathrm{d}\mu(\underline{x}) \,, \quad P_{t,k}(\epsilon; \delta \underline{x}_0) = \int_{\Lambda_k(\epsilon)} \mathrm{d}\mu_t(\underline{x}; \delta \underline{x}_0). \quad (1.48)$$

This leads to the coarse grained invariant density $\rho(\underline{x};\epsilon)$:

$$\rho(\underline{x};\epsilon) = \sum_k \rho_k(\underline{x};\epsilon)\ , \quad \text{with} \quad \rho_k(\underline{x};\epsilon) = \begin{cases} P_k(\epsilon)/\epsilon^d & \text{if } x \in \Lambda_k(\epsilon) \\ 0 & \text{else} \end{cases} \tag{1.49}$$

Let Z_i be the number of bins of form $\left[x_i^{(q)} - \epsilon/2, x_i^{(q)} + \epsilon/2\right)$, $q \in \{1, 2, \ldots, Z_i\}$, in the i-th direction. Then, the marginalization of the coarse grained distribution yields the following set of Z_i probabilities:

$$p_i^{(q)}(\epsilon) = \int_{x_i^{(q)}-\frac{\epsilon}{2}}^{x_i^{(q)}+\frac{\epsilon}{2}} \left\{ \int \rho(\underline{x};\epsilon) \prod_{j\neq i} dx_j \right\} \mathrm{d}x_i = \mathrm{Prob}\left(x_i \in \left[x_i^{(q)} - \frac{\epsilon}{2}, x_i^{(q)} + \frac{\epsilon}{2}\right)\right) \tag{1.50}$$

each of which is the invariant probability that the coordinate x_i lie in one of the Z_i bins. In an analogous way, one may define the marginal of the evolving coarse grained perturbed probability $p_{t,i}^{(q)}(\epsilon;\delta\underline{x}_0)$. In both cases, dividing by ϵ, one obtains the coarse grained marginal probability densities $\rho_i^{(q)}(\epsilon)$ and $\rho_{t,i}^{(q)}(\epsilon;\delta\underline{x}_0)$, as well as the ϵ-coarse grained version of the response function $B_i(x_i, \delta\underline{x}_0, t)$:

$$B_i^{(q)}(x_i, \delta\underline{x}_0, t, \epsilon) = \frac{1}{\epsilon}\left[p_{t,i}^{(q)}(\epsilon, \delta\underline{x}_0) - p_i^{(q)}(\epsilon)\right] = \rho_{t,i}^{(q)}(\epsilon, \delta\underline{x}_0) - \rho_i^{(q)}(\epsilon) \tag{1.51}$$

In the following, we will show that the r.h.s. of (1.51) tends to a regular function of x_i in the $Z_i \to \infty$, $\epsilon \to 0$, limit. Then, in the limit of small perturbations $\delta\underline{x}_0$, $B_i^{(q)}(x_i, \delta\underline{x}_0, t, \epsilon)$ may be expanded as a Taylor series, to yield an expression similar to standard response theory, in the sense that it depends solely on the unperturbed state. The difference, here, is that the invariant measure is singular and represents a nonequilibrium steady state.

To illustrate this fact, we run a set of N trajectories with uniformly distributed initial conditions in the phase spaces of two simple, but substantially different, 2-dimensional maps: a dissipative baker map, and the Henon map.

1.6.1 *The Dissipative Baker Map*

Let us consider one of the simplest examples of dissipative discrete dynamical systems. Let $\mathscr{M} = [0, 1] \times [0, 1]$ be the phase space, and consider the evolution equation

$$\begin{pmatrix} x_{n+1} \\ y_{n+1} \end{pmatrix} = M \begin{pmatrix} x_n \\ y_n \end{pmatrix} = \begin{cases} \begin{pmatrix} x_n/l \\ r y_n \end{pmatrix}, & \text{for } 0 \le x_n < l; \\ \begin{pmatrix} (x_n - l)/r \\ r + l y_n \end{pmatrix}, & \text{for } l \le x_n \le 1. \end{cases} \tag{1.52}$$

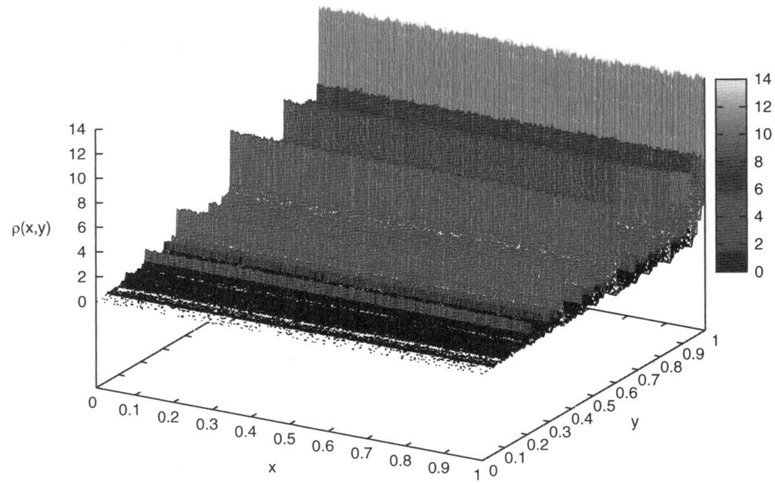

Fig. 1.20 Invariant probability distribution of the map defined by (1.52)

whose Jacobian determinant is given by

$$J_M(\underline{x}) = \begin{cases} J_A = r/l, & \text{for } 0 \le x \le l; \\ J_B = l/r = J_A^{-1}, & \text{for } l \le x \le 1. \end{cases} \quad (1.53)$$

and shows that the M is dissipative for $l < 1/2$. The map M is hyperbolic, since stable and unstable manifolds which intersect each other orthogonally are defined at all points $\underline{x} \in \mathscr{M}$, except in the irrelevant vertical segment at $x = l$. The directions of these manifolds coincide, respectively, with the vertical and horizontal directions. It can also be shown that this dynamical system is endowed with an invariant measure μ which is smooth along the unstable manifold and singular along the stable one, cf. Fig. 1.20. In particular, μ factorizes as $\mathrm{d}\mu(\underline{x}) = \mathrm{d}x \times \mathrm{d}\lambda(y)$, similarly to the case of [54].

In order to verify whether the functions corresponding to the above introduced $B_i^{(q)}(x_i, \delta\underline{x}_0, t, \epsilon)$ become regular functions in the fine graining limit, let us consider first an impulsive perturbation, directed purely along the stable manifold, i.e. $\delta\underline{x}_0 = (0, \delta y_0)$. Ruelle's work on singular measures is clearly relevant, in this case, because the support of the marginal perturbed probability measure, obtained projecting out the y-direction has simply drifted preserving its singular character, while the state may have fallen outside the support of the unperturbed invariant measure, cf. left panel of Fig. 1.21. Consider now an initial impulsive perturbation with one component, no matter how small, along the unstable manifold, $\delta\underline{x}_0 = (\delta x_0, \delta y_0)$ and rotate the vectors of the basis of the 2-dimensional plane, so that the coordinate x lies along the direction of the perturbation. We find that $B_x^{(q)}(x, \delta\underline{x}_0, t, \epsilon)$ is regular as a function of x. Indeed, the projections of μ and of its perturbations onto the direction of $\delta\underline{x}_0$ have a density along all directions except the vertical one. Hence, a small perturbation does not take the state outside the corresponding projected support.

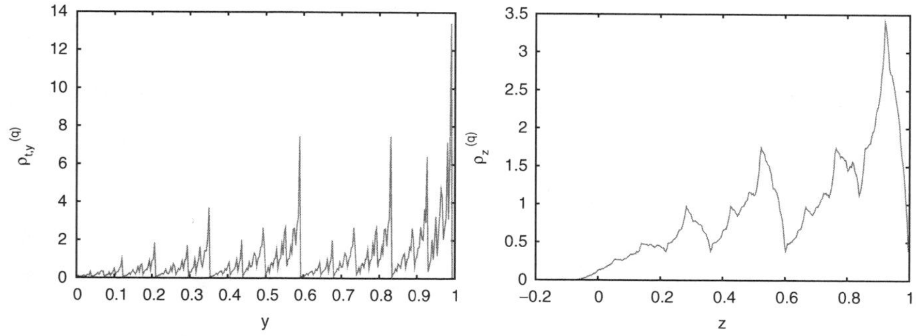

Fig. 1.21 *Left panel*: singular distribution of the unperturbed system, obtained projecting the invariant measure onto the vertical direction y, which is the direction of the stable manifold. *Right panel*: projected invariant density along the direction of the perturbation, which forms an angle $\alpha = \pi/8$ radiants with the y-direction. $z = 1 - (x\cos(\alpha) - y\sin(\alpha) + \sin(\alpha))$ is the coordinate along this direction. Although hardly differentiable, this projected distribution has a density

As already noted in [54], this baker map shows that the response to very carefully selected perturbations, cannot be computed in general from solely the invariant measure. However, similarly to the case of [54], the factorization of μ makes the present case rather peculiar. Indeed, for the overwhelming majority of dynamical systems, it looks impossible to select directions such that the projected measures preserve the same degree of singularity as the full measures. This is a consequence of the fact that stable and unstable manifolds have different orientations in different parts of the phase space, provided they exist. Clearly, the higher the dimensionality of the phase space and the larger the number of projected out dimensions, the more difficult it is to preserve singular characters.

1.6.2 The Henon Map

Consider for instance the Henon map defined by:

$$\begin{pmatrix} x_{n+1} \\ y_{n+1} \end{pmatrix} = M \begin{pmatrix} x_n \\ y_n \end{pmatrix} = \begin{pmatrix} y_n + 1 - a x_n^2 \\ b x_n \end{pmatrix}. \tag{1.54}$$

one the phase space $\mathcal{M} = [-\frac{3}{2}, \frac{3}{2}] \times [\frac{1}{2}, \frac{1}{2}]$, where $a = 1.4$ and $b = 0.3$ imply a chaotic dissipative dynamics, with a fractal invariant measure μ, which is not the product of the marginal measures obtained by projecting onto the horizontal and the vertical directions. These marginals are indeed regular and would yield a regular product. As stable and unstable manifolds wind around, changing orientation, in a very complicated fashion, it seems impossible, here, to disentangle the contributions of one phase space direction from the other.

Then, because no direction appears to be privileged in phase space, an initial perturbation along one of the axis should not lead to any singular perturbed projected

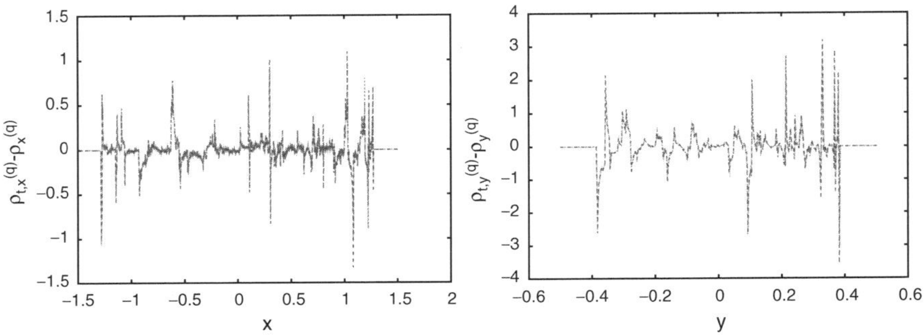

Fig. 1.22 *Left panel*: plot of the function $B_x^{(q)}(x, \delta \underline{x}_0, t, \epsilon)$, introduced by (1.46), for the Henon map, after $t = 2$ iterations, with an initial perturbation along the x direction, $\delta \underline{x}_0 = (2.5 \cdot 10^{-2}, 0)$, and different numbers of bins Z_x on the x-axis, as well as different numbers N of trajectories, to keep the statistics in the $\epsilon \to 0$ limit: $N = 3 \cdot 10^6$, $Z_x = 1320$ (*green curve*), $N = 5 \cdot 10^6$, $Z_x = 1650$ (*purple curve*), $N = 8 \cdot 10^6$, $Z_x = 1980$ (blue curve). *Right panel*: plot of the function $B_y^{(q)}(y, \delta \underline{x}_0, t, \epsilon)$ after $t = 2$ iterations, for the same initial perturbation considered in the *left panel*, and different values of bins Z_y on the y-axis and different values of N: $N = 3 \cdot 10^6$, $Z_y = 336$ (*green curve*), $N = 5 \cdot 10^6$, $Z_y = 420$ (*purple curve*), $N = 8 \cdot 10^6$, $Z_y = 505$ (*blue curve*). The *curves* largely overlap, but the figure does not clarify whether $B_x^{(q)}$, $B_y^{(q)}$ get smoother as Z_x and Z_y increase with N

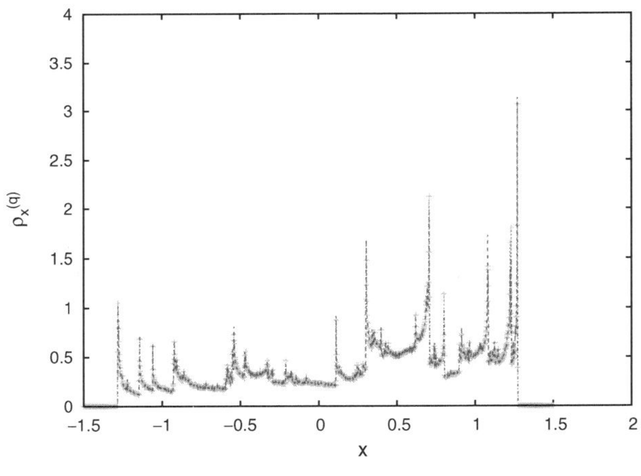

Fig. 1.23 Plot of the projected invariant probability density $\rho_x^{(q)}(\epsilon)$, for the Henon map, with $N = 5 \cdot 10^6$, $Z_x = 1650$ (*purple curve*, with *green errorbars*) and $N = 8 \cdot 10^6$, $Z_x = 1980$ (*blue curve*). The figure does not clarify whether the invariant distribution is singular or not

measure, or irregular response function, see e.g. [28, 29]. Unfortunately, this is not obvious from the histograms constructed with growing numbers of bins, as they seem to be quite irregular and to develop singularities in some parts of the phase space, cf. Figs. 1.22 and 1.23. However, this does not necessarily prevent the projected measures from having a density. Therefore, to clarify whether the

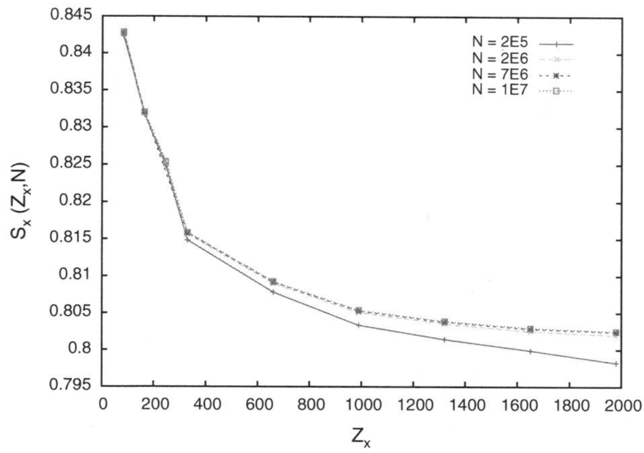

Fig. 1.24 Numerical simulations of the Shannon Entropy $S_x(\epsilon)$ for the Henon map, where Z_x denotes the number of bins considered on the x-axis and N denotes the number of trajectories. The *curves* collapse onto each other, approximating the asymptotic value of S_x associated with the $N, Z_x \to \infty$ limits

projected probability density $\rho_x^{(q)}(\epsilon)$ exists or not in the $\epsilon \to 0$ limit, we have examined the behavior of the Shannon Entropy, defined as

$$S_i(\epsilon) = -\epsilon \sum_{q=1}^{Z_i} \rho_i^{(q)}(\epsilon) \log(\rho_i^{(q)}(\epsilon)) \tag{1.55}$$

with ϵ the size of the bin along the direction of the perturbation. Note that this entropy is often defined differently; our definition is meant to introduce a quantity whose $\epsilon \to 0$ limit is finite if a density exists, while it diverges if the measure is singular. We approximated $S_x(\epsilon)$ by running different sets of trajectories, with different sizes of the coarse graining of the x-axis. Our simulations with $N = 2 \cdot 10^6$, show that S_x has substantially converged to its asymptotic $N \to \infty$ limit, cf. Fig. 1.24. Moreover, for fixed N, S_x decreases as the number Z_x of bins grows, and appears to tend to a constant as $1/Z_x \to 0$, cf. Fig. 1.25.

Figures 1.26 and 1.27 further prove that S_α is always a finite quantity in the Henon case while, in the baker case, it diverges logarithmically only when S_α tends to $\pi/2$, which is the only angle for which the projected invariant measure is singular. Therefore, the response can be obtained from the invariant measure at all perturbation angles in the case of the Henon map, and at all but a single angle for the Baker map. This confirms the applicability of a generalized FDT, which yields the response function in terms of the unperturbed state only, even if supported on a fractal set, except in very special situations, such as a negligible set in cases in which the invariant measure is the product of regular and singular measures. In particular,

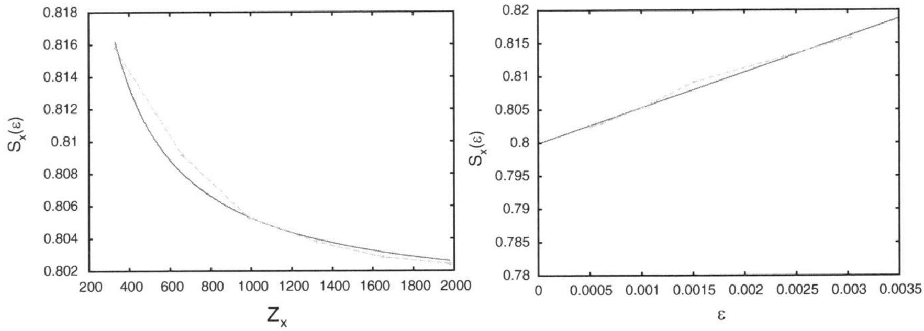

Fig. 1.25 *Left panel*: Fit of the numerical data for S_x corresponding to $N = 7 \cdot 10^6$, with $f(x) = a_0 + \frac{a_1}{Z_x}$, $a_0 = 0.7998$ and $a_1 = 5.3847$. *Right panel*: Same curve as in the *left panel*, plotted vs. the variable $\epsilon = 1/Z_x$, to extrapolate the asymptotic value $S_x \simeq 0.8$

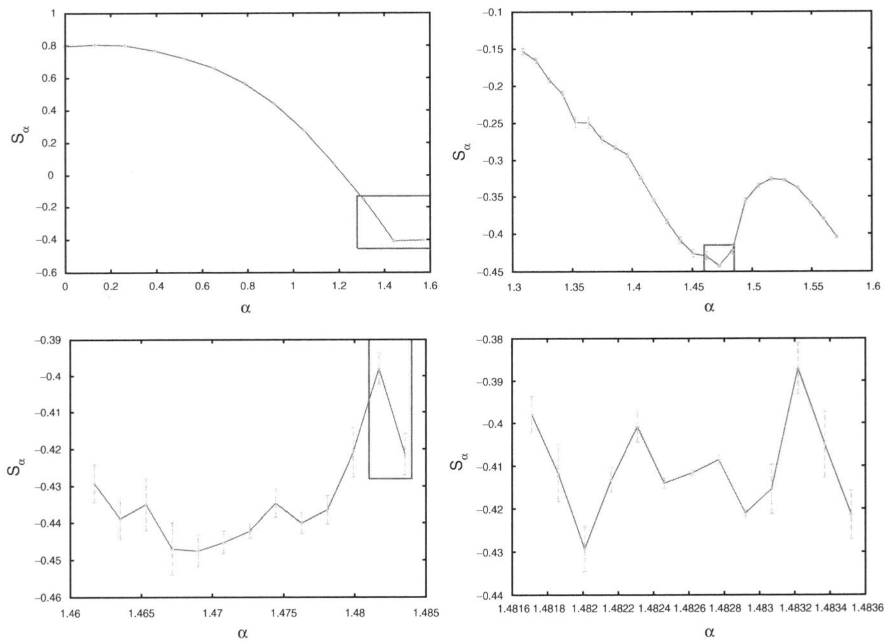

Fig. 1.26 Numerical data with *error bars*, for the Shannon Entropy S_α of the Henon map (1.54), as a function of the angle $\alpha \in [0, \pi/2]$ with respect to the horizontal direction, at increasing orders of magnification. The *top right panel* is a magnification of the framed part of the *top left panel*. The *bottom left panel* is a magnification of the framed part of the *top right panel*. The *bottom right panel* is a magnification of the framed part of the *bottom left panel*. The quantity S_α is quite structured, especially for α close to $\pi/2$, but it is a finite quantity

for the baker map, the response to a perturbation may be expressed just in terms of the smooth projected invariant measure if one does not perturb uniquely the vertical coordinates. For the Henon map, all directions lead to the existence of a projected invariant measure, although very finely structured.

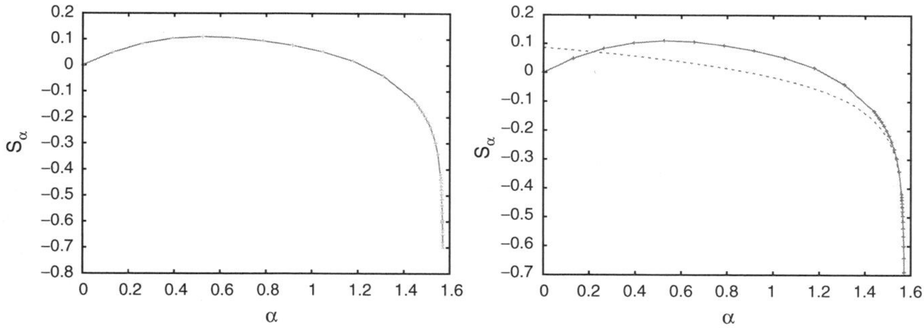

Fig. 1.27 *Left panel*: Numerical data, supplemented by error bars, for the Shannon Entropy S_α of the Baker map (1.52), as a function of the angle $\alpha \in [0, \pi/2]$ with respect to the x-axis. *Right panel*: Fit of S_α for the Baker map, with the curve $f(\alpha) = a \log(\frac{\pi}{2} - \alpha) + b$, $a = -0.101524 \pm 0.002768$, and $b = 0.0411212 \pm 0.01401$. In this case, S_α is a much simpler function of α than in the Henon case, and it diverges in the $\alpha \to \pi/2$ limit, because the projected invariant measure is singular at $\alpha = \pi/2$

1.7 Conclusions

In this work, we reviewed two fundamental tools of nonequilibrium statistical mechanics, the Fluctuation Relations and the Fluctuation-Dissipation Relations, from a novel perspective. We aimed at investigating, in particular, the relevance of the time-reversal symmetry in the derivation of the so-called Λ-FR as well as the extension of the standard FDT to dissipative dynamical systems. We focused on some simple deterministic chaotic maps which are amenable to a mathematical description. The map studied in Sects. 1.2–1.4, is a version of the classical baker model, proposed in [56], whose properties are determined by two parameters, one of which, q, may be suitably tuned in order to fix the distance from "equilibrium". This low-dimensional dynamical system represents just a caricature of a real particle system subjected to an external driving, but it can be studied in detail and is, therefore, useful to understand the role of the projection procedures used to obtain a coarser, stochastic-like, description from a microscopic, deterministic, one. If the specific observable of interest is, e.g., the phase space contraction rate Λ, our model allows one to readily identify the relevant and the irrelevant variables: the phase function Λ depends on the Jacobians of the mapping, and the Jacobians only depend on x, which is, hence, the only relevant variable. This implies that the addition of a source of irreversibility concerning only the "irrelevant" degree of freedom y, affects neither the equilibrium state, defined by $\langle \Lambda \rangle = 0$, nor the validity of the Λ-FR. In spite of the irreversibility of the phase space dynamics, the projection onto the space of the relevant variable not only produces a smooth marginal probability density but, as long as $\langle \Lambda \rangle = 0$, also yields a projected dynamics satisfying detailed balance. We also considered the validity of the Λ-FR when the considered irreversible dynamical system is pulled out of equilibrium. So far, the validity of the Λ-FR has been derived as a property of time-symmetric dynamical systems. Our analysis, also supported by numerical tests, showed that as long as the source of irreversibility affects only the "irrelevant" degrees of freedom, the Λ-FR and the transport laws (linear and nonlinear response) hold indistinguishably for both reversible and irreversible phase space dynamics. Our results extend, hence, to deterministic phase space dynamics some of the considerations reported in [57, 58], which refer to stochastic dynamics.

Next, in Sects. 1.5 and 1.6, we reviewed the methods introduced in [53] and in [6, 27, 55], concerning the derivation of response formulae for systems in nonequilibrium steady states. In particular, we showed that the approach traced in [27], based on the existence of a smooth invariant probability density, can be extended to dissipative deterministic dynamics [55].

The presence of noise, in any physically relevant dynamical system, does contribute to smoothing out the invariant density, but even in the absence of noise, the fact that statistical mechanics is typically interested in projected dynamics allows an approach to FDT which only requires the properties of the unperturbed states, as in standard response theory. Clearly, this is better and better justified as the dimensionality of the phase space grows. In particular, it is appropriate for macroscopic systems in nonequilibrium steady states, because the dynamics of interest take place in a space whose dimensionality is enormously smaller than that of the phase space. Then, as projecting out more and more produces smoother and smoother distributions, one finds that the approach of [27] can be used to obtain the linear response function for nonequilibrium steady states from the unperturbed measure only.

Our results support the idea that the projection procedure makes unnecessary the explicit calculation of the term discovered by Ruelle, which was supposed to forbid the standard approach. This does not mean that Ruelle's term is necessarily negligible [54]. The singularity of the phase space distribution may even have rather dramatic consequences [59]. However, except in very peculiar situations, such as our baker map which has carefully oriented manifolds, and for carefully chosen perturbations, that term does not need to be explicitly computed and the calculation of response may be carried out referring only to the unperturbed dynamics, as in the standard cases.

References

1. Castiglione, P., Falcioni, M., Lesne, A., Vulpiani, A.: Chaos and Coarse Graining in Statistical Mechanics. Cambridge University Press, London (2008)
2. Evans, D.J., Searles, D.J.: The fluctuation theorem. Adv. Phys. **52**, 1529 (2002)
3. Ruelle, D.: Smooth dynamics, new theoretical ideas in nonequilibrium statistical mechanics. J. Stat. Phys. **95**, 393 (1999)
4. Rondoni, L., Mejia-Monasterio, C.: Fluctuations in nonequilibrium statistical mechanics: models, mathematical theory, physical mechanisms. Nonlinearity **20**, R1–R37 (2007)
5. Klages, R.: Microscopic Chaos, Fractals and Transport in Nonequilibrium Statistical Mechanics. World Scientific, Singapore (2007)
6. Marconi, U.M.B., Puglisi, A., Rondoni, L., Vulpiani, A.: Fluctuation-dissipation: response theory in statistical physics. Phys. Rep. **461**, 111 (2008)

7. Jaksic, V., Pillet, C.-A., Rey-Bellet, L.: Entropic fluctuations in statistical mechanics I. Classical dynamical systems. Nonlinearity **24**, 699 (2011)
8. Evans, D.J., Cohen, E.G.D., Morriss, G.P.: Probability of second law violations in nonequilibrium steady states. Phys. Rev. Lett. **71**, 2401 (1993)
9. Gallavotti, G., Cohen, E.G.D.: Dynamical ensembles in nonequilibrium statistical mechanics. Phys. Rev. Lett. **74**, 2694 (1995)
10. Gallavotti, G., Cohen, E.G.D.: Dynamical ensembles in stationary states. J. Stat. Phys. **80**, 931 (1995)
11. Evans, D.J., Searles, D.J.: Equilibrium microstates which generate second law violating steady states. Phys. Rev. E **50**, 1645 (1994)
12. Bonetto, F., Gallavotti, G.: Reversibility, coarse graining and the chaoticity principle. Comm. Math. Phys. **189**, 263 (1997)
13. Lebowitz, J.L., Spohn, H.: A Gallavotti-Cohen-type symmetry in the large deviation functional for stochastic dynamics. J. Stat. Phys. **95**, 333 (1999)
14. Chetrite, R., Gawedzki, K.: Fluctuation relations for diffusion processes. Comm. Math. Phys. **282**(2), 469 (2008)
15. Searles, D.J., Rondoni, L., Evans, D.J.: The steady state fluctuation relation for the dissipation function. J. Stat. Phys. **128**, 1337 (2007)
16. Chechkin, A.V., Klages, R.: Fluctuation relations for anomalous dynamics. J. Stat. Mech. (2009) L03002
17. Jepps, O.G., Rondoni, L.: Deterministic thermostats, theories of nonequilibrium systems and parallels with the ergodic condition. J. Phys. A: Math. Theor. **43**, 1 (2010)
18. Gonnella, G., Pelizzola, A., Rondoni, L., Saracco, G.P.: Nonequilibrium work fluctuations in a driven Ising model. Physica A **388**, 2815 (2009)
19. Bonaldi, M., Conti, L., De Gregorio, P., Rondoni, L., Vedovato, G., Vinante, A., Bignotto, M., Cerdonio, M., Falferi, P., Liguori, N., Longo, S., Mezzena, R., Ortolan, A., Prodi, G.A., Salemi, F., Taffarello, L., Vitale, S., Zendri, J.P.: Nonequilibrium steady-state fluctuations in actively cooled resonators. Phys. Rev. Lett. **103**, 010601 (2009)
20. Rondoni, L., Morriss, G.P.: Large fluctuations and axiom-C structures in deterministically thermostatted systems. Open Syst. Inform. Dynam. **10**, 105 (2003)
21. Mejia-Monasterio, C., Rondoni, L.: On the fluctuation relation for Nos-Hoover boundary thermostated systems. J. Stat. Phys. **133**, 617 (2008)
22. Colangeli, M., Maes, C., Wynants, B.: A meaningful expansion around detailed balance. J. Phys. A: Math. Theor. **44**, 095001 (2011)
23. Lucarini, V., Colangeli, M.: Beyond the fluctuation-dissipation theorem: the role of causality. J. Stat. Mech.: Theor. Exp. (2012) P05013
24. Zwanzig, R.: Nonequilibrium Statistical Mechanics. Oxford University Press, London (2001)
25. Colangeli, M., Karlin, I.V., Kröger, M.: Hyperbolicity of exact hydrodynamics for three-dimensional Grad equations. Phys. Rev. E **76**, 022201 (2007)
26. Colangeli, M., Kröger, M., Öttinger, H.C.: Boltzmann equation and hydrodynamic fluctuations. Phys. Rev. E **80**, 051202 (2009)
27. Boffetta, G., Lacorata, G., Musacchio, S., Vulpiani, A.: Relaxation of finite perturbations: beyond the fluctuation dissipation relation. Chaos **13**, 3 (2003)
28. Evans, D., Rondoni, L.: Comments on the entropy of nonequilibrium steady states. J. Stat. Phys. **109**, 895 (2002)
29. Bonetto, F., Kupiainen, A., Lebowitz, J.L.: Absolute continuity of projected SRB measures of coupled Arnold cat map lattices. Ergod. Theor. Dynam. Syst. **25**, 59 (2005)
30. Colangeli, M., De Gregorio, P.M., Klages, R., Rondoni, L.: Steady state fluctuation relation with discontinuous "time reversibility" and invariant measures. J. Stat. Mech. (2011) P04021
31. Lasota, A., Mackey, M.C.: Chaos, Fractals, and Noise. Stochastic Aspects of Dynamics. Cambridge University Press, London (1985)
32. Roberts, J.A.G., Quispel, G.R.W.: Chaos and time-reversal symmetry. Order and chaos in reversible dynamical systems. Phys. Rep. **216**, 63 (1992)

33. Harris, R.J., Schütz, G.M.: Fluctuation theorems for stochastic dynamics. J. Stat. Mech. (2007) P07020
34. Morriss, G.P., Rondoni, L.: Equivalence of "nonequilibrium" ensembles for simple maps. Physica A **233**, 767 (1996)
35. Tolman, R.C.: The Principles of Statistical Mechanics. Oxford University Press, London (1938)
36. Falcioni, M., Palatella, L., Pigolotti, S., Rondoni, L., Vulpiani, A.: Initial growth of Boltzmann entropy and chaos in a large assembly of weakly interacting systems. Physica A **385**, 170–184 (2007)
37. Onsager, L.: Reciprocal relations in irreversible processes I. Phys. Rev. **37**, 405 (1931); Reciprocal relations in irreversible processes II. Phys. Rev. **38**, 2265 (1931)
38. Ellis, R.S.: An overview of the theory of large deviations and applications to statistical mechanics. Scand. Actuarial J. **1**, 97 (1995)
39. Rondoni, L., Morriss, G.P.: Applications of periodic orbit theory to N-particle systems. J. Stat. Phys. **86**, 991 (1997)
40. Evans, D.J., Searles, S.J., Rondoni, L.: On the application of the Gallavotti-Cohen fluctuation relation to thermostatted steady states near equilibrium. Phys. Rev. E **71**, 056120 (2005)
41. Gallavotti, G.: Extension of Onsager's reciprocity to large fields and the chaotic hypothesis. Phys. Rev. Lett. **77**, 4334 (1996)
42. Rondoni, L., Cohen, E.G.D.: Orbital measures in non-equilibrium statistical mechanics: the Onsager relations. Nonlinearity **11**, 1395 (1998)
43. Rondoni, L.: Deterministic thermostats and fluctuation relations. In: Dynamics of Dissipation. Lecture Notes in Physics, vol. 597, p. 35. Springer, Berlin (2002)
44. Rondoni, L., Cohen, E.G.D.: Gibbs entropy and irreversible thermodynamics. Nonlinearity **13**, 1905 (2000)
45. Rondoni, L., Tel, T., Vollmer, J.: Fluctuation theorems for entropy production in open systems. Phys. Rev. E **61**, R4679–R4682 (2000)
46. Cohen, E.G.D., Rondoni, L.: Particles, maps and irreversible thermodynamics. Physica A **306**, 117 (2002)
47. Gallavotti, G.: Breakdown and regeneration of time reversal symmetry in nonequilibrium statistical mechanics. Physica D **112**, 250–257 (1998)
48. Rondoni, L., Cohen, E.G.D.: On some derivations of irreversible thermodynamics from dynamical systems theory. Physica D **168–169**, 341 (2002)
49. Green, M.S.: Markoff random processes and the statistical mechanics of time-dependent phenomena. J. Chem. Phys. **20**, 1281 (1952)
50. Green, M.S.: Markoff random processes and the statistical mechanics of time-dependent phenomena: II. Irreversible processes in fluids. J. Chem. Phys. **22**, 398 (1954)
51. Kubo, R.: Statistical-mechanical theory of irreversible processes: I. General theory and simple applications to magnetic and conduction problems. J. Phys. Soc. Jpn. **12**, 570 (1957)
52. Cencini, M., Cecconi, F., Vulpiani, A.: Chaos: From Simple Models to Complex Systems. World Scientific, Singapore (2009)
53. Ruelle, D.: General linear response formula in statistical mechanics, and the fluctuation-dissipation theorem far from equilibrium. Phys. Lett. A **245**, 220 (1998)
54. Cessac, B., Sepulchre, J.-A.: Linear response, susceptibility and resonances in chaotic toy models. Physica D **225**, 13 (2007)
55. Colangeli, M., Rondoni, L., Vulpiani, A.: Fluctuation-dissipation relation for chaotic non-Hamiltonian systems. J. Stat. Mech. (2012) L04002
56. Colangeli, M., Rondoni, L.: Equilibrium, fluctuation relations and transport for irreversibile deterministic dynamics. Physica D **241**, 681 (2011)
57. Gabrielli, D., Jona-Lasinio, G., Landim, C.: Onsager reciprocity relations without microscopic reversibility. Phys. Rev. Lett. **77**, 1202–1205 (1996)
58. Gabrielli, D., Jona-Lasinio, G., Landim, C.: Onsager Symmetry from Microscopic TP Invariance. J. Stat. Phys. **96**, 639 (1999)
59. Cessac, B.: Does the complex susceptibility of the Hénon map have a pole in the upper-half plane? A numerical investigation. Nonlinearity **20**, 2883 (2007)

Part II
Monsoon Chaos and Wind Turbine System

Chapter 2
A Bond Graph Approach to Modeling and Simulation of Nonlinear Wind Turbine System

Tore Bakka and Hamid Reza Karimi

Abstract This chapter addresses the problem of bond graph methodology as a graphical approach for modeling of wind turbine generating systems. The purpose of this chapter is to show some of the benefits of the bond graph approach in contributing a model for wind turbine systems. We will present a nonlinear model of a wind turbine generating system, containing pitch, drive train, tower motion and generator. All which will be modeled by means of bond graph. We will especially focus on the drive train, and show the difference between modeling with a classical mechanical method and by using bond graph. The model consists of realistic parameters, but we are not trying to validate a specific wind turbine generating system. Simulations are carried out in the bond graph simulation software 20-sim [Kleijn, "20-sim 4.1 Reference Manual" Enschede, Controllab Products B.V. (2009). ISBN 978-90-79499-05-2].

2.1 Introduction

The demand for energy world wide is increasing every day. And in these green times renewable energy is a hot topic all over the world. Wind energy is currently the most popular energy sector. The growth in wind power industry has been tremendous over the last decade, its been increasing every year and it is nowadays one of the most promising sources for renewable energy. Since the early 1990s wind power

T. Bakka · H.R. Karimi (✉)
Department of Engineering, Faculty of Engineering and Science, University of Agder, 4879 Grimstad, Norway
e-mail: tore.bakka@uia.no; hamid.r.karimi@uia.no

S. Banerjee and L. Rondoni (eds.), *Applications of Chaos and Nonlinear Dynamics in Science and Engineering 3*, Understanding Complex Systems,
DOI 10.1007/978-3-642-34017-8_2, © Springer-Verlag Berlin Heidelberg 2013

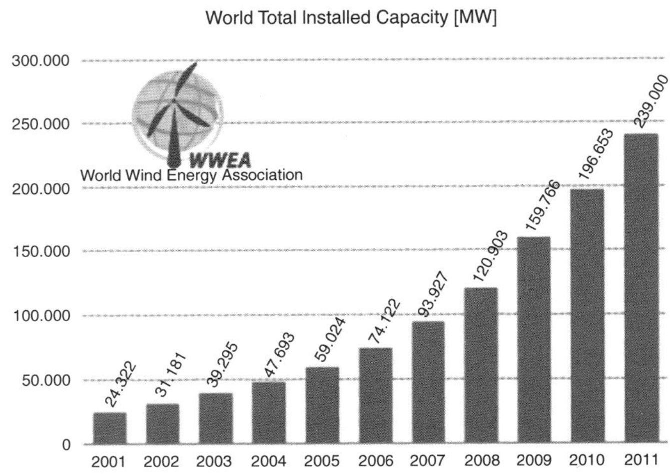

Fig. 2.1 World total installed capacity [MW]

has enjoyed a renewed interest, particularly in the European Union where the annual growth rate is about 20 %. The growth in the wind energy sector is illustrated in Fig. 2.1.

Whenever we are talking about models of wind turbine systems, the turbine model becomes a critical part of the discussion. Over the years it has been some discussion about how to model the wind turbine accurately. In [1–3] they perform dynamic analysis on a one-mass-model, in [4–7] they examine a two-mass-model. In [8] they use actual measured data from a wind turbine and compare it with both a one-mass and a two-mass-model. They validate the model using a recorded case obtained in a fixed speed, stall regulated wind turbine. In [9] a six-, three- and a two-mass model are compared. They argue that a six-mass-model is needed for the precise transient analysis of the wind turbine system, and they develop a way to transform a six-mass-model into a two-mass-model. The goal of that paper is not to use the model in the control scheme, but in the use of transient stability analysis of grid connected system.

The pitching of the blades are usually executed by means of a hydraulic system, but for system modeling purposes it is often considered as a first or second order system.

We are here dealing with variable speed generating system, therfore a wound machine or a double fed induction generator is needed. These can be modeled in different ways, ranging from complex electric equivalent circuits to a first order system.

Several advanced wind turbine simulation softwares has emerged during the last decade. HAWC2 [10], Cp-Lambda [11] and FAST [12] are a few examples. They are developed at RISØ in Denmark, POLI-Wind in Italy and NREL in the US, respectively. In these codes the turbine and structure is considered as complex flexible mechanisms, and uses the finite-element-method (FEM) multibody approach. An aero-servo-elastic model is introduced, which consists of aerodynamic forces from the wind, the servo dynamics from the different actuators and the elasticity in the different joints and the structure. Both FAST and HAWC2 can simulate offshore and onshore cases while Cp-Lambda is limited to the onshore case.

As seen above there are many ways to model a wind turbine generating system, some are simple and some are very complex. In a simulation point of view it is desirable that the model is as simple as possible and can capture as much of the dynamics as appear in reality. This is an absolute demand, another important issue is to keep the central processing unit (CPU) labor to a minimum. For example if we are dealing with hardware in the loop (HIL) simulation, then it is necessary to download the model to a programmable logic controller (PLC). This argues in favor of the importance in having a fast C-code. Things that can potentially have a negative effect on the execution of our C-code are for example algebraic loops and differential causality on the different elements in the system. These topics bring us to the use of the bond graph methodology. This is a unified approach to model all types of physical systems, producing both linear and nonlinear mathematical models. Engineers must work and interact in many different disciplines. An understanding of the intersections of these different disciplines is a valuable asset for any engineer. Using the language of bond graphs, one may construct models of electrical-, magnetic-, mechanical-, hydraulic-, pneumatic- as well as thermal-systems. It is a systematic way to model these dynamic systems, and there are standard ways to translate them into differential equations or computer simulation schemes. After constructing the bond graph one can easily spot algebraic loops and whether you have integral causality on the dynamic elements by inspecting the graph. There are various ways to spot these things in typical simulation software such as MATLAB, but it is beneficial to spot them before the implementation. It is a quite intuitive way in setting up the bonds and connecting the elements, this will be discussed in a later section. The outcome from the bond graph model is a set of first order differential equations, which afterwards can be used for systems response or for example controller design. After constructing the bond graph one gets a better understanding of what actually happens in the system. In an educational point of view one can easily understand which element decides what element in the system. For example in a simple mass-spring-damper system, one can easily see which component decides the speed and which component decides the force.

The wind turbine generating system can be divided into several subsystems, see Fig. 2.2.

The system setup is adopted from [13], where V_w is the wind speed, V_a is the wind speed for power production, $\dot{z}$ is the tower speed, F_t is the thrust force acting on the tower, β_{ref} is the pitch angle reference, β is the actual pitch angle, T_a is the aerodynamic torque, Ω_H is the hub speed, Ω_G is the generator speed, T_{EMref} is the generator torque reference and T_{EM} is the actual generator torque.

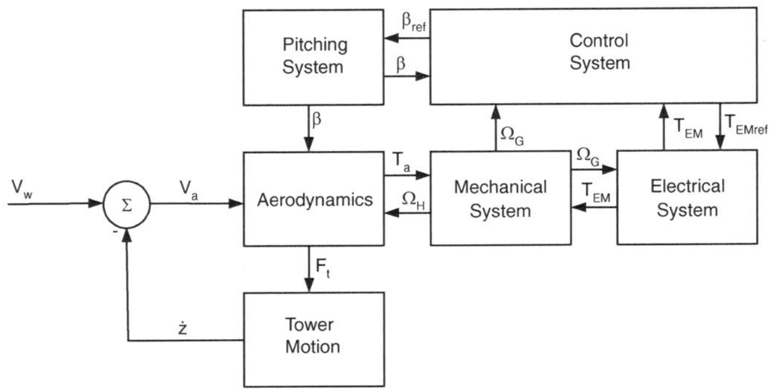

Fig. 2.2 Setup for wind turbine generating system

The expression for power produced by the wind is given by [14]:

$$P_a = \frac{1}{2}\rho\pi R^2 v^3 C_p(\lambda, \beta) \tag{2.1}$$

The dimensionless tip-speed ratio (TSR) λ is defined as:

$$\lambda = \frac{v_b}{v} \tag{2.2}$$

where v_b is the tip speed of the blade and v is the wind speed. From (2.1) we can find the aerodynamic torque and the thrust force acting on the tower:

$$T_a = \frac{1}{2}\rho\pi R^3 v^2 C_p(\lambda, \beta) \tag{2.3}$$

$$F_t = \frac{1}{2}\rho\pi R^2 v^2 C_T(\lambda, \beta) \tag{2.4}$$

where P_a is the aerodynamic power, ρ is the air density, R is the blade radius. C_p gives the relation between how much power is available in the wind and how much can be converted to electrical power. Not all the available power can be converted, this is due to the fact that the wind cannot be completely drained of energy, otherwise the wind speed at the rotor front would reduce to zero and the rotation of the rotor would stop. It can be proven that the theoretical upper limit of C_p is $16/27 \approx 0.59$, this is known as the Betz limit. A general modern wind turbine has a maximum power coefficient of about 0.5. C_t is the thrust force coefficient, both these coefficients are dependent on the TSR λ and the pitch angle β.

This chapter starts with an introduction to wind turbine generating system modeling. Section 2.2 gives a short overview on the bond graph methodology and its different elements. Section 2.3 describes the different parts of our system model; aerodynamics, pitch, drive train, tower motion and generator. Section 2.4 states the simulation results and Section 2.5 gives the conclusion and states some suggestions regarding future work.

2.2 Introduction to Band Graph

Bond graph is a graphical way of modeling physical systems. All these physical systems have in common the conservation laws for mass and energy. Bond graph, originated by Paynter [15] in 1961, deals with the conservation of energy. This gives a unified approach to modeling physical systems. Further follows a short introduction to this modeling tool, more information can be found in [16, 17]. The bond graph based modelling has several advantages over conventional simulation methods as follows: (1) providing a visual representation of the design; (2) controling the consistency of the topological settings of the design; (3) providing the hierarchical modelling of designs; (4) extracting the system equations symbolically in a structured way.

Within physical systems, energy is transported form one item to another. This energy is either stored or converted to other forms. But the important thing is that it does not dissipate. If the energy is changing in one place, it also changes in an opposite way at another location. The definition of power is the change in energy (E) with respect to time:

$$P = \frac{d}{dt}(E) \tag{2.5}$$

This power is transferred between the different parts in bond graph model with the use of power bonds, see Fig. 2.3. Power can be expressed as the product of an effort and a flow variable, thus the general expression:

$$P(t) = e(t)f(t) \tag{2.6}$$

The symbols $e(t)$ and $f(t)$ are used to denote effect and flow quantities as functions of time. Table 2.1 shows what the effort and flow quantities can be in some familiar domains.

2.2.1 *System Elements*

In bond graph modeling there are a total amount of nine different elements. We will also here introduce the causality assignments, but first we have to explore the cause and effect for each of the basic bond graph elements. Only elements with its preferred causality will be discussed. The importance of causality will be dealt with later in the chapter.

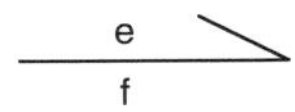

Fig. 2.3 Power bond with effort and flow

Table 2.1 Effort and flow variables in different domains

Domain	Effort, $e(t)$	Flow, $f(t)$
Mechanical translation	Force, $F(t)$	Velocity, $v(t)$
Mechanical rotation	Torque, $\tau(t)$	Angular velocity, $\omega(t)$
Hydraulic	Pressure, $P(t)$	Volume flow rate, $Q(t)$
Electric	Voltage, $e(t)$	Current, $i(t)$

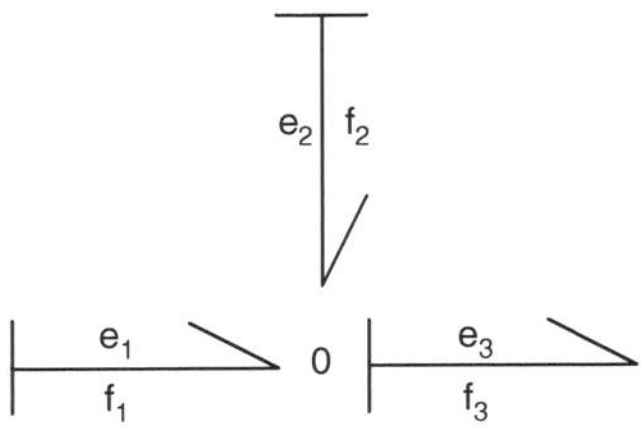

$$\begin{aligned} e_1 &= e_2 = e_3 \\ f_3 &= f_1 + f_2 \end{aligned} \tag{2.7}$$

Fig. 2.4 0-junction

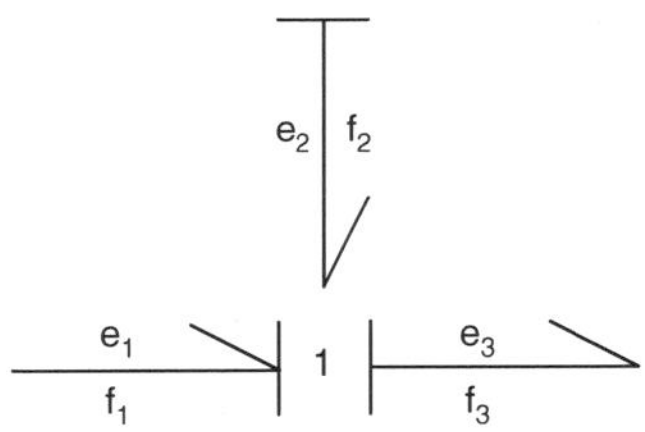

$$\begin{aligned} f_1 &= f_2 = f_3 \\ e_3 &= e_1 + e_2 \end{aligned} \tag{2.8}$$

Fig. 2.5 1-junction

2.2.1.1 Junctions

There are two different types of junctions that connects the different parts in a bond graph model, the 0-junction and the 1-junction. The 0-junction is an effort equalizing connection, see Fig. 2.4 and its corresponding equation in (2.7). Since the efforts are the same, only one bond can decide what it is. The 1-junction is a flow equalizing connection, see Fig. 2.5 and its corresponding equation in (2.8). Since the flows are the same, only one bond can decide what it is. Which bond decides the flow and which one decides the effort is indicated with the vertical causality stroke. If the vertical line is closest to the junction, then this element decides the effort, furthest away from the junction decides the flow.

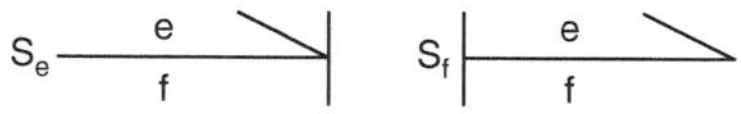

Fig. 2.6 Effort and flow source with their causality assignment

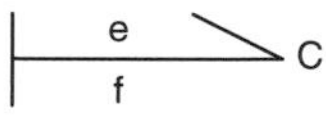

Fig. 2.7 Example of a compliance element with integral causality

$$
\begin{aligned}
e &= \frac{1}{C}\int f dt \\
&= \frac{q}{C}
\end{aligned}
\tag{2.9}
$$

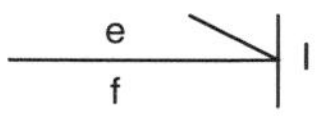

Fig. 2.8 Example of an inertia element with integral causality

$$
\begin{aligned}
f &= \frac{1}{I}\int e dt \\
&= \frac{p}{I}
\end{aligned}
\tag{2.10}
$$

2.2.1.2 Source Element

We can divide the source elements into two different kinds, effort- and flow-source. The effort source gives an effort into the system, then it is up to the system to decide the flow. This is what is meant with cause and effect, and its vice versa for the flow source. Figure 2.6 shows how the causality is indicated on the graphical elements. For the source elements these causality assignments are fixed.

2.2.1.3 Compliance Element

The causality assignment for the C-element has two possibilities, but one is preferred in contrast to the other. This is discussed at the end of this section. The preferred case is seen in Fig. 2.7 and its corresponding equation in (2.9). We see from both the equation and the figure that flow is given to the element/equation and it gives the effort in return.

The variable q is called *the generalized displacement*. For example, this can be rotational position of the rotor in a wind turbine.

2.2.1.4 Inertia Element

There are two choices for the causality assignment for the I-element, also here one is preferred in contrast to other. The preferred case is seen in Fig. 2.8 and its corresponding equation in (2.10).

The variable p is called *the generalized momentum*. For example, this can be rotor inertia times rotor velocity in a wind turbine.

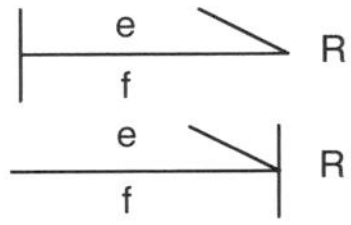

Fig. 2.9 Example of resistive elements

$$e = Rf$$
$$f = \frac{1}{R}e \tag{2.11}$$

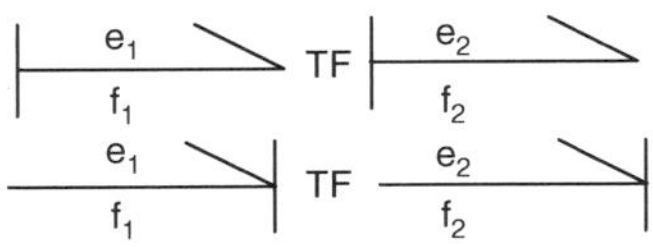

Fig. 2.10 Example of the two transformers

$$e_1 = me_2$$
$$f_2 = mf_2 \tag{2.12}$$

$$e_2 = \frac{1}{m}e_1$$
$$f_1 = \frac{1}{m}f_2 \tag{2.13}$$

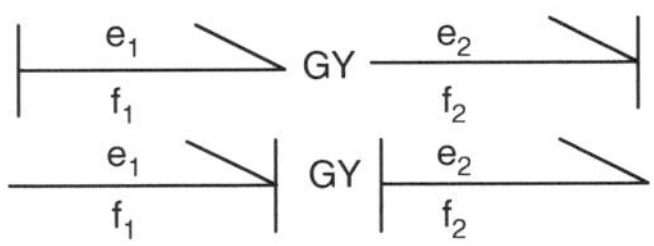

Fig. 2.11 Example of the two gyrators

$$e_1 = rf_2$$
$$e_2 = rf_1 \tag{2.14}$$

$$f_1 = \frac{1}{r}e_2$$
$$f_2 = \frac{1}{r}e_1 \tag{2.15}$$

2.2.1.5 Resistive Element

It is a bit more freedom when it comes to the causality assignment for the R-element. Its equation do not include any dynamics, it is only an algebraic expression. The two causality choices are shown in Fig. 2.9 and its corresponding equation in (2.11).

2.2.1.6 Transformer

The transformer element can work in two ways; either it transforms a flow into another flow or it transforms an effort into another effort. Figure 2.10 corresponds to (2.12) and (2.13), where m is the transformation ratio.

For example, this can represent a mechanical gearing or an electric transformer.

2.2.1.7 Gyrator

The gyrator can also work in two ways; either it transform a flow into an effort or it transform an effort into a flow. Figure 2.11 corresponds to (2.14) and (2.15), where r is the gyrator ratio.

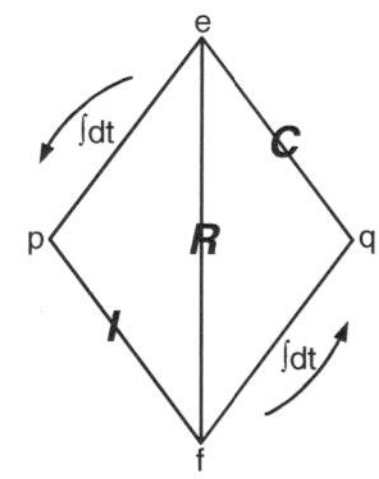

Fig. 2.12 The tetrahedron of state

This can for example be an electric motor, where you have voltage as input and a rotational speed as output. The importance of integral causality is nicely explained in [18]. First imagine a step in effort is imposed on a C-element, then the causality assignment will be opposite of what is shown in Fig. 2.7. This means the flow output is proportional to the derivative of the input effort. From calculus we know that the derivative of the step function at the beginning is infinite, i.e. this do not give any physical meaning. We can imagine a simple electric circuit containing a voltage source coupled with a capacitor, if a step input were to be imposed on the voltage source, the capacitor would experience a very high current and it would blow up. From this we can conclude that nature integrates and only mathematicians differentiate!

On the other hand, the ability to spot algebraic loops is one of the benefits with the use of bond graph as a modeling tool. These loops can be spotted simply by inspection of bond graph representation, if the causality assignment on the R-elements are different from each other, then we have algebraic loops in the system. If they have the same causality, there are no algebraic loops. These loops occur for example if you have two resistors in series. In this circuit both resistors will try to decide on the current, i.e. they depend on each other. This will not necessarily cause problems to the simulation, but it might. Especially if the resistors are nonlinear, then the simulation could easily crash. The simulation program will also be forced to spend time to solve this algebraic loop. If we can easily spot these loops early in the modeling process, then we can try to fix them by simply adding an element. For example, regarding our circuit with two resistors in series, we can add an inductive element to the circuit. Then it would be the inductive element who decides on the current and not the resistive elements, the resistive elements would simply have to take what current the inductive element lets through. We can give the inductive element a value such that the voltage drop over the element is very low, i.e. it does not play any major role in the circuit. Now when our model has no algebraic loops and all the dynamic elements have integral causality, the simulation should go smooth. If we have a large set of equations or a Simulink model it is not easy to spot these things right away, but with a bond graph representation of our model we can spot them simply by inspection.

To simple remember the aforementioned relations we can use what Paynter called *the tetrahedron of state*, shown in Fig. 2.12.

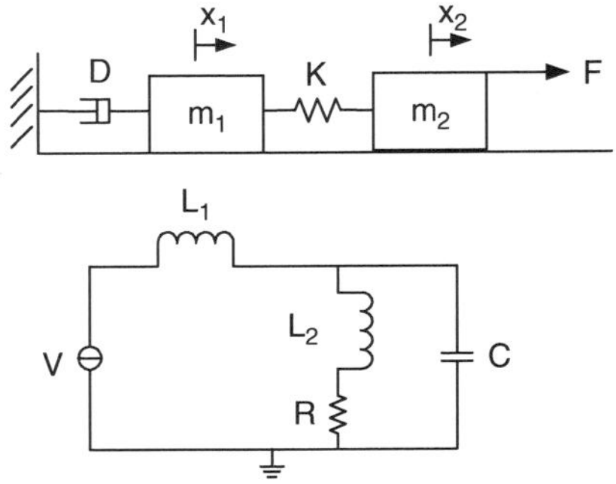

Fig. 2.13 Two equivalent circuits

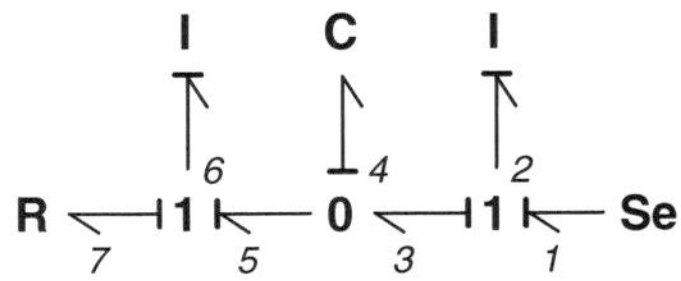

Fig. 2.14 Bond graph of the two equivalent circuits

The procedure of how to extract the algebraic and dynamic equations from a bond graph model is not included in this short overview, but it can be done in a very systematic way and it will partly be shown in the next section.

We will end this section with a small example. We want to show how to set up a bond graph of a simple system, and also show the difference in relation to block diagrams. Figure 2.13 shows two equivalent circuits in two different domains, and they have exactly the same governing equations. The corresponding bond graph is shown in Fig. 2.14. The easiest way to set up a bond graph when having a mechanical system, is to start with setting up 1-junctions. One junction for each mass, this gives two 1-junctions in our example. We add a 0-junction in between, because we know the speed is different but the force is the same. Force is transfered through the C-element (spring). The right side of the damper has the same speed as m_1, R-element and I-element is therefor connected to the left 1-junction.

Regarding the electric circuit, we know that the source and L_1 have the same current i_1. We know that L_2 and R have the same i_2, and we know that the parallel branches have the same voltage. In this way we end up with the exact same bond graph. We also note that the graph has integral causality. The two I-elements receive effort and give flow in return, the C-element receives flow and gives effort in return.

We will now find the governing equations. First we find $\dot{p}_i$, second we find $\dot{q}_i$. In mechanical terms this is $m\ddot{x}_i$ and $\dot{x}$, respectively. Subscript i corresponds to in which bond we are at.

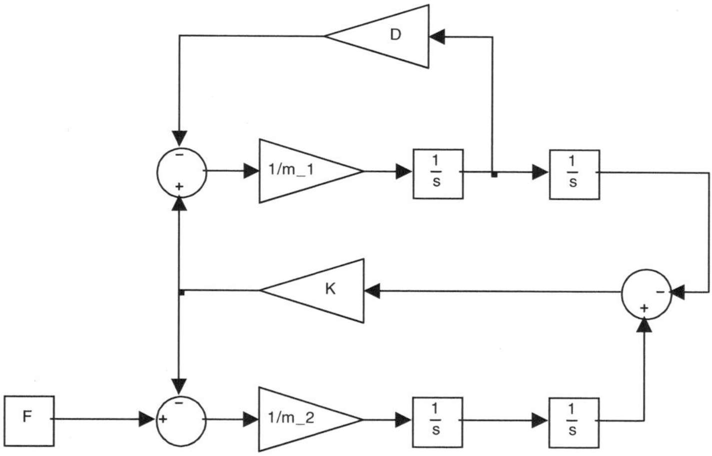

Fig. 2.15 Block diagram of mechanical example

$$\dot{p}_2 = e_2 = e_1 - e_3 = S_e - e_4 = S_e - \frac{q_4}{C_4} \tag{2.16a}$$

$$\dot{p}_6 = e_6 = e_5 - e_7 = e_4 - R_7 f_7 = \frac{q_4}{C_4} - R_7 f_6 \tag{2.16b}$$

$$= \frac{q_4}{C_4} - R_7 \frac{p_6}{I_6}$$

$$\dot{q}_4 = f_4 = f_3 - f_5 = f_2 - f_6 = \frac{p_2}{I_2} - \frac{p_6}{I_6} \tag{2.16c}$$

In mechanical domain terms, (2.16) correspond to (2.17).

$$m_2 \ddot{x}_2 = F - K x_0 \tag{2.17a}$$

$$m_1 \ddot{x}_1 = k x_0 - D x_1 \tag{2.17b}$$

$$x_0 = x_2 - x_1 \tag{2.17c}$$

These are exactly the same equations we will end up with if we do it in the classical Newtons 2nd law approach. The block diagram for these equations are shown in Fig. 2.15. Block diagrams represent the structure of the mathematical model and displays which variables must be known in order to compute others. They do not reflect the physical structure. The reason is that feedback is represented in separate feedback loops.

By using bond graph as the modeling tool we get a good overview of the model's physical structure and we can do simulations in one step, instead of first deriving the equations and then drawing the block diagram.

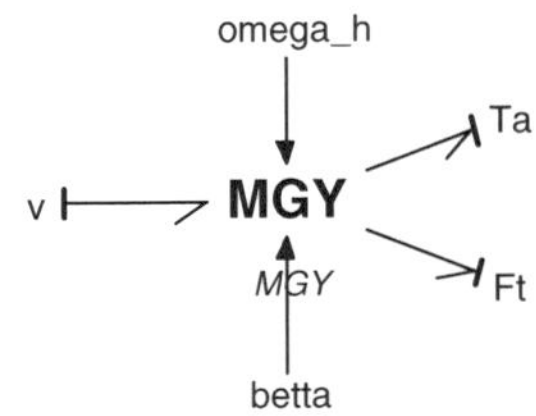

Fig. 2.16 Modulated gyrator transforming wind speed into aerodynamic torque and thrust force

2.3 Model Description

In the following section, the bond graph based modelling for the different subsystems, shown in Fig. 2.2, will be presented. It is shown that the bond graph method provides a hierarchical modeling for the entire wind turbine generating system as well as the system equations can be extracted symbolically in a structured way.

2.3.1 Aerodynamics

In the aerodynamics part we need to find a way to convert the wind into torque and thrust force, i.e. transform a flow into efforts. This is done by means of a modulated gyrator. We use the torque and thrust equations given in (2.3) and (2.4). The only difference between a MGY and a GY is that the gyrator ratio is not a constant parameter, but it is a varying parameter. In this case the transformation is dependent on two varying parameters, the pitch angle β and the rotor rotational speed ω_r (Fig. 2.16).

A generic equation is used to model C_p. This equation, based on the modeling turbine characteristics of [19], is shown in (2.18). The power coefficient used in the calculation of the torque is given in (2.18). A plot of the C_p curve is shown in Fig. 2.17, the plot is made with different pitch- and λ- values. Similar formulas can be found regarding the thrust force coefficient C_T, in our calculations only a simple relation is used.

$$\lambda = \frac{\omega_r R}{v} \tag{2.18a}$$

$$\lambda_i = \frac{1}{\frac{1}{\lambda + 0.08\beta} - \frac{0.035}{\beta^2 + 1}} \tag{2.18b}$$

$$C_p = c_1 \left(\frac{c_2}{\lambda_i} - c_3\beta - c_4 \right) e^{-\frac{c_5}{\lambda_i}} + c_6\lambda \tag{2.18c}$$

where $c_1 = 0.5176, c_2 = 116, c_3 = 0.4, c_4 =, c_5 = 21, c_6 = 0.0068$.

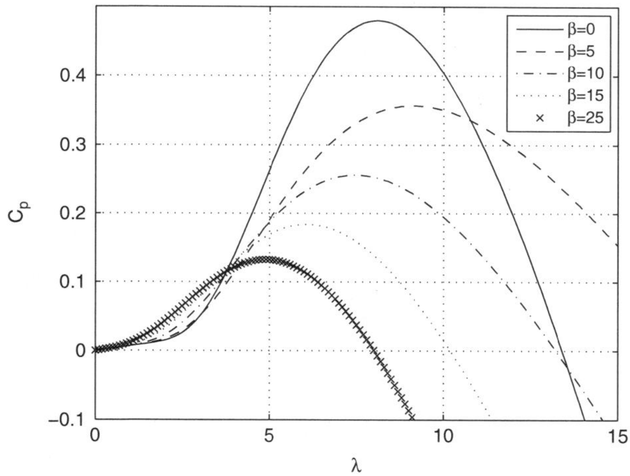

Fig. 2.17 C_p curve

Fig. 2.18 Mass spring damper

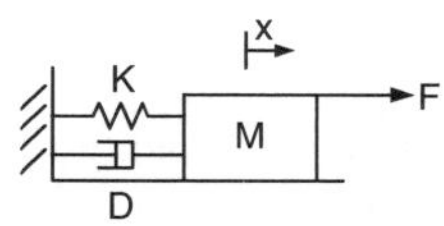

2.3.2 Pitching System

The pitching mechanism can be modeled as a second order system;

$$\omega_n^2 \theta_{ref} = \ddot{\theta} + 2\zeta\dot{\theta} + \omega_n^2 \theta \tag{2.19}$$

where θ_{ref} is the reference pitch angle, ω_n is the natural frequency and ζ is the damping ratio. By setting up the dynamic equation of the mass spring damper system in Fig. 2.18, we can compare the elements in the equation with (2.19). In this way we can set up the bond graph in Fig. 2.19 with appropriate coefficients.

$$F = \theta_{ref} \ , \ M = \frac{1}{\omega_n^2} \ , \ D = \frac{2\zeta}{\omega_n^2} \ , \ K = 1$$

2.3.3 Drive Train

A sketch of a two-mass drive train model is seen in Fig. 2.20. As discussed in the introduction there are many types of drive train models, ranging from for example one- to six-mass models. For simplicity we will assume a two-mass-model is enough. To derive the governing equations out of a two-mass-model is not too

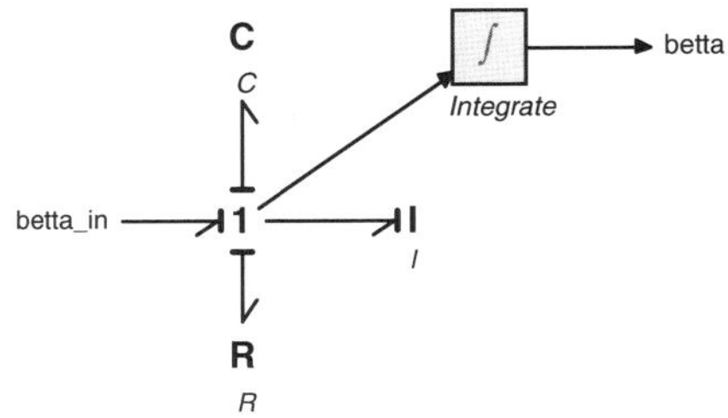

Fig. 2.19 Bond graph of pitching system

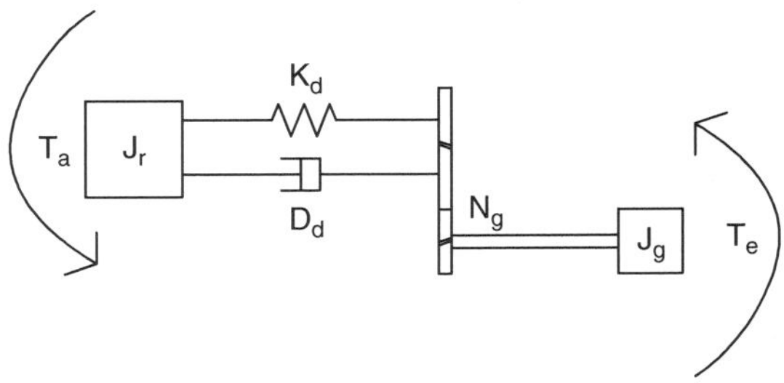

Fig. 2.20 Sketch of wind turbine

hard, but if we are talking about a six-mass-model the work can be quite extensive, and the possibility of making a mistake in the process is high. This is one of the reasons bond graph is a safer choice. As the complexity of the mechanical system grows, our work as modelers stays about the same. If we have a six-mass-model with many springs and dampers, this gives us many equations and to translate this into a block diagram can take quite some time. As for dealing with bond graph the work is to set up the graphical representation, and if we want to see the equations, these can be derived in a very specific way. Or, off course we can choose to get them from our bond graph simulation program 20-sim.

By utilizing Newton's second law on rotational form of the wind turbine sketch in Fig. 2.20, we end up with the following differential equations:

$$T_r = I_r\dot{\omega}_r + \dot{\phi}_\Delta D_d + \phi_\Delta K_d \tag{2.20a}$$

$$-T_g N_g = I_g N_g^2 \frac{\dot{\omega}_g}{N_g} - \dot{\phi}_\Delta D_d - \phi_\Delta K_d \tag{2.20b}$$

where

$$\phi_\Delta = \phi_r - \frac{\phi_g}{N_g}, \; \dot{\phi}_\Delta = \omega_r - \frac{\omega_g}{N_g}$$

In a quite intuitive way we can translate the mechanical system in Fig. 2.20 into a bond graph representation, as shown in Fig. 2.21. This can again be simplified a bit in order to make a minimal bond graph representation, see Fig. 2.22. The bond graph model consists of two 1-junctions and one 0-junction. The 1-junction connected to the rotor inertia describes the rotor rotational speed. Since there are

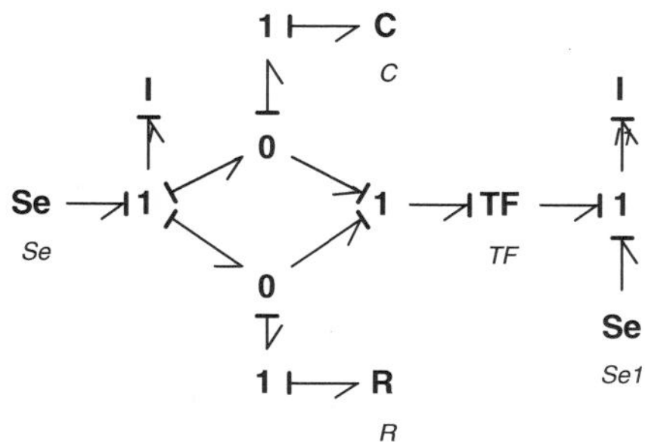

Fig. 2.21 Bond graph of drive train

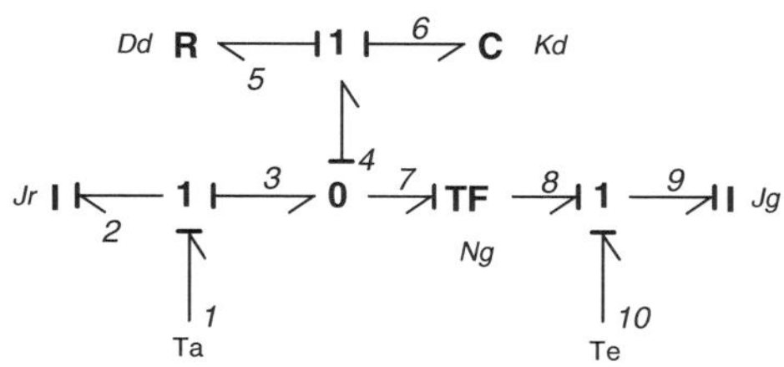

Fig. 2.22 Minimal bond graph of drive train

dynamics in between the rotor inertia and the generator inertia they do not have the same speed. This is the reason for the 0-junction, because we know the transferred torque is the same (no loss included in the drive train). The 1-junction connected to the resistive- and the compliance- element indicates the rotational speed difference between the two inertias. This connection also indicates that the compliance- and resistive- element have the same rotational speed (flow), but different torque (effort).

Once the bond graph representation is made, the procedure for extracting its governing equations is quite straight forward. One have to follow some certain rules, and at the end the equations will be the outcome. We can also choose to get the equations from the simulation software. The equations can be derived as follows.

From the bond graph representation we see there are three dynamic elements, two inertias and one spring, i.e. three dynamic equations must exist. These first order differential equations are given in (2.21).

$$\dot{p}_2 = e_2 = e_1 - e_3$$

$$= T_a - \frac{q_5}{C_6} - R_5 f_5 \tag{2.21a}$$

$$\dot{q}_5 = f_5 = f_3 - f_7$$

$$= \frac{p_2}{I_2} - N_g \frac{p_9}{I_9} \tag{2.21b}$$

$$\dot{p}_9 = e_9 = e_8 + e_{10}$$

$$= -T_e + \frac{1}{N_g}\left(\frac{q_5}{C_6} + R_5 f_5\right) \tag{2.21c}$$

With some manipulations this is exactly the same as in (2.20).

Fig. 2.23 Sketch of wind turbine structure

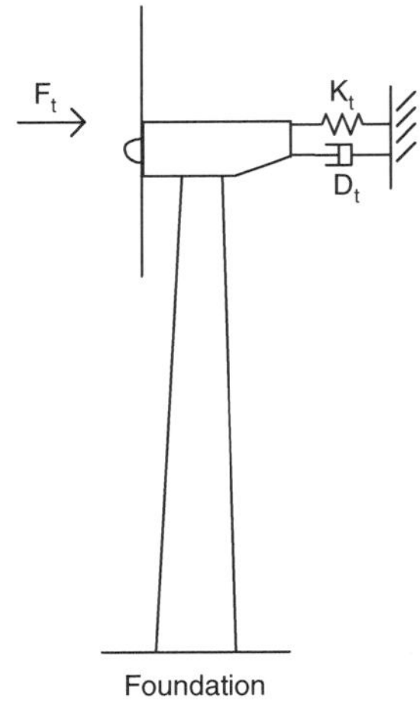

2.3.4 Generator

There are many ways to model the generator dynamics. One of the recurring ways is with equivalent circuit. In our system we assume a first order transfer function will capture its dynamics. We do this in the same way as for the pitching system, but since it is first order we do not include the spring,

$$T_{ref} = \tau \dot{T}_e + T_e \tag{2.22}$$

where T_{ref} is the reference torque and τ is the time constant. In this way we can set up the bond graph similar to Fig. 2.19 with appropriate coefficients.

$$F = T_{ref} \ , \ M = \tau \ , \ D = 1$$

2.3.5 Tower

It is assumed that the tower movement do not influence the mechanical system, it only affects its input, i.e. the wind speed (Fig, 2.23). The bond graph model of the tower can be seen in Fig. 2.24. Since the deflections of the tower are assumed to be small, we assume tower movement only in horizontal direction.

The dynamic equation from the bond graph model, shown in Fig. 2.24, is given in (2.23).

$$\dot{p}_2 = S_e - R\frac{p_2}{I} - \frac{q_3}{C} \tag{2.23a}$$

$$\dot{q}_3 = \frac{p_2}{I} \tag{2.23b}$$

Fig. 2.24 Bond graph of tower motion

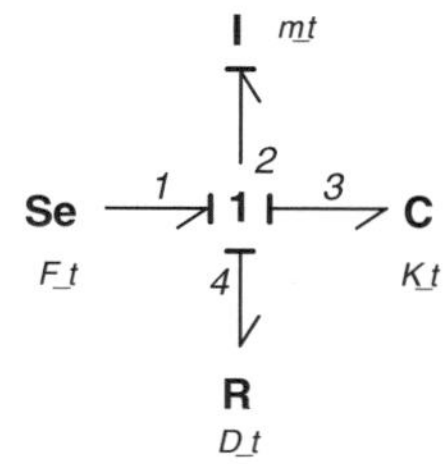

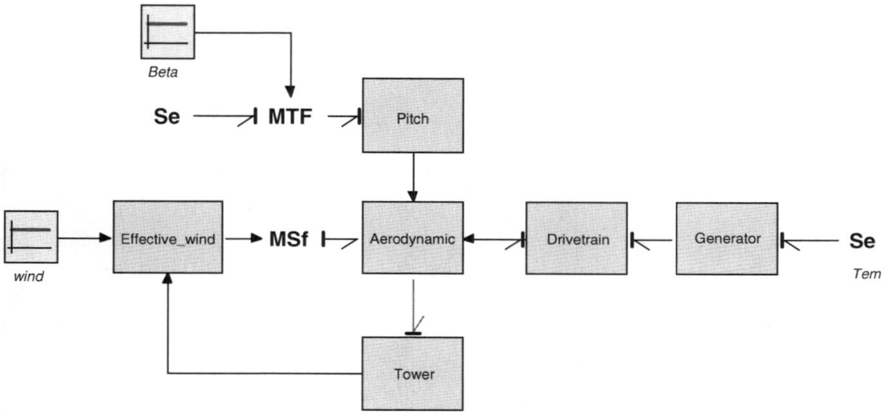

Fig. 2.25 Bond graph of wind turbine generating system

We can rewrite (2.23) in a non bond graph notation:

$$m_t \ddot{z} = F_t - D_t \dot{z} - K_t z \tag{2.24}$$

where m_t is the tower mass, F_t is the thrust force acting on the tower, D_t is the tower damping and K_t is the tower stiffness.

2.4 Simulation Results

In this section we want to validate the bond graph design. The procedure is to first connect together all the subsystems from Section III. Second, the same model is implemented as block diagrams in MATLAB/Simulink. This software is widely established throughout the academic community and the result from MATLAB/Simulink will act as a reference output for validation purpose.

The bond graph representation of the system setup in Fig. 2.2 is shown in Fig. 2.25. The inputs to the systems are pitch angle, reference power and wind speed. The simulations are made with maximum pitch angle, maximum wind condition, maximum power and with initial conditions on the rotor and generator. All wind turbine parameters used in the simulations are found in [20]. Once the simulations

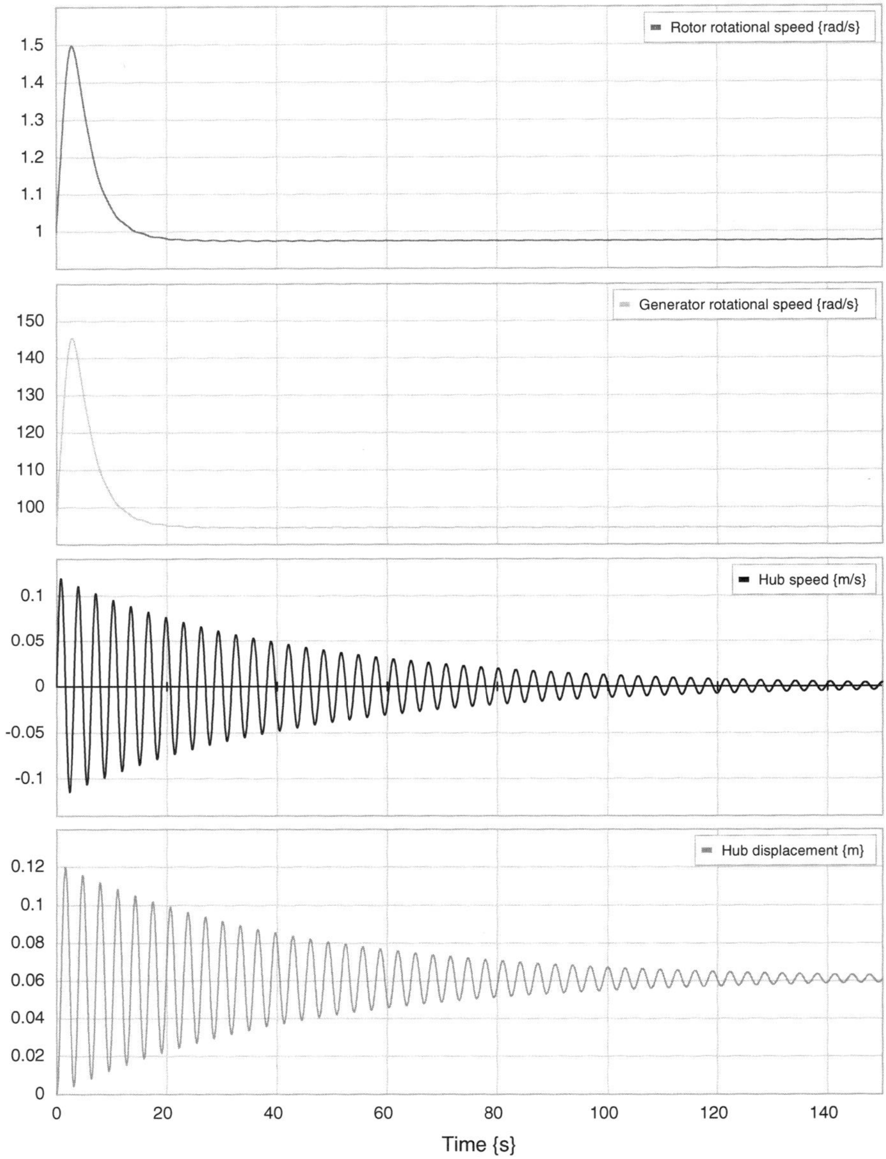

Fig. 2.26 Time behavior of the selected signals from 20-sim

are carried out in the two softwares, time behavior of the most important dynamics are inspected. As seen in Figs. 2.26 and 2.27, the behavior of the two systems are identical. This confirms the fact that we eventually end up with the same governing equations whether one uses the classical Newton's $2nd$ law or the bond graph approach.

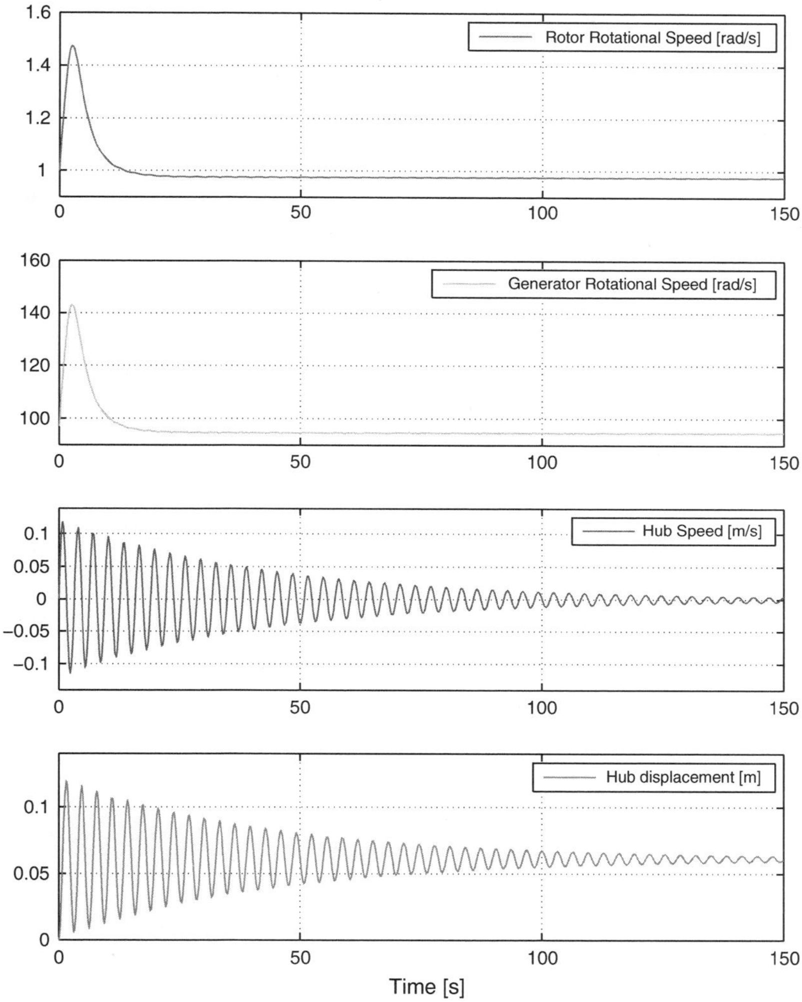

Fig. 2.27 Time behavior of the selected signals from MATLAB/Simulink

2.5 Conclusions

The purpose of this chapter is to make a nonlinear model of a wind turbine generating system by using the bond graph approach. We are not looking to validate a specific turbine system, but we want to show a simple and suitable way to model it. The nonlinear wind turbine consists of drive train, pitching system, tower and generator. Modeling dynamic systems in the classical way and the bond graph way is quite different, but the outcome is exactly the same governing equations. We have tried to emphasize that using the bond graph approach will give a better understanding of what actually happens in the system. It is a unified approach

Table 2.2 Wind turbine generating system parameters

Pitch	
Natural frequency	$\omega_n = 0.88\ [\frac{rad}{s}]$
Damping ratio	$\zeta = 0.9[-]$
Maximum pitch angle	$\beta_{max} = 25\ [°]$
Minimum pitch angle	$\beta_{min} = -5\ [°]$
Drive train	
Nominel power	$P_{nom} = 5e6\ [W]$
Rotor inertia	$I_r = 5.9154e7\ [Kg \cdot m^2]$
Generator inertia	$I_g = 500\ [Kg \cdot m^2]$
Drive train stiffness	$K_d = 8.7354e8\ [\frac{N}{rad}]$
Drive train damping	$D_d = 8.3478e7\ [\frac{N}{rad \cdot s}]$
Gear ratio	$N_g = 97[-]$
Generator	
Time constant	$\tau = 0.1\ [s]$
Nominal generator speed	$\omega_{gmax} = 122.91\ [\frac{rad}{s}]$
Minimum generator speed	$\omega_{gmin} = 70.16\ [\frac{rad}{s}]$
Structure/Tower	
Rotor radius	$R = 63\ [m]$
Hub height	$h = 90\ [m]$
Tower mass	$m_t = 4.2278e5\ [Kg]$
Tower stiffness	$K_t = 1.6547e6\ [\frac{N}{m}]$
Tower damping	$D_t = 2.0213e3\ [\frac{N}{m \cdot s}]$

to model all types of physical systems. Most engineers work in many different disciplines, an understanding of the intersection of these disciplines is a valuable asset for any engineer. Based on the results in the chapter, interesting future research is to design controller based on the bond graph modeling. This can be done in the bond graph simulations program 20-sim, but it is also possible to export the model to MATLAB via S-function.

2.6 Appendix

The wind turbine parameters used for this study in the model system are given in Table 2.2.

References

1. Tamura, J., Yamajaki, T., Ueno, M., Matsumura, Y., Kimoto, S.: Transient stability simulation of power system including wind generator by PSCAD/EMTDC. In: IEEE Porto Power Tech Proceedings, vol. 4, EMT-108. IEEE, Porto, Portugal (2001)
2. Zubia, I., Ostolaza, X., Tapia, G., Tapia, A., Saenz, J.R.: Electrical fault simulation and dynamic response of a wind farm. In: Proceedings of IASTED international conference on power and energy system, No. 337–095, p. 595. Acta Press, Rhodes, Greece (2001)
3. Abdin, E.S., Xu, W.: Control design and dynamic performance analysis of a wind turbine induction generator unit. IEEE Trans. Energ. Convers. **15**(1), 91 (2000)
4. Ledesma, P., Usaola, J., Rodriíguez, J.L.: Transient stability of a fixed speed wind farm. Renewable Energ. **28/9**, 1341–1355 (2003)
5. Carrillo, C., Feijoo, A.E., Cidras, J., Gonzalez, J.: Power fluctuations in an isolated wind plant. IEEE Trans. Energ. Convers. **19**(1), 217–221 (2004)
6. Petru, T., Thiringer, T.: Modeling of wind turbines for power system studies. IEEE Trans. Power Syst. **17**(4), 1132–1139 (2002)
7. Akhmatov, V., Knudsen, H.: Modeling of windmill induction generators in dynamic simulation programs. In: International conference on electric power engineering, PowerTech Budapest 1999, 29 August to 2 September 1999. p. 108. IEEE, Budapest. Hungary. doi:10.1109/PTC.1999.826539, ISBN: 0-7803-5836-8
8. Martins, M., Perdana, A., Ledesma, P., Agneholm, E., Carlson, O.: Validation of fixed speed wind turbine dynamic models with measured data. Renewable Energ. **32**, 1301–1316 (2007)
9. Muyeen, S.M., Hasan Ali, Md., Takahashi, R., Murata, T., Tamura, J., Tomaki, Y., Sakahara, A., Sasano, E.: Comparative study on transient stability analysis of wind turbine generator system using different drive train models. IET Renew. Power Gener. **1**(2), 131–141 (2007)
10. Larsen, T.J.: How 2 HAWC2, the user's manual. Ris-R-1597(ver. 3–9)(EN), Sept 2009
11. Bottasso, C.L., Croce, A.: Cp-Lambda user manual. Dipartimento di Ingnegneria Aerospaziale, Politecnico di Milano, Milano, Italy (2009)
12. Jonkman, J., Buhl, M.L. Jr.: *FAST User's Guide.* Technical Report NREL/EL-500-38230, National Renewable Energy Laboratory, August 2005
13. Hammerum, K., Brath, P., Poulsen, N.K.: A fatigue approach to wind turbine control. J. Phys. Conf. Ser. **75**, 012081 (2007)
14. Eggleston, D.M., Stoddard, F.S.: Wind Turbine Engineering Design. Van Nostrand Reinhold Co., New York (1987)
15. Paynter, H.M.: Analysis and Design of Engineering Systems. MIT, Cambridge (1961)
16. Karnopp, D.C., Margolis, D.L., Rosenberg, R.C.: System Dynamics: Modeling and Simulation of Mechatronic Systems, 4th edn. Wiley, New York (2006)
17. Borutzky, W.: Bond Graph Methodology: Development and Analysis of Multidisciplinary Dynamic System Models, 1st edn. Springer, Berlin (2010)
18. Pedersen, E., Engja, H.: Mathematical Modeling and Simulation of Physical Systems. Lecture Notes in Course TMR4275 Modeling, Simulation and Analysis of Dynamic Systems at NTNU. Trondheim, Norway (2003)
19. Heier, S.: Grid Integration of Wind Energy Conversion Systems. Wiley, New York (1998)
20. Henriksen, L.C.: Model Predictive Control of a Wind Turbine. Master Thesis, Technical University of Denmark (2007)
21. Kleijn, C.: "20-sim 4.1 Reference Manual" Enschede, Controllab Products B.V. (2009). ISBN 978-90-79499-05-2

Chapter 3
A General Circulation Model en Route to Intraseasonal Monsoon Chaos

P. Carl

Abstract Since the seminal work of Lorenz, chaotic behaviour of truncated atmospheric equations bears conceptual grounds in the issue of weather predictability. As for climatic scales, low-order models of the El Niño–Southern Oscillation system show chaotic motions as well. The gap to General Circulation Models (GCMs) has been bridged in conceptual studies using a coarse spatial resolution—but temporally and physically resolved—tropospheric GCM. Cross sections of its attractor set across the boreal summer hint at an inverse period doubling route ('out of chaos') in the active-break cycle of the global monsoon system. These dynamics, best visible in integrals of motion, represent a distinct sub-regime of the seasonal cycle, borne in topological changes between spring and autumn bifurcations. Computational problems of first-generation simulations notwithstanding, the paper summarizes the GCM's 'geometry of behaviour' and its observational analogues. The concluding conceptual discussion updates a 'monsoon hypothesis' and addresses hints at the dynamical status and potential evolution of the climate system.

3.1 Introduction

For half a century since Lorenz' celebrated paper [62], conceptions of dynamic systems theory have fundamentally influenced our qualitative understanding of atmospheric and climate dynamics (e.g., [40, 42]). Midlatitude vacillation phenomena and their rotating-annulus experimental analogues stood godfather to momentous progress in the theory of weather regimes and their transitions (e.g., [30, 39, 63]), leant upon idealized equations of atmospheric flow (e.g., [61, 64]). Climate dynamics were mainly addressed during the first decades from a paleoclimatic perspective,

P. Carl (✉)
ASWEX-Applied Water Research (formerly IAMARIS e.V.), Hausvogteiplatz 5–7, 10117 Berlin, Germany
e-mail: pcarl@wias-berlin.de

S. Banerjee and L. Rondoni (eds.), *Applications of Chaos and Nonlinear Dynamics in Science and Engineering 3*, Understanding Complex Systems,
DOI 10.1007/978-3-642-34017-8_3,

using simple energy and (ice sheet) mass balance models (e.g., [9, 38, 48, 75]). As alternative to the evolving class of General Circulation Models (GCMs; e.g., [7, 35]; cf. also [73]), either approach stands for the power of conceptual modelling.

Based on 'primitive' transport equations of mass, momentum and energy, and on parameterized subgrid scale processes, GCMs are the physically most comprehensive members of a hierarchy of climate models. Their complexity goes with spatial resolution and grid scale detail of sources and sinks, and with functional coupling of subsystems. Since atmospheric GCMs (AGCMs) grew up from numerical weather prediction (NWP) models to spatial and physical completion, whereas the Lorenz attractor was borne in an adverse effort toward maximum simplification of the NWP equations [61], there is a notion (calling GCMs with their nonlinear equations to witness) of low-dimensional behaviour and related conceptions of nonlinear science belonging to a world quite apart from climate reality. 'Perpetual' GCM climates were run until the 1990s and beyond (annual or seasonal mean boundary forcing [68, e.g.]), but a systematic GCM based verification of qualitative dynamics as addressed by lower-order models is largely missing.

The evolution of computational capabilities since the mid-1980s posed into relief a class of models coined with the poorly defined, floating term of 'intermediate' complexity (ICMs). ICMs soon became charged with the expectation of substantial contributions, by 'exhaustive' computational experimentation [41, e.g.], to the understanding of climate dynamics. Justification of this hope, and intriguing insights into the potential role of nonlinear dynamics at timescales much below paleoclimatic ones, emerged from a series of studies on the origin and performance of the El Niño–Southern Oscillation (ENSO) system of the tropical Pacific. Based on extensive simulations using intermediate coupled atmosphere–ocean models [27, 28, 51, 86, 87], the nature of the observed irregularity and phase preference to the seasonal cycle (SC) of an ENSO cycle as hypothesized by the delay oscillator conception [8, 77] has successfully been addressed until the mid-1990s.

Induced by a set of AGCM projections on the emerging 1991 Kuwait oil well fire problem showing sensitive, oscillatory response to lower tropospheric smoke loads ([11, 12]; Fig. 3.1, left panel), a 'civilian' study using the same GCM led to a hypothesis on the dynamic constitution of the global monsoon [13, 24]. As a 'natural' (not astronomically driven) oscillator in the 40–60 days band with interacting regional branches, this GCM's boreal summer monsoon rules a distinct global climate regime of the season [14, 84]. Its dynamic 'skeleton', probed in extensive intraseasonal attractor soundings, became most clearly visible in variables that relate to global integrals of motion [23]: an inverse route to (planetary scale!) "monsoon chaos".

Here an extended presentation is given of the work [17], together with a more comprehensive list of references. Section 3.2 introduces the GCM, Sect. 3.3 outlines the 'architecture' of its boreal summer monsoon system, and Sect. 3.4 discusses monsoon retreat dynamics and their interannual aspects. Effects of changing boundary conditions are addressed in Sect. 3.5, and a new Sect. 3.6 presents a completely updated monsoon hypothesis of the 1990s. Section 3.7 offers a range of conclusions.

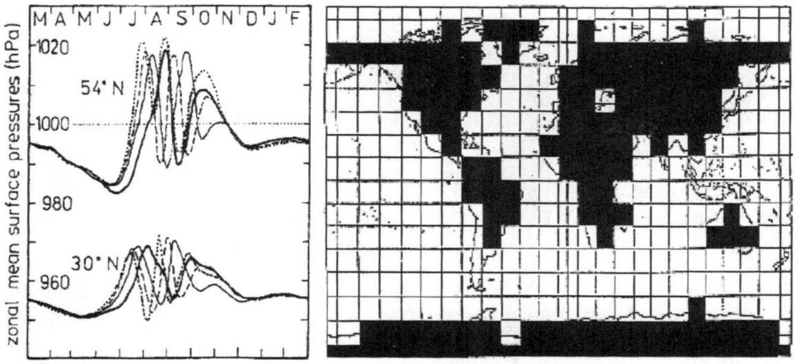

Fig. 3.1 *Left panel*: GCM response to lower tropospheric smoke immission scenarios of the Kuwait oil well fire study ([11, 12]; constant immission rates start March 1 each—simulations done before the fact): Control run (no smoke; *thick line*), lower edge scenario ('nominal' smoke load, corresponding to Kuwait's pre-war oil production of ∼220 kt/d; *thin line*), reference case (doubled smoke; *dashed*), upper edge scenario (fourfold smoke; *dotted*); shown are zonally averaged evolutions of surface pressure at latitudes 30°N and 54°N; *right panel*: 12°lat × 15°lon land–sea mask of the coarse resolution CCAS GCMs

3.2 A (Very) Coarse Resolution Mintz–Arakawa AGCM

A family of AGCMs, coined "Mintz–Arakawa" after their original authors [7], emerged from early efforts at the University of California at Los Angeles (UCLA; [6, 56]) and spread around the world to seed a number of climate modelling activities, both in development and training (cf. also [73]). It includes the RAND Corporation [35] and Oregon State University (OSU; [37]) models which found their way to Moscow for use at the Computing Centre of the USSR Academy of Sciences (CCAS) with their Earth system model GEA [69]. To this end, a coarse resolution version of the OSU AGCM had been developed in cooperation [2] and became equipped with a simple model of the oceanic upper (mixed) layer [74] to form the 'CCAS model' [3].

A similar RAND version (coined CCAS-M here) was used in Moscow for first GCM simulations of climatic effects of nuclear war [80], and then in Berlin within a related cooperative endeavour [20, 25, 76]. After in-depth regeneration in order to make of it a fexible tool of dynamic systems analysis for research and education [10], this 'CCAS-B' AGCM unveiled intriguing monsoon dynamics [13] not known before from GCMs, when used in diurnal and seasonal cycle mode in the Kuwait oil well fire study [11, 12]—in contrast to annual mean forcing in the 'nuclear winter' case (e.g. [20, 76]).

3.2.1 Brief Model Description

The AGCM is a hydrostatic ('shallow water'; SW) model with a 'rigid lid' top of the dynamic atmosphere, at mean tropopause pressure $p_T = 200$ hPa. For two tropospheric layers of 400 hPa 'thickness' each, wind speeds in zonal and meridional direction (u, v), temperatures T, water vapour q (mixing ratio) and surface pressure p_s (via $\pi := p_s - p_T$) are computed at a 90 min-timestep, composed of five 18-min substeps. At standard horizontal resolution of 12°lat × 15°lon, the grid consists of $16 \times 24 = 384$ meshes (Fig. 3.1, right panel; polar latitude extensions are halved and single polar cells emulated by zonal averaging of prognostic variables). A vertical 'sigma' coordinate, $\sigma := (p - p_T)/\pi$, represents atmospheric pressure p in hydrostatic approximation.

As a grid model, the CCAS GCM solves the set of transport equations on its fixed spatial mesh. Spectral GCMs, in contrast, advance their solutions in terms of a zonal wavenumber hierarchy of spherical harmonics. Numerical schemes of RAND and OSU versions of the GCM are different; CCAS-B applies an alternating-direction predictor–corrector scheme due to Arakawa [35]. Dynamic layers are labelled 1 (upper troposphere) and 3 (lower one), the interface between both 2, the tropopause 0, near-surface layer 4, and ground level s. The (purely radiative) stratosphere is parameterized.

The RAND model [35] has eight prognostic variables $(u_i, v_i, T_i, q_3, \pi; i = 1, 3)$, the OSU GCM [37] ten, including snow cover s and upper tropospheric water vapour q_1. In either case, both stratospheric and near-surface layer variables are locally determined once a timestep—as are the 'sigma velocity' at level 2, $\dot{\sigma}_2$ (and related vertical exchange processes), the transport of radiation across the layers, elements of the atmospheric water cycle (ground wetness, evaporation, condensation, convection, cloud formation, rainfall generation etc.) and other source and sink terms. Atmospheric tracers (CO_2 or smoke, for example) may be transported in analogy to water vapour q.

3.2.2 Reference AGCM Climates: Seasonal Cycle(s)

The present paper focuses on the AGCM solution, which bears the dynamics of interest. Sea surface temperatures (SSTs) and seaice distributions thus are not interactively computed in general, but prescribed in their seasonal march by daily interpolation from monthly mean climatology [4]. Non-climatological lower boundary forcing is applied in Sect. 3.5.2 to highlight both sensitivity and structural robustness of this model's atmospheric water cycle—and thus climate—dynamics. The mixed-layer ocean (MLO) model [74] is 'switched on' for reference in Sect. 3.5.3.

Three qualitatively different types of seasonal cycle (SC) have been simulated so far: *(i)* a rather smooth seasonal march (a sort of 'mean' SC), *(ii)* an SC that develops a regime of very active tropical–extratropical interactions in boreal summer ("summer monsoon solution"—the one found in the Kuwait oil well fire study [11]; cf. Figure 3.1), and *(iii)* a solution that exhibits high intraseasonal monsoon activity in both summer and winter [13]. Whereas solution *(i)* is computationally stable and robust, the other two are more difficult to get and turned

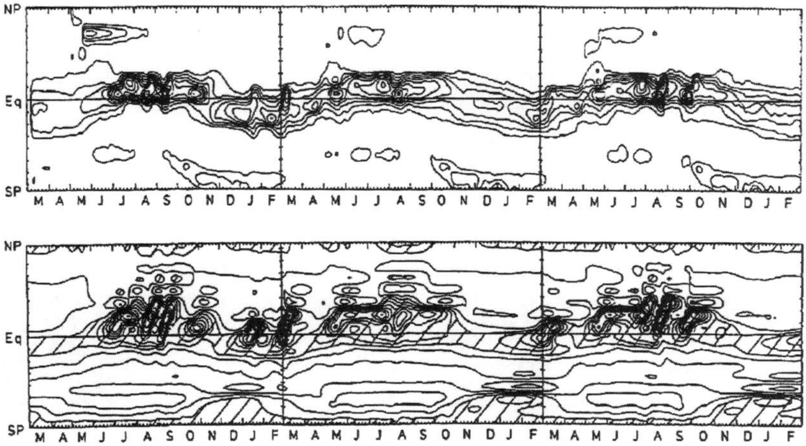

Fig. 3.2 "Summer–winter–monsoon solution" of the coarse resolution CCAS-B GCM [13]: Zonally averaged evolution of pentad mean precipitation (*top*; level spacing 2.5 mm/d) and 400 hPa zonal wind snapshots (*bottom*; level spacing 10 m/s; hatched: easterlies)

out to bear the risk of 'spontaneous' computational instability. Most tricky to find in the GCM's multidimensional parameter space is type *(iii)*, the "summer–winter–monsoon solution" (Fig. 3.2).

Change between types *(ii)* and *(iii)* has first been detected when introducing time-dependence in the estimate of q_4 (and q_4^*, the saturation water vapour pressure) and 'playing' with the corresponding numerical scheme (semi-implicit vs. explicit) [13]. Such type change in the intraseasonal monsoon activity may also have caused differing GCM climates, with substantial distinctions in the hydrological cycle, as presented in a study [47] that tested gridpoint vs. spectral numerics in a 'perpetual July' simulation, using the physics of the two-layer OSU model at horizontal resolution 4°lat × 5°lon. Concerns aside about spurious, numerically induced planetary waves in SW models [59, e.g.], the 'monsoon problem' with present-day GCMs might arise from their insufficient excitation or coupling (Sect. 3.7).

Qualitatively correct GCM dynamics, to which intraseasonal monsoon performance holds a key, are not a question of fine tuning but of an adequate 'operating regime' [88, e.g.]. Sensitive dependence on details of numerical scheme or physical parameterization of hitting or failing when searching for reasonable intraseasonal monsoon performance, hints at existence of critical transitions between climate regimes; that is, at their nonlinear construction.

3.2.3 *On Monsoon Definition: A Biased Note*

Major attributes of the classical monsoon definition are the seasonally reversing wind regimes and the wet–dry transition [93, e.g.]. Elementary mechanisms in the back are borne in the Earth's rotation, the distribution of land and sea, and in the orography of the landmasses—notably the location and elevation of the Tibetan and Mexican plateaus [90]. Differential heating between land and sea in boreal spring, reinforced by the emerging elevated heat sources of the plateaus, drives a huge sea breeze circulation, with a dominant meridional component that merges with the southern Hadley cell as the season advances and the meteorological equator shifts northward. Rising air masses over the northern continents 'suck in' low-level airflow from the oceans and set up this way a strong positive feedback cycle of the wet atmosphere. The two 'conventional' hypotheses on the causation of monsoons, namely differential heating and seasonal migration of the intertropical concergence zone [32, e.g.], are thus not necessarily competing.

The winds collect latent energy in form of water vapour from the ocean surface, which is released to the atmosphere in the cloud formation process. Since the tropical atmosphere is unstable in general above the condensation level, it may form huge precipitating convective towers often reaching the tropopause level and forming the typical anvil-shaped cap. In combination with synoptic and planetary scale wave activity, collective processes may be initiated that generate organized cloud clustering at a hierarchy of spatial scales [71]. (In a general sense, this calls to mind the classical phenomenon of *Bénard* convection in a liquid layer heated from below.)

The self-enhancing supply with energy blows up the southern Hadley cell into the monsoon cell which strengthens and extends further northward thereby, fans the subtropical zonal jets and shifts them to the north until they finally 'tip over' to pass the plateaus at their northern side (active monsoon at planetary scale). A major break monsoon commences if this northward shove has built up restoring forces (e.g., pressure gradients) which cannot any longer be balanced by tropical supply. The system 'falls back' and starts its activity anew in collecting excess budgets of integrals of motion in the tropics to prepare their export to the extratropics. Whereas onset is an abrupt process, however, monsoon revival proceeds more smooth. Besides this 'lateral' monsoon, zonal circulation cells extend from the ascending air masses, in boreal summer over South Asia, both eastward and westward [93].

The active-break cycle, a broadband phenomenon in the "40–60 days" range of period, is a salient feature of the present-day monsoon (e.g., [53, 95, 96]) but has not entered any monsoon definition [93, e.g.]. The same holds for abrupt onset and fast retreat which extend the spatial range of the system beyond that of its classical definition [98]: Monsoons are circulation regimes from pole to pole [52, e.g.]. Indispensable ingredients (including the role of the 'Madden–Julian wave' [66]), and even the very nature of the system and its dynamics, are under persistent debate [44, 49, 85, 89, 99].

3.2.4 Boreal Summer Monsoon Solution: Rainfall Patterns

A qualitatively correct spring-to-summer evolution of the GCM's global rainfall fields is illustrated in the six pentad-averaged panels of Fig. 3.3 [13] (top left to

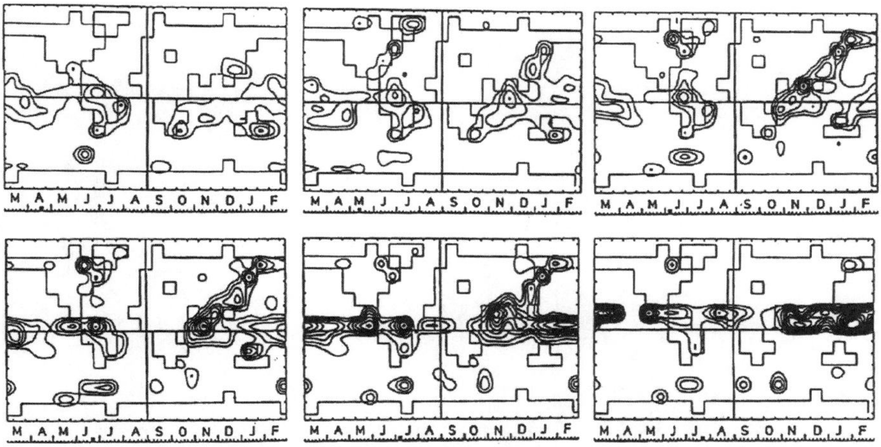

Fig. 3.3 Pentad-mean evolution of rainfall patterns of the coarse resolution CCAS-B GCM [13], from the beginning 'Mei-Yu' rainy period in China to the first break of the Indian Southwest monsoon (ISM); *left-to-right, top*: pentads 22, 27, 32; *bottom*: pentads 35, 38, 41 (cf. mark at the ruler below each individual panel); level spacing 5 mm/d

bottom right: pentads 22 (April 16–20), 27 (May 11–15), 32 (June 5–9), 35 (June 20–24), 38 (July 5–9), and 41 (July 20–24)). Pre-monsoon rains over the Bay of Bengal (pentad 27) develop together with a rainband along the East Asian shoreline ('Mei-Yu'). Northward shift and strengthening of the convective center over the Bay and its merger with the East Asian frontal zone in early June (pentad 32) are close to observation, as is the northwest shift of the merged rainbelt in the Indian Ocean–Tibetan sector. Thereby, the center of action moves around India until it jumps to the subcontinent in early July (Indian summer monsoon (ISM) onset; pentad 38).

The GCM's seasonal march of rainfall for grid cells representing South and Southeast Asia [14] is shown in comparison to observations [53, 58] in Fig. 3.4 (bottom). The latter area comprises the Yangtse basin which repeatedly suffers from catastrophic inundation. Computational access to intraseasonal monsoon dynamics there, as demonstrated here in terms of timing and structure of major rainfall episodes, is of high societal relevance.

Though simulated East Asian rainfall in Fig. 3.4 amounts to about half the climatological level, whereas the Indian monsoon rains are clearly stronger than those observed around 1970, the structural features of both are striking. Note also that another such type East Asian monsoon simulation, though showing lesser timing accuracy and structurally less pronounced rainfall spells after 'Mei-Yu' termination (and a much weaker rainfall amplitude) [60], has been obtained from an advanced version of the two-layer Mintz–Arakawa AGCM of the Beijing Institute of Atmospheric Physics (IAP) of the Chinese Academy of Sciences—applied at higher (4°lat × 5°lon) resolution.

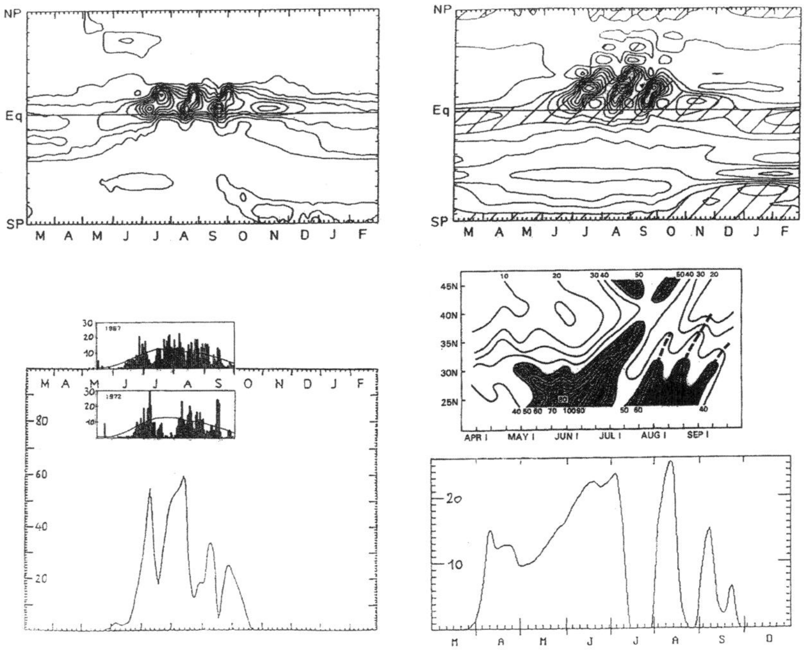

Fig. 3.4 "Summer monsoon solution" of the coarse resolution CCAS-B AGCM under climatological forcing [13]: Seasonal march of global rainfall and upper troposphere zonal winds (*top*; cf. also Fig. 3.2) and regional rainfall over the Indian subcontinent (*bottom left*) and Southeast Asia (*bottom right*) [14] vs. observation [53, 58]

When Asian climate dynamics run into the first break monsoon (early in July here; Fig. 3.3), monsoonal intensification over the western Pacific to the north of the equator prepares a marked structural change that activates the west African and Carribean/Mexican monsoons. It terminates 'Mei-Yu' over Southeast Asia and is about to develop into break over India (second half of July; pentad 41), with typical zonally streched rainbelts and rainfall maxima over the seas, notably the Indian Ocean and Western Pacific [33, e.g.].

This overall dynamic picture of the planetary scale boreal summer monsoon and its individual branches reflects a proper model climate regime of the season. Figure 3.5 displays a 100 year segment of the GCM's rainfall under climatological boundary forcing, exhibiting variability in the tropics and subtropics close to observation. In contrast, the weak rainfall variability found in [43] to hold for all participating GCMs was largely driven by variable SSTs (1979–1991 observed), i.e. the GCMs' hydrological cycle fundamentally differed from observation (cf. also the related comment in [44]).

It has early been understood [12] that global organization of the real-world monsoon system—the nowadays 'emerging' concept of "global monsoon" [29, 89] is not really new [52, e.g.]—must be blamed for the fact that a GCM with extremely coarse spatial resolution may show its dynamic essentials. Intraseasonal

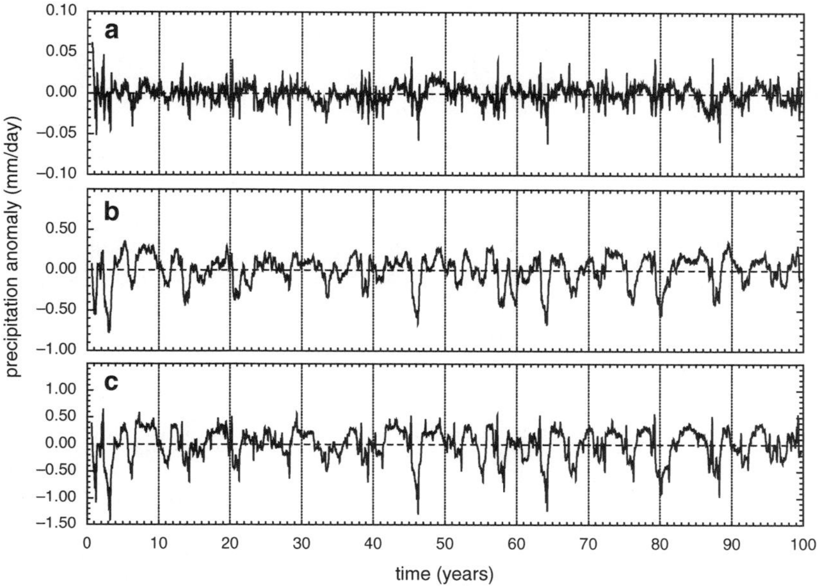

Fig. 3.5 100 year segment of total precipitation (mm/d) in the coarse resolution CCAS-B AGCM's "summer monsoon solution" under climatological boundary forcing (365 days moving averages): (**a**) global (90° S–90° N; mean 2.76 mm/d, standard deviation 0.01 mm/d); (**b**) deep tropics (12° S–12° N; mean 10.46 mm/d, standard deviation 0.19 mm/d); (**c**) tropics (24° S–24° N; mean 20.53 mm/d, standard deviation 0.28 mm/d)

monsoon dynamics are difficult to simulate, now as before [49, e.g.]; neither high spatial resolution nor dynamic atmosphere–ocean coupling provided the missing ingredients (e.g., [15, 44]).

3.3 'Architecture' of the GCM's Boreal Summer Monsoon

Lorenz' suggestion on the potential role of (intra-) seasonal attractor sets in shaping climate variability [65] appears to have died away largely unlistened in the community he addressed. Climate model intercomparison focuses on simulated trajectories, even if taking ensemble means or other statistics, without considering the 'systems-dynamical core' in the rear. Topological structures (or the "geometry of behaviour" [1]) that define the dynamical status of the system, remain unknown. It makes thus sense to recall a contribution [23] to the First Scientific Conference of the Atmospheric Model Intercomparison Project (AMIP [34]). It demonstrated the conceptual use of a GCM in order to unveil its attractor sets and suggested those objects and the method as appropriate supplementary approach to climate model intercomparison.

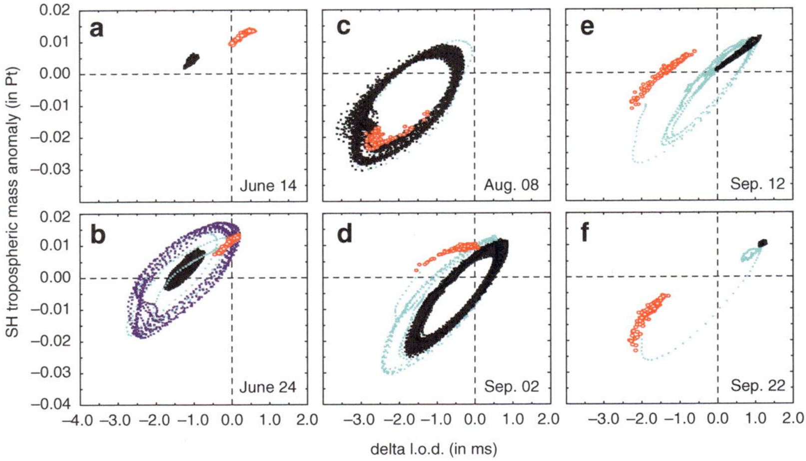

Fig. 3.6 Cross sections of the coarse-grid CCAS-B AGCM's attractor set from boreal spring to autumn [23] (other than in Fig. 3.4, a "summer monsoon solution" with only two activity spells is used; cf. also [22]): Anomalies of SH tropospheric mass (in Pt) vs. relative angular momentum of the troposphere, scaled to length-of-day fluctuations (Δl.o.d.; in ms); *black/violet/turquoise* diurnal mean trajectories represent computationally stable/unstable perpetual-day simulations and their transient motions, respectively; *red circles* are snapshots at the respective day from a 198 year seasonal run; (**a**) spring fixpoint (June 14), (**b**) subcritical Hopf bifurcation around monsoon onset (June 24), (**c**) period-2 monsoon activity cycle (August 8), (**d**) period-1 cycle (September 2), (**e**) slow irregular wander between two unstable fixpoints, representing an 'intraseasonal' Southern oscillation (SO; September 12), (**f**) autumn fixpoint (September 22)

3.3.1 Overall Perspective

Figure 3.6 exemplifies attractor soundings using perpetual-day runs (orbital revolution stopped at specified day, Earth's rotation maintained) with the CCAS-B AGCM, from boreal spring to autumn. Variables closely related to global integrals of motion turned out to best unveil the very nature of this model's boreal summer 'architecture' [23] that controls its intraseasonal dynamics: The seasonal cycle, if understood as a forced limit cycle, blows up into a torus segment during the season, of which the minor circumference is made up of the (dynamically stable) 40–60 day monsoon activity cycle.

Coordinates in Fig. 3.6 include southern hemisphere (SH) tropospheric mass anomaly, representing interhemispheric mass exchange, and atmospheric angular momentum (AAM) relative to the solid Earth, scaled to length-of-day (l.o.d.) variations [5, e.g.]. Whereas studies on interhemispheric mass exchanges at subseasonal scales focus on boreal winter [26, e.g.], the fact of 40–60 day co-variations in AAM and monsoon activity in boreal summer is well established [54, e.g.]—as is the interhemispheric organization of the Asian monsoon systems [53, e.g.]. The Δl.o.d. amplitude in Fig. 3.6 (~2.5–3 ms) exceeds the observed one by a factor of ~2–3 [5] (recall the GCM's coarse resolution—and enhanced zonal flow over mean, not envelope, orography).

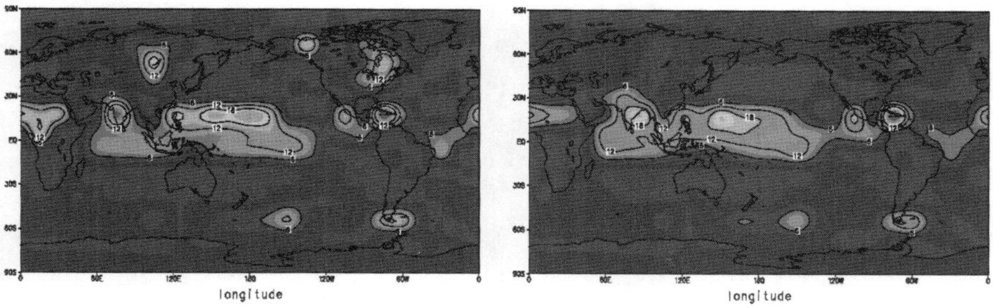

Fig. 3.7 Mean global precipitation fields (mm/day) at June 24 on the 'pre-monsoon' (*left panel*) and 'interactive' summer monsoon attractors (*right panel*), respectively, as found in the coarse resolution, two-level CCAS-B AGCM (cf. Fig. 3.6b)

The model's boreal summer torus segment bears a non-trivial structure. Embedded between spring and autumn fixpoints (Fig. 3.6a, f), it starts with coexisting oscillatory states (Fig. 3.6b) that bear hard transition into a chaotic July regime (cf. Sect. 3.3.3), runs than backward as the season advances via period-2 (Fig. 3.6c) and period-1 (Fig. 3.6d) cycles, and ends up in a wrinkled shape of torus. The "40–60 day" cycle shows pronounced frequency drift across the season, which contributes to its broadband nature [84]. Off the late-summer torus wrinkle, an 'intraseasonal' Southern Oscillation (SO) remains as degenerate monsoon cycle (Fig. 3.6e; cf. also [22])—another hint at the kinship between both subsystems [92].

Abrupt seasonal transitions have early been observed over the Eurasian landmass [98]. Elementary monsoon mechanisms [90] and related models [91, e.g.] do not explain these large-scale, fast dynamics, and GCMs do not simulate them in general. Using an idealized potential vorticity equation to address midlatitude oscillations in the 30–60 days band [50], regimes of "jet shifting oscillations" were found in the flow over topography, which come into effect via Hopf bifurcation and may show period multiplication. Two-level stability analysis [57] provides mechanisms of quickly growing unstable waves in boreal summer and makes their absence during the winter season plausible.

Given clear signatures of monsoon activity in the dynamics of tropospheric jets [78, e.g.], such mechanisms may participate indeed in tropical–extratropical interactions inherent to the monsoons [97, e.g.] (cf. Figs. 3.2–3.4), and might be blamed for the observed abruptness of monsoon onset. Other than with the summer monsoon, difficulties to stably find the "summer–winter–monsoon solution" may hint at a problem of two-layer models in representing unstable processes of generic multilevel character [57].

3.3.2 *Subcritical Monsoon Onset*

Coexistence of GCM solutions in early boreal summer hints at a subcritical transition into the 'interactive' (tropical–extratropical [67]) global monsoon. Figure 3.7 shows mean precipitation fields on either June 24 attractor of Fig. 3.6b. The spring fixpoint (Fig. 3.6a) evolves supercritically into a 'breathing' oscillation which may computationally be followed into the range of coexistence when settling on this attractor and gradually advancing the boundary conditions. The last day of coexistence, after which the inner attractor ceases to exist, was experimentally found this way to be July 3 (for the AGCM version used, and under climatological boundary conditions).

Distinctions between the two attractors in their precipitation fields might bear paleoclimatic relevanve. In northern middle and higher latitudes three centers of substantial terrestrial rainfall are found at the 'pre-monsoon' attractor (Fig. 3.7, left panel): to the north of Tibet (where a large paleo lake existed), to the north of the Gulf of Alaska, and the larger area of the Great Lakes. Over the Indian subcontinent a spatially stagnant (yet 'breathing') rainfall center resides, whereas the 'interactive' monsoon (right panel) shows two weaker centers over the head Bay of Bengal and the Arabian Sea off the Somali coast, indicating a moving center in contrast. The West African monsoon is active at that date in either solution, but the inner attractor shows much more extended African rainfall—calling to mind the "green Sahara".

Due to 'screening' of the inner solution when the outer one exists (cf. seasonal snapshots in Fig. 3.6b), the pre-monsoon attractor may be consulted only by 'tunneling' trajectories—or if the outer solution does not exist due to changed boundary conditions. A real-world analogue of tunneling might have happened in 2003, when the major intraseasonal activity cycle was missed, at least in midlatitudes where it is normally felt by changes between blocking and zonal circulation regimes. Figure 3.8 shows for a perpetual June 19 run an initial transient, both in elements of the Indian monsoon system (left column) and of the global circulation (right), that 'feels' either monsoon attractor.

Rainfall over India (Fig. 3.8c) most clearly reflects structural changes in the Indian summer monsoon branch after about day 180, when the activity cycle extends meridionally to grasp both the 'Somali jet' (Fig. 3.8d) and the Tibetan plateau (Fig. 3.8a). As may be seen from Hovmoeller plots of zonal winds, for example ([14, 22]; not shown here), the stagnant yet 'breathing' pre-monsoon oscillation is orographically excited. Δl.o.d. fluctuations (Fig. 3.8j) are not subrotational in the real system, which has more contributors to the AAM balance than a tropospheric GCM, and surface friction (Fig. 3.8f) is only extrapolated from the lower troposphere here.

3.3.3 Monsoon Chaos

Hidden in a mine field of computational instability the first generation of simulations stumbled across, the attractor set of the model version used for Fig. 3.6 is not sufficiently uncovered from mid-June until the beginning of August. A regime with various types of coexisting attractors (fixpoints, cycles, chaotic oscillations) has

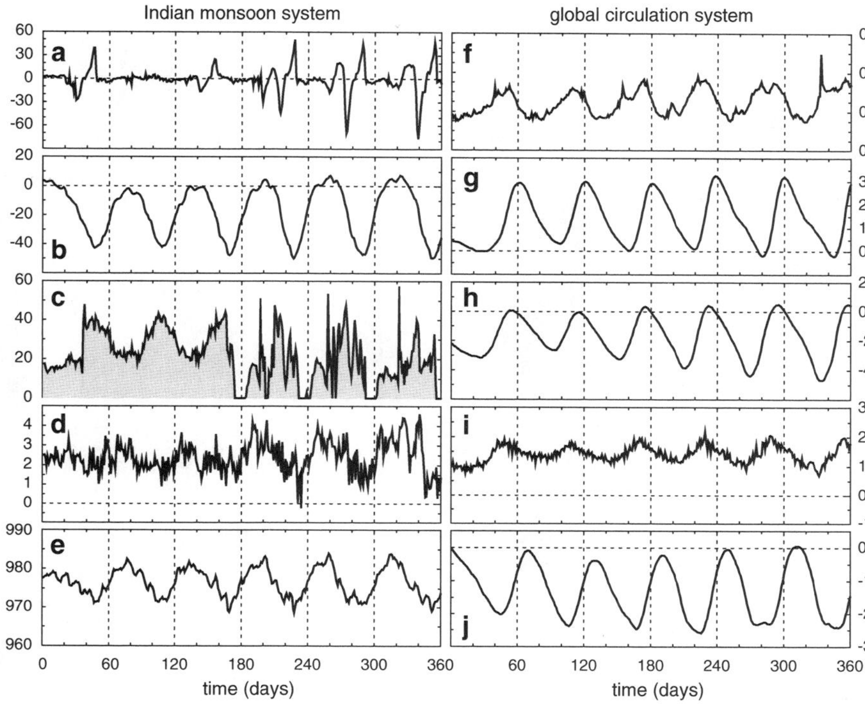

Fig. 3.8 Initial transient 360 days of the June 19 perpetual-day simulation [14, 22], reflecting both 'pre-monsoon' and 'interactive' monsoon behaviour in elements of the ISM system (*left column*) and the global circulation system (*right column*)—(**a**) Tibetan surface heat balance (W/m^2), (**b**) Gangetan plane 800 hPa zonal winds (m/s), (**c**) Central India rainfall (mm/d), (**d**) 800 hPa equator-crossing winds (m/s) over the Arabian sea ('Somali jet'), (**e**) 'Madagascar' surface pressure (hPa), (**f**) globally averaged surface friction (N/m^2), (**g**) 48°N zonally averaged 400 hPa zonal winds (m/s), (**h**) 12°N zonally averaged 400 hPa zonal winds (m/s), (**i**) zonally averaged 800 hPa equator-crossing meridional winds (m/s), (**j**) changes in the length-of-day (Δl.o.d.; ms)

been found there, however, in a slightly different version [21]. Figure 3.9 shows those chaotic oscillations of the global circulation system, with a basic period of the 40–60 days monsoon activity cycle, which appear to be borne in irregular changes between coexisting attractors. The physics of monsoon clearly allows for substantial amplitude and frequency modulation (AM, FM) of the major activity cycle, and in contrast to common wisdom, chaotic dynamics of the (wet) atmosphere may be preserved when expanding the spatial scale. Global integrals of motion even most clearly unveil the system's dynamic 'architecture' here.

3.3.4 Out of Chaos: Inverse Period Doubling

The 'post-chaotic' August 8 attractor (Fig. 3.6c) exhibits the basic structure of period-2. Aside of tropospheric mass and relative angular momentum, the rotational

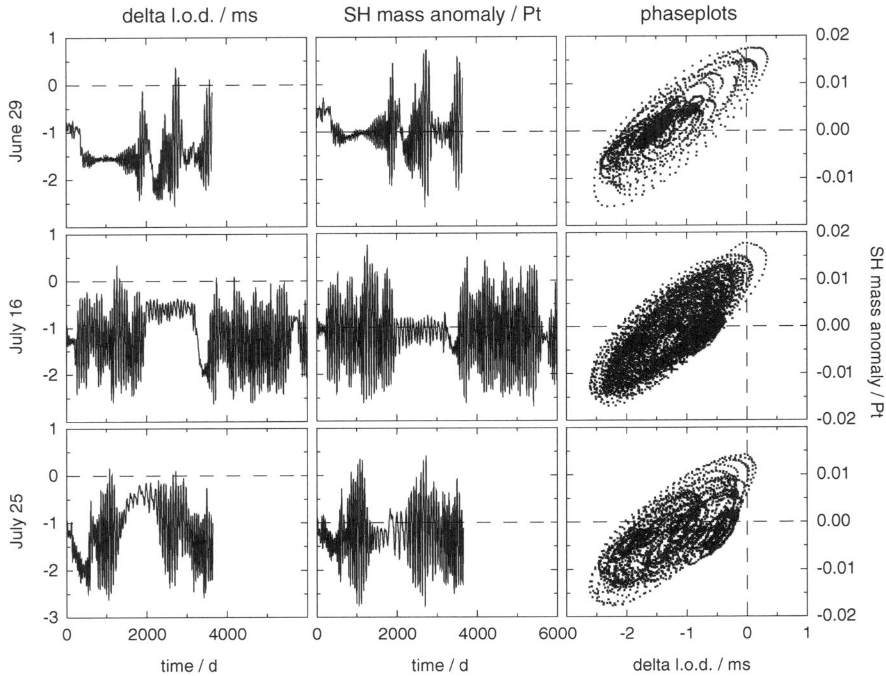

Fig. 3.9 Planetary scale chaotic motions on the GCM's intraseasonal attractor set; *top to bottom*: Perpetual-day simulations at June 29, July 16, and July 21; Δl.o.d. (*left column*), SH tropospheric mass anomaly (*center*), phaseplot of both variables (*right*; cf. Fig. 3.6)

(non-divergent) part of atmospheric motion, measured in terms of global enstrophy, may be used as a third dimension of attractor representation—into which the two 'leaves' of the period-2 cycle split in fact (Fig. 3.10). This midsummer state shows highest 'attractivity': The seasonal trajectory most closely approaches the attractor set there (Fig. 3.6), and within the pronounced frequency drift across the season [14, 84], the major activity cycle reaches its shortest period ($\sim$52.5 days).

That highest attractivity means highest internal synchronization, may be inferred from Fig. 3.11 showing transition from period-2 to period-1 in a long transient at the border between midsummer and post-midsummer attractors. Accompanied by a swing in the SO index (Fig. 3.11d), the transition bears loss of internal synchrony among a hierarchy of planetary waves, as seen in the Fourier spectra of Fig. 3.11b. A set of $\sim$12/20/30/60 day oscillators (frequency relationships $\sim$1/2, 2/3, 1/3, 3/5, 2/5, 1/5) turns into one that loses both the 1/5 relationship and a sharp peak in the mediating $\sim$20 day period. Low rational frequency relationships, at top of the Farey tree of rational numbers, may characterize well-organized dynamic systems [55].

Whereas the midsummer attractor (Fig. 3.6c) completely resides in the range of atmospheric subrotation, Δl.o.d. < 0, the system enters a regime of superrotation during break-monsoon conditions at transition to the post-midsummer state

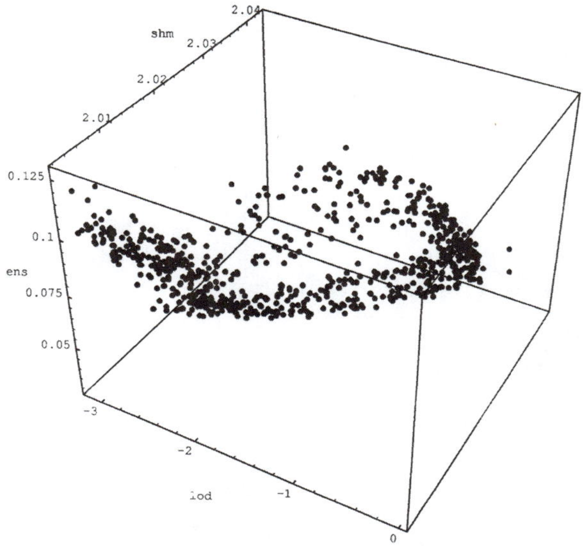

Fig. 3.10 Period-2 attractor in 3D projection: SH tropospheric mass (Pt), Δl.o.d. (ms), enstrophy (in $10^{-7}s^{-}2$), showing separation of its 'leaves' in the enstrophy direction

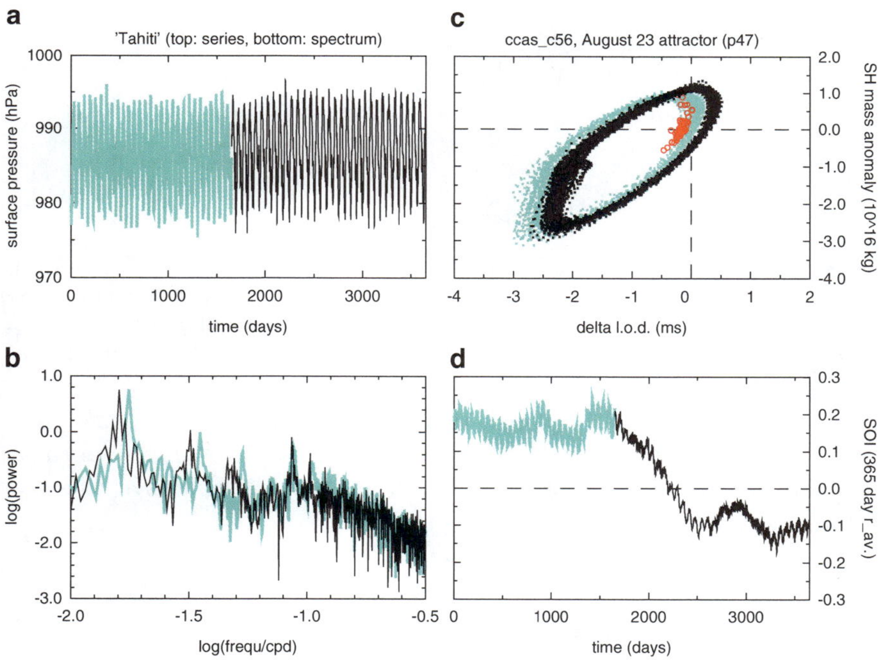

Fig. 3.11 Spontaneous, slow transition between midsummer and post-midsummer attractors [23], initiated by a common jump of three planetary waves with low rational frequency relationships (periods ~20/30/60 days), which breaks a fourth (1/5) synchrony; (**a**) 'Tahiti' surface pressure (hPa) and (**b**) corresponding Fourier spectra, (**c**) August 23 perpetual-day trajectory (cf. Fig. 3.6), (**d**) SO index (SOI; 365-days running average)

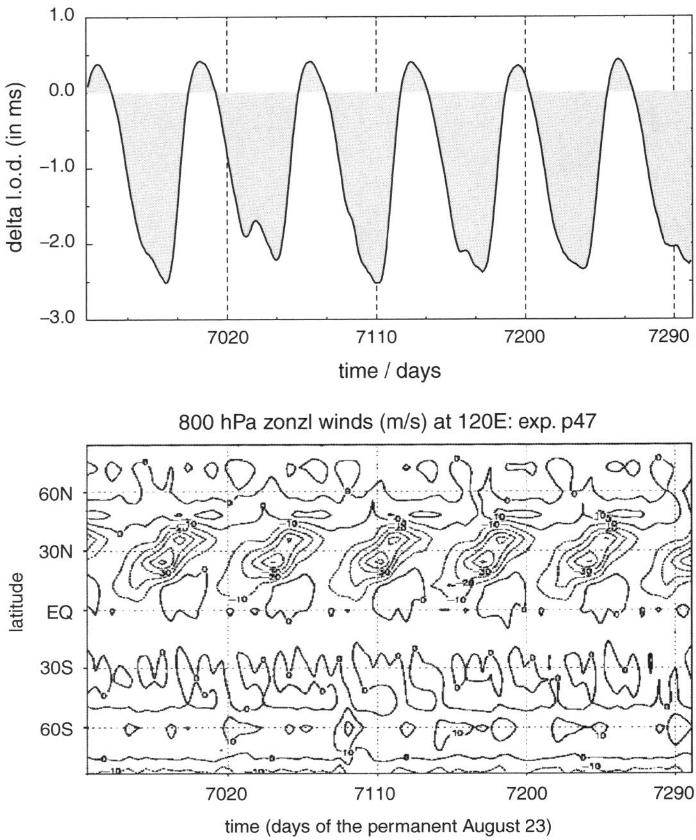

Fig. 3.12 365 day segment of the monsoon activity cycle at perpetual August 23 conditions; *upper panel*: Δl.o.d. fluctuation, entering superrotation at breaks; *lower panel*: North–South Hovmoeller plot of 800 hPa zonal winds at 120°E, showing patches of near-equatorial westerlies toward the dormant state each of the monsoon cycle (cf. Fig. 3.11c; [23])

(Fig. 3.12, upper panel). This dormant phase of monsoon activity, which bears a *memento* of the winter circulation [91], is announced by "westerly wind bursts" (WWBs) over the Western Pacific (lower panel) which are potential real-world precursors of an evolving El Niño [93, e.g.].

3.3.5 *Post-Midsummer Torus Wrinkling*

When the separate period-2 surfaces past together, a 'scar' is left as signature of their former split (Fig. 3.13). The re-united, period-1 cycle slows down, continues to shift into the range of superrotation, and clearly loses 'attractivity' until it degenerates—just at the Julian day when its center passes atmospheric co-rotation (Δl.o.d. $\sim$0).

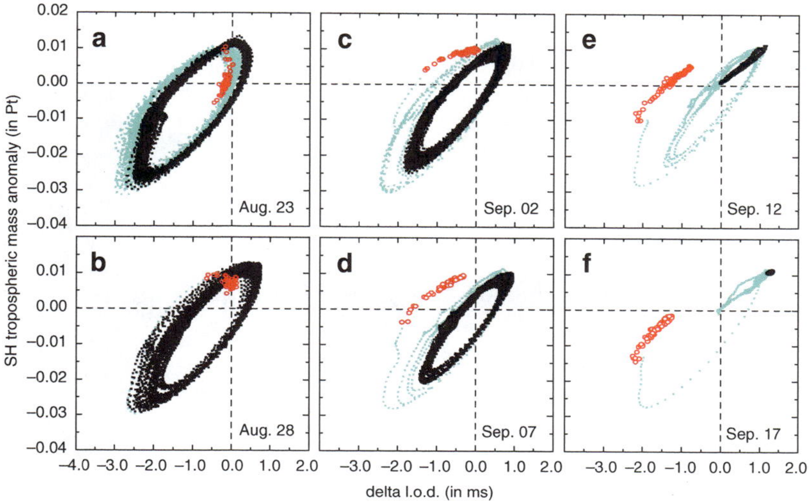

Fig. 3.13 As in Fig. 3.6, but post-midsummer evolution of the attractor set into an 'intraseasonal' SO [82]; (**a**) August 23 (Fig. 3.11c), (**b**) August 28, (**c**) September 2 (Fig. 3.6d), (**d**) September 7, (**e**) September 12 (Fig. 3.6e), (**f**) September 17

The remnant 'scar' of the former period-2 split develops into a 'wrinkle' on the torus where the activity cycle tends to stay longer and longer—as if the system has to collect 'power' before returning to active conditions. Recall, collection of excess budgets of integrals of motion prepares the export phase in the reviving monsoon [67].

3.3.6 An 'Intraseasonal' Southern Oscillation

Having passed the state of co-rotation by its center, the monsoon activity cycle is extinguished and the center itself shows up (for a few Julian days only) as an unstable fixpoint of the fading boreal summer. Figure 3.14 displays the corresponding slow SO index fluctuation, roughly in concert with the evolution of the global enstrophy budget, which together reflect the system's irregular wander between unstable fixpoints of both summer and winter circulation (Figs. 3.6e, 3.13e). The period of this generically intraseasonal SO [16] amounts to ∼500 days [14].

Five Julian days later (September 17), the unstable summer fixpoint is only temporarily visited by a long transient which lately settles on the winter fixpoint (Fig. 3.13f). Remarkably, the latter became stable (and the 'intraseasonal' SO extinguished) just when the unstable summer fixpoint reached the center of the plot, i.e. a state of simultaneous atmospheric co-rotation and balanced mass distribution between the hemispheres.

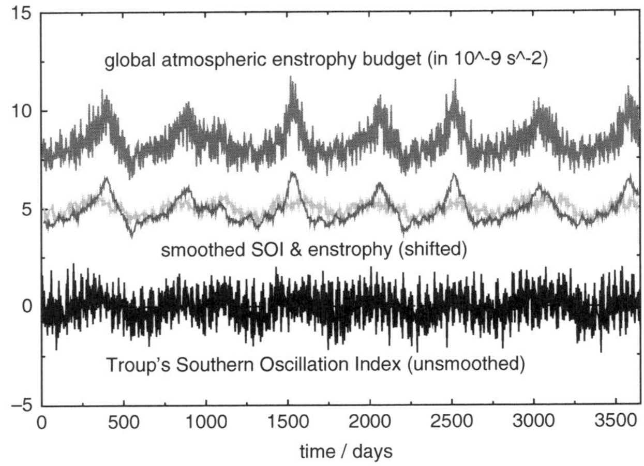

Fig. 3.14 3,650 day segment of a perpetual September 12 simulation using the coarse resolution tropospheric AGCM's "summer monsoon solution": SO Index (*bottom*; cf. Fig. 3.13e), global tropospheric enstrophy budget (*top*; in 10^{-9} s^{-2}), and 365 day moving averages of both (*center*; shifted each)

'Warm' and 'cold' phases of the SO (which normally refer to Eastern Pacific SSTs during the ENSO cycle), are inferred from weaker and stronger zonal circulations here, respectively [83]. Both differ in their divergent circulations as expected. Figure 3.15 (upper panel) shows the difference ('cold' minus 'warm') of the velocity potentials, indicating enhanced convection over the Western Pacific and India during the 'cold' phase. Enhanced Western Pacific convection and upper tropospheric outflow over the region belong to the ENSO 'cold' phase indeed. The lower panel displays slow changes between the two phases for a 1,825 day segment of vertical air flow over the equator; corresponding global circulation patterns [16] confirm correct assignment.

Existence in the real climate system of an SO that results from the dynamic organization of the monsoon system would be conceptually relevant with a view on the two types of ENSO that have recently been 're-discovered' (e.g., [45, 46, 81]). Moreover, it is of interest with respect to forecasting since the 'monsoon borne' SO described here has precursor states during the season which might hint at the course the autumn climate will take.

3.4 Monsoon Retreat Behaviour and Interannual Effects

As with a roulette, the phase of the monsoon cycle when the system passes the retreat bifurcation controls the autumn circulation in a sensitive manner (Fig. 3.16). Three general retreat types may roughly be distinguished among the scatter of trajectories provided by a long seasonal simulation (Fig. 3.16, left panel): from

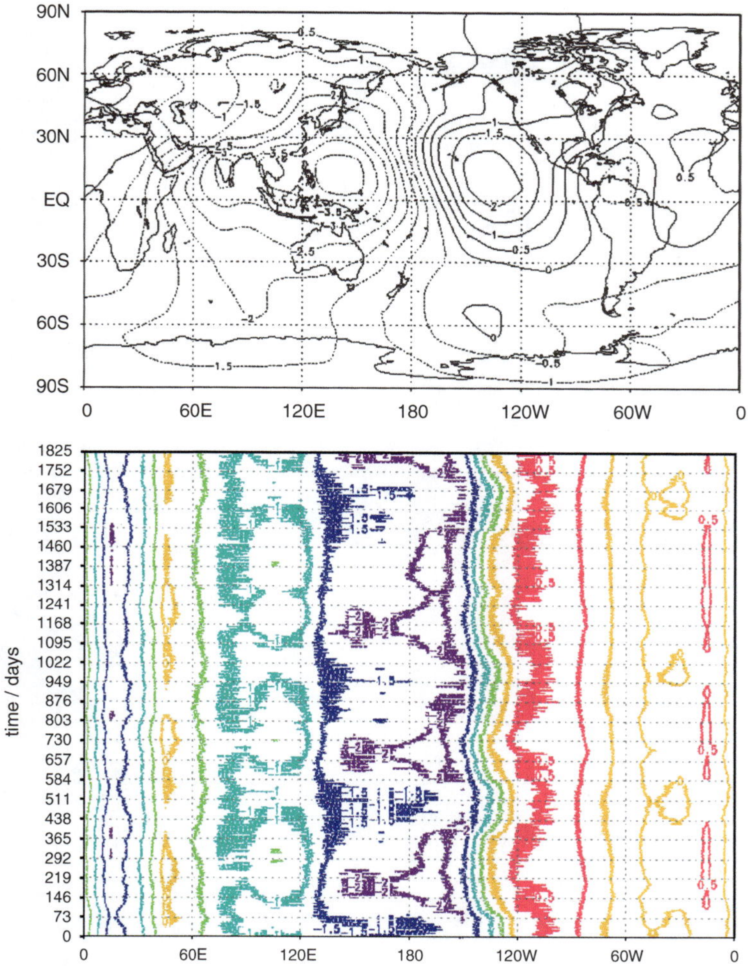

Fig. 3.15 *Upper panel*: Difference of the upper troposphere velocity potential (in $10^7\,m^2/s$) between 'cold' and 'warm' phases (c-w) of the GCM's 'intraseasonal' SO at September 12 (cf. Fig. 3.14); *lower panel*: 1,825 days of equatorial vertical mass flux, indicating phases of enhanced deep convection over the Western Pacific as part of this SO (15-days running average)

dormant, active, or revival phases. As can be inferred from the right panel (cf. also Fig. 3.17), this topologically conditioned situation may be blamed for the observed difference between spring and autumn circulations [31]. Moreover, 'breathing' of the seasonal cycle due to variable monsoon retreat as shown directly translates into interannual variability.

(*i*) Dormant phase retreat (*La Niña prone*): Export to NH extratropics of excess budgets of integrals of motion is completed, their 'mining' by internally organized backward transport is similarly efficient as at mid-season—and is

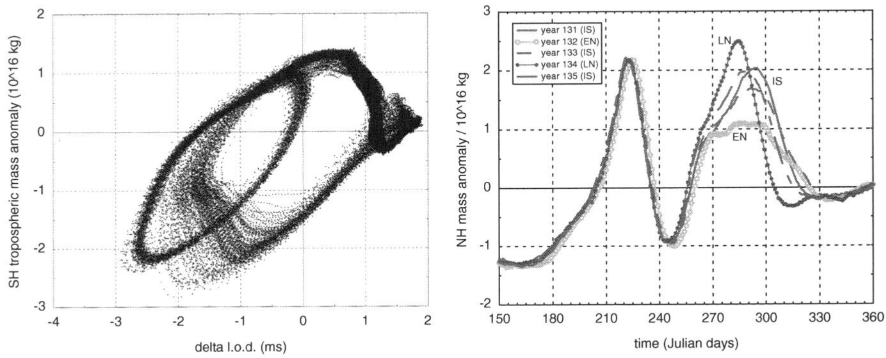

Fig. 3.16 *Left panel*: Physical phase space projections (cf. Fig. 3.6); 198 year AGCM simulation under climatological boundary conditions ("summer monsoon solution"); *right panel*: Monsoon retreat behaviour in terms of northern hemisphere (NH) tropospheric mass anomaly (in 10^{16} kg), showing three types of autumn circulation: *La Niña prone* (LN), *Indian summer type* (IS), and *El Niño trigger* (EN); displayed are seasonal trajectories of years 131–135 of the 198 years shown in the left panel

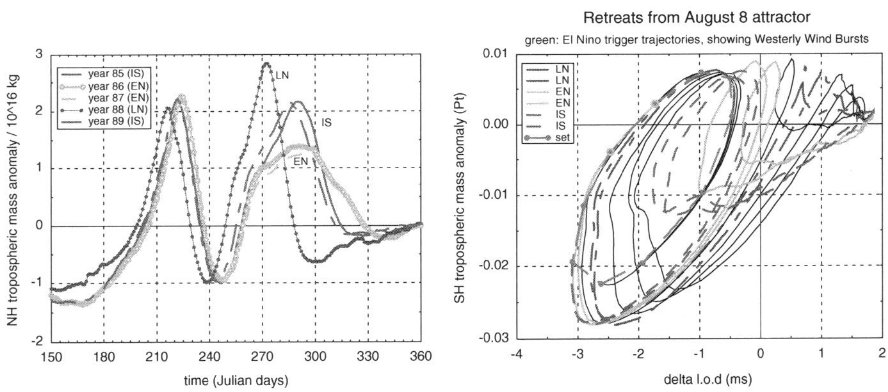

Fig. 3.17 *Left panel*: as in Fig. 3.16 (right), but for years 85–89, showing two El Niño trigger autumn trajectories, immediately followed by an exceptionally early La Niña prone one; *right panel*: Monsoon retreat ensemble started from different phases of the activity cycle on the August 8 attractor [23] (*long dashed*), exhibiting all three retreat types, including two El Niño trigger trajectories (*thick full lines*; starting points emphasized)

supported by the coinciding seasonal transition. This results in a fast backward swing, i.e. early commencement of NH winter conditions. The situation is symmetric with respect to the monsoon season [84]: Both (sub)tropical and extratropical planetary circulations retreat from the intraseasonal cycle roughly along their spring trajectories.

(ii) Active phase retreat (*Indian summer type*): During its export phase, the monsoon cycle is interrupted by the externally conditioned seasonal transition, leaving excess budgets of integrals of motion in both (sub)tropics and

extratropics. Their dissipation does no longer take the break-monsoon path. The situation is completely asymmetric with respect to the season [84]: Not only (sub)tropical but also extratropical circulations experience hystereses at retreat with respect to their spring trajectories.

(iii) Revival phase retreat (*El Niño trigger*): The system has collected excess budgets of integrals of motion in the (sub)tropics but is hindered to export them to the extratropics via its lateral branch. The system adopts a non-lateral dissipation path via zonal, predominantly eastward, transport. The situation is partially asymmetric with respect to the season [84]: Whereas extratropical circulations follow their spring trajectories backward, the (sub)tropics exhibit more or less marked hysteresis.

Systematic study of the effects of diverse phase relationships at monsoon retreat between the intraseasonal active-break cycle and the seasonal cycle has been conducted with a focus on the *El Niño trigger* type of monsoon retreat. In its right panel, Fig. 3.17 shows an ensemble of ten trajectories, started from a set of roughly equidistant phases of the midsummer attractor. Two of these trajectories temporarily enter the range of tropospheric superrotation. Their mass displacements correspond to type *(iii)* retreat—including WWBs (as demonstrated in Fig. 3.12) and induced eastward propagating large-scale convection cells in the sequel [21, 82].

In a highly excited model solution [14, 84], with a more detailed parameterization of the planetary boundary layer, an eastwards propagating, zonally oriented equatorial ("Walker") circulation cell was released at monsoon retreat and traversed the Pacific in the course of 3–4 months—resembling the real-world 1982/1983 ENSO event. Note, this 'atmospheric ENSO' has been found in an AGCM with climatological SST forcing, without interactive ocean. Close to structural retreat, the same GCM bears a regional ISM attractor which controls the global circulation and releases a very narrow ensemble of retreat trajectories [14, 84], in contrast to Fig. 3.17 (right).

3.5 Effects of Changing Boundary Conditions

Occupying problems of the AGCM's computational tractability ruled out by itself extended studies beyond the type of solution displayed. Exploratory excursions into the conceptual, parametric, and functional 'environment' include varied insolation (solar constant) under maintenance of the present-day lower boundary climatology (Sect. 3.5.1), non-climatological SST forcing for 1982–1988 as proposed by the AMIP programme [34] (Sect. 3.5.2), and coupling with the thermodynamic mixed layer ocean model [74] (Sect. 3.5.3).

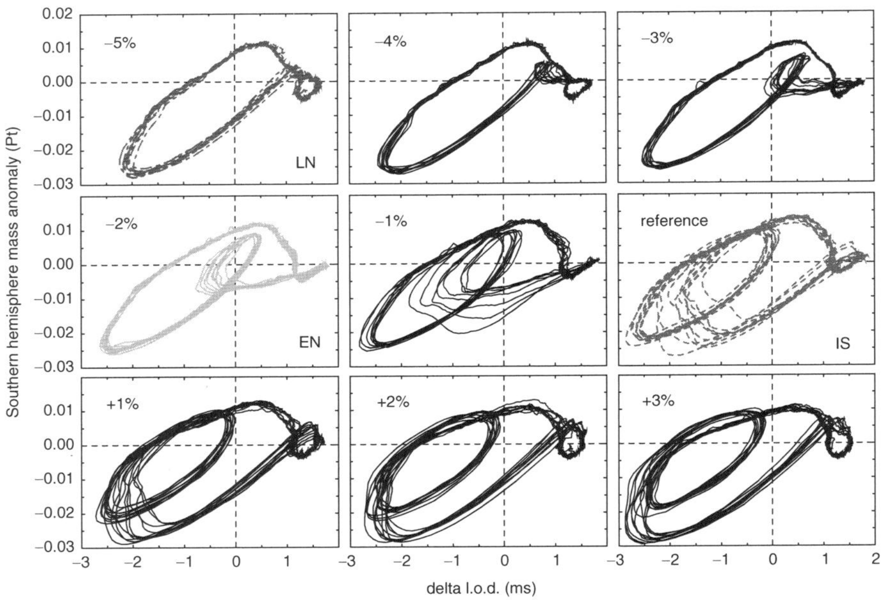

Fig. 3.18 Physical phase space projections (cf. Fig. 3.6) of the initial 10 year segments of a set of AGCM simulation under climatological lower boundary conditions ("summer monsoon solutions") at varying insolation, from −5 % to +3 % of present-day ("reference")

3.5.1 *Varied Insolation*

It goes without saying that a structurally sensitive dynamic object like this AGCM's boreal summer monsoon, which acts as a 'modulation unit' of the global climate system's seasonal cycle in fact and controls the worldwide patterns of seasonal drought and flood, bears paleoclimatic relevance. When present-day climatological lower boundary forcing is maintained, the 'interactive' monsoon ceases to show up during the seasonal march at reduced insolation by about 11 % [22] (reference case used here throughout: 1,396 W/m^2). Even though such an attractor continues to exist (at least down to −13 %; not shown here), it may no longer capture the seasonal trajectory. The same effect appears with varied model parameterizations. Increasing atmospheric background friction, e.g., may result in a complete loss of the target of study, the 'interactive' boreal summer monsoon [22].

Figure 3.18 shows structurally changing seasonal trajectories under varied solar constant (1 % steps, from −5 % to +3 %; 10 years each):

(*i*) With increasing (decreasing) insolation, the intraseasonal 'rotation number', i.e. the (non-integer, in general) number of active-break sequences per season, increases (decreases).

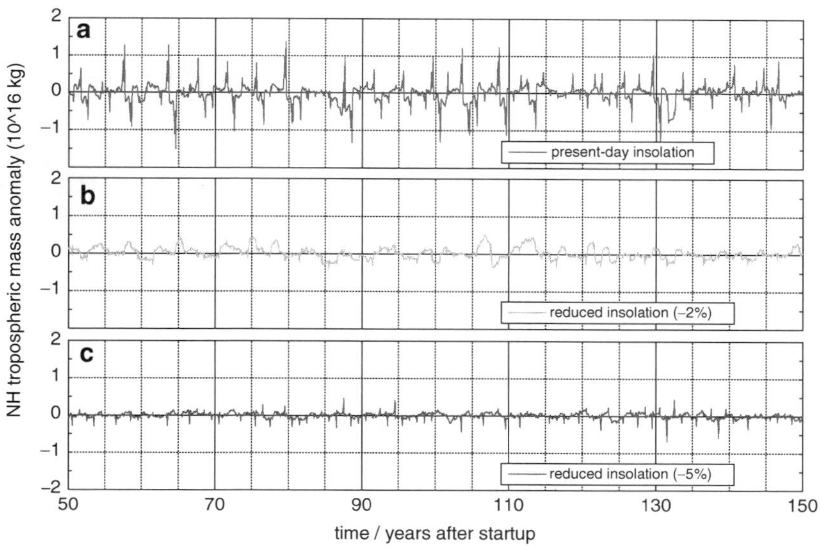

Fig. 3.19 NH tropospheric mass anomalies for centennial seasonal AGCM runs ("summer monsoon solutions" each) under (**a**) present-day and reduced insolation of (**b**) $-2\,\%$ and (**c**) $-5\,\%$ (daily data, 365 day moving averages); lower boundary conditions: present-day climatology

(ii) With higher excitation, i.e. increasing 'rotation number', the system loses retreat trajectories of the El Niño trigger type—but these appear again at even higher insolation ($+6\,\%$; [21]).

(iii) At insolation below 'present-day', regimes are found of a substantially higher number ($-1\,\%$), or even permanent ($-2\,\%$) monsoon retreat trajectories of this type—recalling paleoclimatic evidence of "Super El Niños".

(iv) A simple case of permanent "La Niña prone" trajectories, in contrast, shows up at further reduced insolation ($-5\,\%$).

(v) Highest model climate variability is found in the vicinity of the reference state (rotation number $\sim 1.5 \cdots < 2$), whereas evolution into integer ("resonant") rotation numbers in either direction reduces interannual variability; its change does not parallel the direction of changes in forcing.

The AGCM's "non-resonant" or "resonant" variability types develop at only gradually changing topological background. Time series of NH tropospheric mass anomaly for the three cases emphasized in Fig. 3.18 are given in Fig. 3.19, two examples each of five consecutive years are shown in detail in Fig. 3.20 (cf. also Fig. 3.16 and 3.17).

The structural evolution of the midsummer attractor with varied insolation (Fig. 3.21) shows maintenance of (weakening) midsummer period doubling down to $\sim -2\,\%$ insolation. The drift into the range of subrotation with increasing insolation is interrupted at $+4\,\%$ (and $+7\%$; not shown) where the attractor shifts back toward co-rotational break-monsoon conditions. Figure 3.22 (left panel) summarizes these

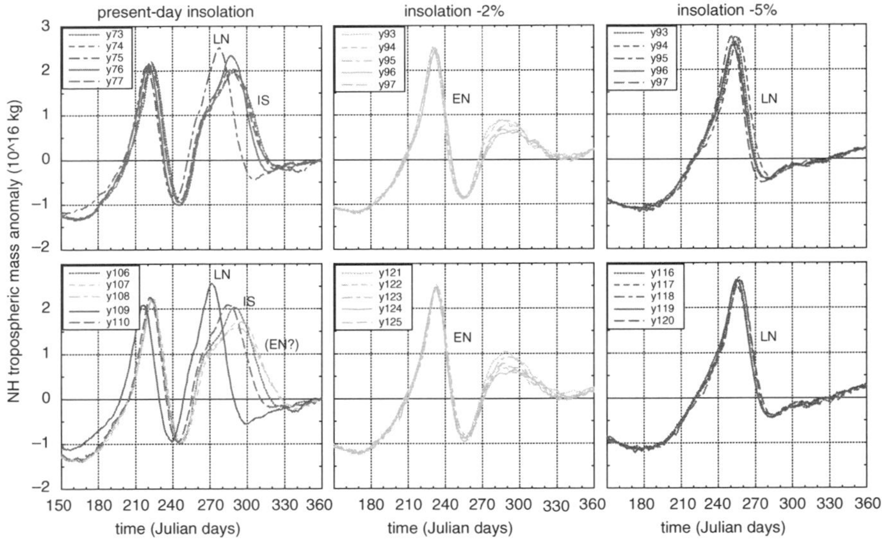

Fig. 3.20 NH tropospheric mass anomalies for two 5-year segments each of seasonal runs for present-day and reduced insolation (cf. Fig. 3.19, also for years given in the insets)

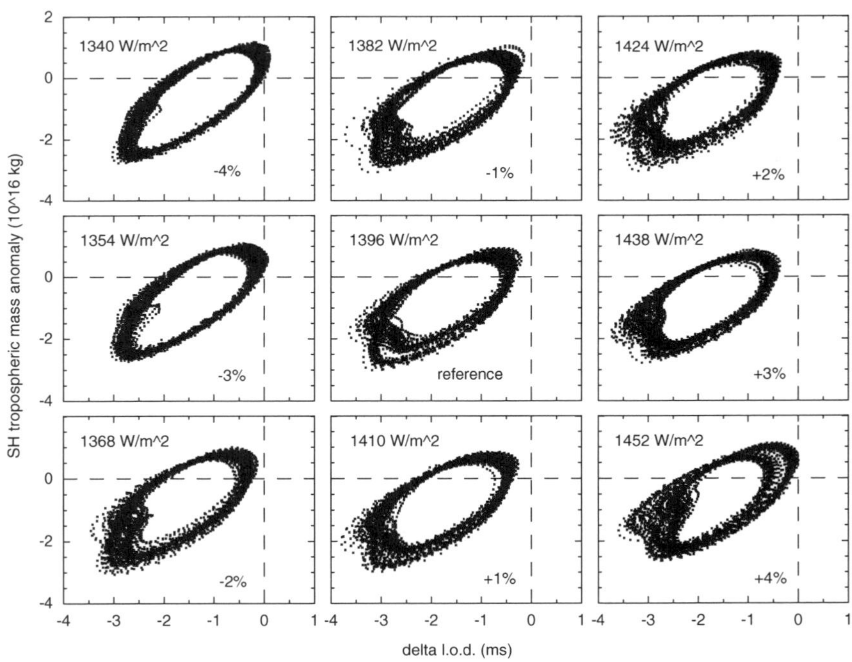

Fig. 3.21 Phase space projections (cf. Fig. 3.6) of 3,650 day segments of a set of perpetual August 8 AGCM simulations under climatological lower boundary forcing ("summer monsoon solution") but varying insolation [82]; −4 % to +4 % of 'present-day' (reference)

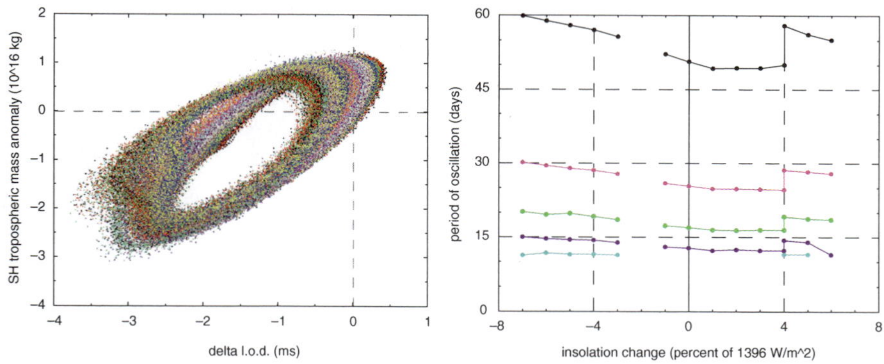

Fig. 3.22 Midsummer study as in Fig. 3.21; *left panel*: Torus segment for insolations from −13 % to +7 %; *right panel*: Major Fourier modes of the Δl.o.d. coordinate show preserved internal synchrony (at −2 % a broadband spectrum without clear modal structure is found)

evolutions over the range from −13 % to +7 % insolation. The right panel shows a result of paramount interest with regard to the dynamic organization of the boreal summer monsoon: preservation of high internal synchrony among the hierarchy of planetary waves over a substantial range of insolation (−7 % to +6 %). The gap at −2 % is due to a broadband spectrum without clearly dominating peaks, pointing to weakly chaotic behaviour (irregular alternation between period-1 and period-2). The jump at +4 % is due to a long transient, very similar to Fig. 3.11.

3.5.2 *Non-climatological Lower Boundary Forcing*

AMIP [34] SSTs and seaice distributions for 1982–1988 have been used for comparison with the AGCM's solution under climatological lower boundary conditions. Figure 3.23 shows such a 7-year trajectory. Though the GCM's spinoff must be considered and year 1981 is missing for sound conclusion, the strong 1982–1983 El Niño is reflected in shifted monsoon retreat loops, but these are not of the *El Niño trigger* type, in contrast to the three trajectories before and during the 1987/1988 ENSO with its marked monsoon signature. This calls to mind again the two types of El Niño [45, e.g.]. Whereas the 1982/1983 case has been exceptional due to a potential volcanic trigger (El Chichon eruption), the El Niño that started 1986 belongs to the "Central Pacific" (in contrast to "Eastern Pacific") type [46]—and might have to do with the monsoon retreat mechanism outlined (Sects. 3.3.6 and 3.4).

Though the AGCM version used here differs from the one in [84], weaker excitation of its boreal summer monsoon for 1982–1988 lower boundary forcing is confirmed. The 1987 drought over India had been traced back there to a mid-season 'tunneling' of the interactive monsoon by a sort of blown-up 'pre-monsoon'

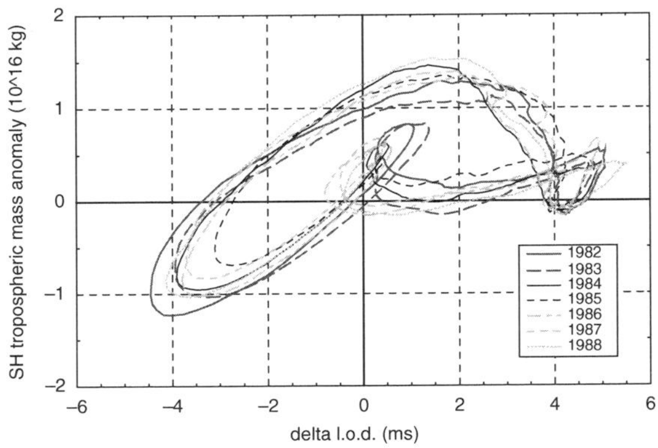

Fig. 3.23 Physical phase space projection (as in Fig. 3.16, but changed coordinate range): Seasonal AGCM simulation under non-climatological SST and seaice forcing, 1982–1988

attractor [21, 82, 84]. The stronger zonal wind component of Fig. 3.23 appears to 'blow away' part of the meridional wind structures, thus weakening the 'lateral' monsoon in the model version used here (the mean orography in the CCAS GCMs makes them generally inclined to solutions with stronger zonal winds—which may also be triggered by other factors, however).

3.5.3 Mixed-Layer Ocean

The coupled model preserves a reminder of the dynamics the AGCM alone shows up, but the intraseasonal motion becomes substantially attenuated when 'switching on' the thermodynamic MLO model. Figure 3.24 (left panel) illustrates this for a highly excited "summer monsoon solution" from a series of experiments aimed to find an appropriate coupling strength that does not extinguish the dynamic features of the AGCM but shows some ocean response to leave the climatological lower boundary forcing. In contrast to this seasonal monsoon response, the perpetual, 'intraseasonal' SO of the AGCM not only remains unattenuated, but even becomes amplified in the coupled model (right panel).

To finally return 'back to the roots', a new Kuwait oil well fire experiment was started from the stabilized solution of the coupled model. It exhibits strongly enhanced intraseasonal response during the 'smoky season' (i.e. to the additional persistent lower tropospheric heat source over the region) that approaches the AGCM-only solution (Fig. 3.25). South Asian lower tropospheric forcing thus surmounts the attenuating effect of the global ocean on the GCM's intraseasonal monsoon activity. This is a remarkable result in view of mechanisms that drive the global boreal summer monsoon, of its sensitivity and structural robustness. Note also that even highly excited monsoon dynamics of the AGCM are extinguished by an upper tropospheric smoke load, whereas they are kept alive, though strongly attenuated, in the coupled case (upper troposphere smoke scenarios not shown here).

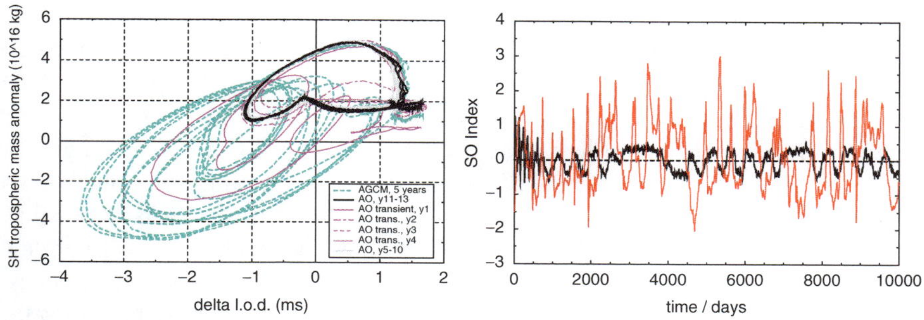

Fig. 3.24 *Left panel*: Highly excited "summer monsoon solution" of the AGCM and its response to coupling to the thermodynamic mixed layer ocean (MLO) of the CCAS-B GCM; shown are 5 years of AGCM solution (*bluish green*), 10 years of transient response after MLO coupling (*pink/grey*), and 3 years of stabilized AGCM-MLO (AOGCM) motion (*black*); *right panel*: SO Index of two perpetual day September 12 GCM simulations, with climatological SST forcing (*black*) and with interactive, thermodynamic mixed-layer ocean model (*red*); 10 days moving average applied twice

Fig. 3.25 Relative AAM (scaled to Δl.o.d.) of the disturbed AOGCM climate in another Kuwait oil well fire experiment; shown are four scenario calculations of seasonal smoke load during the first year, and the corresponding control run

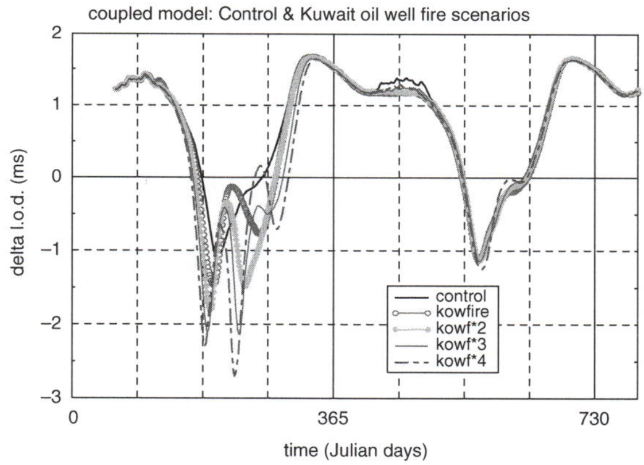

3.6 Hypothesis on the Global Monsoon: An Update

The basic question that emerged when discovering the sensitive oscillatory GCM response to forcing scenarios [11] read: externally forced or internally generated? The hypothesis given in [12, 13] adopts the topological view of self-maintained oscillations. Abruptness of monsoon onset, nearly 'explosive' evolutions of (observed and simulated) wind regimes, an activity cycle within the observed period band, etc., called for in-depth analyses of the CCAS-B AGCM's phase space trajectories. In conjunction with theoretical results (e.g., [50, 92]), detailed conclusions on the topology in the rear of observed monsoon dynamics could be drawn [22], including the "roulette" situation at monsoon retreat [12]. Computational stability problems led to speculation about a chaotic July regime [24] which later became confirmed [21].

The geometrical interpretation of boreal summer monsoon dynamics is summarized first (Sect. 3.6.1), based on the rigid-body Earth, atmosphere–land model system under climatological boundary conditions (insolation, SSTs, seaice, snowlines). The major observable consequence of low-dimensional behaviour of the high-dimensional climate system rests in its internal organization which by necessity bears synchronous motions [18]. Non-observability of the system's attractor sets and higher complexity of the real world notwithstanding, basic features of the observed climate may appear in a new conceptual light when accepting this topological view (Sect. 3.6.2). Key monsoonal aspects of the Earth's climatic evolution have been found to be borne in the coupled atmosphere–land system (Sect. 3.6.3).

3.6.1 Topological Perspective

(i) *The boreal summer monsoon constitutes a regime of planetary scale oscillations in the 40–60 days band that is separated from the seasonal cycle by Hopf bifurcations; the climate system's intraseasonal attractor set thus forms a torus segment in phase space with minor (poloidal) circumference made up by the major active-break monsoon cycle.*

(ii) *This type of monsoon dynamics emerges in a two-stage process that starts with a supercritically evolving, orographically excited standing planetary 'pre-monsoon' wave, followed by passage of a subcritical Hopf bifurcation of the global circulation system at abrupt onset of the 'interactive' monsoon with its inherent tropical–extratropical interactions.*

(iii) *Subcriticality of the monsoon onset bifurcation may also be blamed for the observed 'predictability barrier' in boreal spring [92].*

(iv) *Hard transition at onset of the major monsoon activity cycle kicks the system into a regime of chaotic oscillations, followed by a backward running period multiplication route ('out of chaos') across the season.*

(v) *The 'interactive' monsoon is organized by synchronized planetary waves, with lowest rational frequency relationships at top of the* Farey tree *of rational numbers, i.e. at strongest possible synchronous motion known from nonlinear dynamic systems which are made up of a hierarchy of oscillators [55].*

(vi) *Highest synchrony and shortest period of the major monsoon activity cycle are reached at midsummer, and are gradually lost again as the season advances toward monsoon retreat, where the period becomes substantially stretched: the cycle thus shows a marked intraseasonal frequency drift which contributes to its broadband nature [84].*

(vii) *The subrotational spring fixpoint evolves into the 'pre-monsoon' oscillation when the NH/SH atmospheric mass distribution approaches its seasonal mean; the 'interactive' monsoon commences when this balance is being broken in favour of the NH; it evolves chaotically as long as both the subrotational center of the cycle remains near that state of mass balance and monsoon activity may enter atmospheric superrotation during break conditions.*

(viii) *The period-2 structure is bound to an 'enstrophy stage' located in the subrotational, NH mass excess phase of oscillation; it is lost in favour of period-1 when mass balance is approached by the phase space location of this stage and when the cycle enters superrotation at break monsoon conditions.*

(ix) *As the season advances, the geometrical center of the activity cycle diverts from its 'center of gravity' due to an increasing density of states along an evolving 'wrinkle' on the torus mantle; borne in the enstrophy stage, the wrinkle runs along a mantle line that ends in a point of perfect relative AAM and hemispheric mass balance, where the monsoon oscillation is extinguished, leaving an unstable fixpoint of the boreal summer circulation.*

(x) *As a 'remnant' of the torus wrinkle, for a short period after structural monsoon retreat a slow irregular wander remains between unstable summer and winter fixpoints, showing essentials of the Southern Oscillation—until the (superrotational, SH mass excess) winter fixpoint stabilizes.*

(xi) *Monsoon retreat, which is governed by the relative phase between intraseasonal activity cycle and seasonal cycle, determines the state of autumn climate by selecting (in a kind of "roulette") one of roughly three retreat classes: from (a) active, (b) dormant, or (c) reviving monsoon phases.*

(xii) *Retreat from an active monsoon phase results in (a) 'Indian summer' type autumn circulation, retreat from dormant monsoon supports (b) fast, 'La Niña prone' transition into the winter circulation, and retreat from reviving monsoon activity generates (c) 'El Niño trigger' autumn trajectories [84].*

3.6.2 Conceptual Reasoning

(xiii) *The phenomenon of "global monsoon" with individual, but interacting branches around the globe (which follow a common intraseasonal pace at different phases of their respective activity cycles) is a consequence of strong internal synchronization of the hierarchy of planetary waves that organizes global teleconnections; topologically, the "global monsoon" reflects the motion of individual monsoons across the basin of a common global attractor set, or 'monsoon manifold'.*

(xiv) *Whereas there may be some channeling effect on the spring trajectory due to the repellor surface of the subcritical monsoon onset bifurcation, which tends to control the date of monsoon onset, there is no such topologically conditioned focusing at monsoon retreat; this may not only explain observed differences between spring and autumn circulations [31] but also those in the variability of monsoon onset and retreat dates [70].*

(xv) *Monsoon definitions based on seasonally changing wet/dry and wind regimes alone do not grasp essentials of the system, notably salient features including the intraseasonal pulse, the dynamic organization as a separate climate regime, abruptness of planetary reorganization, and intrinsic variability; a topologically founded definition of the system is thus advocated.*

(xvi) *The present-day climate regime on Earth is a "monsoon climate" in the sense that the interactive, planetary scale monsoon governs the boreal summer in general; occasional monsoon failure, however, might hint at a dynamical status in the vicinity of a general "monsoon climate" bifurcation of paleoclimatic dimension.*

3.6.3 Paleoclimatic Relevance

(xvii) *Interannual climate variability borne in atmospheric dynamics may undergo long-term structural change due to intraseasonal changes of monsoon excitation; these include emergence or not of the regime of tropical–extratropical interactions in boreal summer, with structural consequences in the worldwide distribution of drought and flood, and transitions between integer and non-integer number of activity cycles per season leading to systematic, structural changes of the regime of boreal autumn climate.*

(xviii) *The present-day monsoon climate regime of dominant Indian summer type boreal autumn circulations (due to a non-integer rotation number in general) may be superseded by regimes of lower interannual variability; no matter what direction of monsoon excitation the system adopts, its atmospheric contribution will probably develop toward more 'resonant' conditions in boreal summer (integer rotation number).*

(xix) *Paleoclimatic "Super El Niño" or "Super La Niña" regimes may be borne in dominant resonant dynamics of the (interactive) boreal summer monsoon, notably if the boreal winter monsoon does not enter an own interactive state—which would modulate the interannual climatic effects of the system's dynamical status in summer, focused upon here.*

(xx) *Lasting absence or return of interactive summer monsoon dynamics may play a role as amplifier of glaciation cycles; the system's operation in the vicinity of its monsoon bifurcation in boreal summer might explain both abruptness of climatic transitions in the Earth's history and the very fact that glaciations did not necessarily coincide (or even appear) at either hemisphere.*

(xxi) *Persistence of 'pre-monsoon' dynamics in boreal summer if the 'interactive' monsoon ceases to exist might bear a "green Sahara" regime, the existence of a large paleo-lake to the north of Tibet, and paleoclimatic teleconnections between the Indian monsoon region and the North Atlantic.*

3.7 Conclusions

Given the obstinate issue of monsoon simulation in GCMs and the persistent challenge of monsoon forecasting, a dynamic-systems perspective as outlined helps to understand the very nature of these problems and to develop ideas of solution. In view of the reviving concept of "global monsoon", it was time to retell the tale of exploration of a qualitatively correct and quantitatively reasonable GCM solution that allows to pose well-known dynamic essentials of the boreal summer monsoon into a geometrical context.

A major topic was the nature of computational problems encountered. It may hardly be due to a coding bug if a long-term "summer monsoon solution" runs into computational instability during an exceptionally early monsoon onset of its 199st year. The chaotic July regime found is just one aspect that might be blamed. Occasional extreme excursions of purely dynamic origin near the turn of the seasons, reminding of homoclinic orbits, also may drive such a model beyond the limits of validity in any one of its parameterizations. One such excursion has been traced back, for example, to passage during the seasonal march of a state of simultaneous balance in the interhemispheric distribution of tropospheric mass and in relative AAM, or Δl.o.d. (that is, tropospheric co-rotation).

Due to unfortunate data losses over the more than 15 years of 'frozen-in' GCM based conceptual monsoon studies, efforts to reconstruct earlier results are indicated. This includes the "summer–winter–monsoon" solution, a specific code version which ran stably during attractor soundings across the chaotic July regime (but turned out to be unstable in the seasonal run), and an 'atmospheric El Niño' solution coincidentally found in a highly excited model version that also exhibits a strong regional attractor of the Indian monsoon system close to monsoon retreat (though source texts and log files are lost in part, compiled model versions have been saved in general).

Having reached firm footing again with the coarse resolution AGCM, the range of a second generation of computational exploration has to be substantially extended into a variety of directions, including

(1) Detailed characterization of the chaotic July regime,
(2) In-depth analysis of the monsoon retreat bifurcation,
(3) Attractor sounding under non-climatological forcings,
(4) Retreat ensembles under non-climatological forcings,
(5) Study of the planetary wave hierarchy at varied horizontal resolution,
(6) Systematic exploration of the coupled AGCM/MLO model,
(7) IPCC and other scenario simulations.
Tracing of the monsoon solutions found into higher spatial model resolution comprises the task as well to
(8) Refine the vertical model dimension (multi-level GCM).
To this end, it is of paramount importance to understand the monsoon mechanisms at the coarse resolution model world in much greater detail via
(9) In-depth diagnostics of mechanism and feedbacks.

At the turn of the 1980s, the CCAS-B AGCM had been equipped with an advanced diagnostic system which calls for certain regeneration today in order to realize opportunities that result from present-day computational capabilities and capacities. Finally, following original intentions and earlier developments in the field of software engineering [10], the model should be kept in a technical state so as to be applicable as a

(10) "Climate model for research and training" [36].

Confirmation of the monsoon dynamics found in related integrals of motion of the real system is of fundamental importance since low-dimensional behaviour, which emerges from the small scales, is not usually assumed to prevail at global scale—and even to dominate there. Low-dimensional organization of the global climate system via extensive internal synchronization of planetary waves might bear a key to successful monsoon simulation. Whereas observational evidence exists for AAM signatures of monsoon activity, interhemispheric exchanges of atmospheric mass are apparently not measured in sufficient detail so as to follow these motions into intraseasonal scales. There is hope to gain this access via the GRACE project [79, 94].

Though existence of a chaotic July regime might lead one to believe that Lorenz' idea of "chaos and intransitivity" as fundamental source of interannual variability [65] is at work in the model, this may be posed into question. Passage of the July regime in a seasonal simulation might proceed via some (intraseasonal) unstable periodic orbit (UPO), without leaving a footprint of "chaos" in the system's trajectory. More important in this respect is the "complexity" issue, namely the attractor surface of variable shape and 'attractivity' across the season and the 'rotation number', which together control monsoon retreat and may govern interannual variability. Whether the system's trajectories show irregularity due to a complex generating process or a chaotic one, cannot be decided by the argument of 'sensitivity to initial conditions'—which is a symptom, but not the essence of chaos. More reliable is the search for UPOs, as part of the 'dynamic skeleton' of a chaotic system. Related aspects of interannual data analysis are touched in [19].

Due to the bare existence of critical transitions within the annual cycle, the climate system is a sensitive receiver (and amplifier) of signals. The oceans provide a huge 'collector surface' to the monsoon circulation which acts to imprint solar or thermal signals onto the water cycle. The monsoon hypothesis anticipates a relatively simple basic structure in boreal summer, the torus segment, of nevertheless rather complex fine structure. The system's evolution proceeds via changes that these (intra-) seasonal features may experience due to changed forcing. Effects of coupled subsystems, notably atmosphere–ocean interactions, are an integral part of these dynamics, but the monsoon system itself is known for long to play an active role in shaping the climate system's variability and evolution [72]. Out of its intraseasonal scales, interannual to millennial climatic regimes may emerge. Robustness of the present-day "monsoon climate" on Earth may crucially depend on the degree of internal synchronization in boreal summer—which might even root in the shortest scales at the very base of the hierarchy of planetary waves.

Acknowledgements This paper is cordially devoted to the former Climate Dynamics Group, Berlin, with special gratitude to Ina Tschentscher, Klaus Worbs, Judith Perlwitz and Stefan Kroschk. Our report series started with results on the CCAS-B AGCM's global monsoon and its individual branches in boreal summer [22] as presented at the 1993 EGS General Assembly—shortly after formation of the team, assigned to study the GCM's behaviour in detail and to further work out the monsoon hypothesis [13]. The first visualization of a GCM's monsoon dynamics was due to Andreas Hantzschk there (also 1994 at the WCRP monsoon conference). As a 'wilding' born in the winds of change in Germany, the group did not survive 'subcritical' funding. In a final act for 15 years, selected material was distributed at the 1997 IWM-1 workshop [21]. A *post mortem* collection of papers [24], including the 1994 update of the monsoon hypothesis, marks the transition to signal analysis aimed to verify conceptual insights as gained from these studies [18, 19].

References

1. Abraham, R.H., Shaw, C.D.: Dynamics. The Geometry of Behavior. 2nd edn. Addison-Wesley, Redwood City (1992)
2. Aleksandrov, V.V., Gates, W.L.: The performance of a coarse-grid version of the OSU two-level atmospheric GCM. Report No. 24. Climatic Research Institute, Oregon State University, Corvallis (1981)
3. Aleksandrov, V.V., Arkhipov, P.L., Parkhomenko, V.P., Stenchikov, G.L.: A global model of the ocean–atmosphere system, and a study of its sensitivity to changes in CO_2 concentration (in Russian). Izv. AN SSSR, Fiz. Atmosf. Okeana **19**, 451–458 (1983)
4. Alexander, R.C., Mobley, R.L.: Monthly average sea-surface temperatures and ice-pack limits on the 1° global grid. Mon. Weather Rev. **104**(2), 143–148 (1976)
5. Anderson, J.R., Rosen, R.D.: The latitude–height structure of 40–50 day variations in atmospheric angular momentum. J. Atmos. Sci. **40**(6), 1584–1591 (1983)
6. Arakawa, A., Lamb, V.R.: Computational design of the basic dynamical processes of the UCLA General Circulation Model. Meth. Comp. Phys. **17**, 173–265 (1977)
7. Arakawa, A., Katayama, A., Mintz, Y.: Numerical simulation of the general circulation of the atmosphere. Reprint No. 4. Department of Meteorology, University of California, Los Angeles (1968)
8. Battisti, D.S., Hirst, A.C.: Interannual variability in a tropical atmosphere–ocean model: Influence of the basic state, ocean geometry and nonlinearity. J. Atmos. Sci. **46**(12), 1687–1712 (1989)
9. Budyko, M.I.: The effect of solar radiation variations on the climate of the Earth. Tellus **21**(5), 611–619 (1969)
10. Carl, P.: Software engineering aspects of computational systems analysis in physics. In: Sydow, A., Tzafestas, S.G., Vichnevetsky, R. (eds.) Systems Analysis and Simulation 1988, vol. II. Math. Res., Akademie-Verlag Berlin. vol. 47, pp. 375–378 (1988)
11. Carl, P.: Notes on the climatic response in the aftermath of Gulf War II. Z. Meteorol. **41**(6), 476–480 (1991)
12. Carl, P.: Persistent localized lower troposphere smoke within the planetary monsoon system. Available from the author; submitted to Nature Nov. 21, Ms.-No. C11408 (1991)
13. Carl, P.: Zur dynamischen Struktur des planetaren Monsuns. Wiss. Z. Humboldt-Univ. Berlin, R. Math. Nat.wiss. **41**(2), 29–35 (1992)
14. Carl, P.: Monsoon dynamics in a low-dimensional GCM. WCRP-84, WMO/TD-No. 619(II), 773–780. WMO, Geneva (1994)
15. Carl, P.: Summer monsoon, MJO, annual cycle, QBO, Southern Oscillation ... On the dynamic architecture of the atmosphere. Preprint volume, 25–26. Second International Scientific Conference on the Global Energy and Water Cycle, Washington. vol.(46) (1996)

16. Carl, P.: Eine zweite – generisch intrasaisonale – Südliche Oszillation aus dem Atmosphäre–Land–System? Ann. Meteorol. (Offenbach) **37**(2), 351–352 (1998)
17. Carl, P.: On the dynamical status of the climate system – I: A general circulation model en route to chaos. In: Stavrinides, S.G., Banerjee, S., Caglar, H., Ozer, M. (eds.) Chaos and Complex Systems, Proceedings. Springer, Berlin Heidelberg (in print, 2012)
18. Carl, P.: On the dynamical status of the climate system – II: Synchronous motions galore across the records. In: Stavrinides, S.G., Banerjee, S., Caglar, H., Ozer, M. (eds.) Chaos and Complex Systems, Proceedings. Springer, Berlin Heidelberg (in print, 2012)
19. Carl, P.: Synchronous motions across the instrumental climate record. In: Banerjee, S., Rondini, L. (eds.) Applications of Nonlinear Dynamics and Chaos in Science and Engineering, vol. 4. Springer, Berlin Heidelberg (scheduled for 2013) (to be submitted)
20. Carl, P., Stenchikov, G.L.: Structural analysis of the climatic response to a nuclear war. In: Sydow, A., Tzafestas, S.G., Vichnevetsky, R. (eds.) Systems Analysis and Simulation 1988, vol. II. Math. Res., Adademic-Verlag Berlin. vol. 47, pp. 33–36 (1988)
21. Carl, P., Tschentscher, I.: A nonlinear atmosphere–land–system with intraseasonal to interannual scales. Draft of a paper presented at the First WMO International Workshop on Monsoon Studies (IMW-1). Denpasar, Bali, 24–28 February 1997
22. Carl, P., Grell, J., Kroschk, S., Perlwitz, J., Tschentscher, I., Worbs, K.D.: Studies on the dynamic structure of a GCM's boreal summer monsoon. Report No. 1. Climate Dynamics Group, Berlin, Germany (1993)
23. Carl, P., Worbs, K.D., Tschentscher, I.: On a dynamic systems approach to atmospheric model intercomparison. WCRP-92, WMO/TD-No. 732, pp. 445–450. WMO, Geneva (1995)
24. Carl, P., Eichler, T., Kroschk, S., Lönhardt, H., Schimmel, E., Tschentscher, I., Worbs, K.D.: Konzeptionelle Studien zur Dynamik des Atmosphäre–Land–Systems. Report No. 7. Climate Dynamics Group, Berlin, Germany (1998)
25. Carl, P., Svirezhev, Y., Stenchikov, G.: Environmental and biospheric impacts of nuclear war. In: Jørgensen, S.E., Fath, B.D. (eds.) Global Ecology. Encyclopedia of Ecology, vol. 2, pp. 1314–1321. Elsevier, Oxford (2008)
26. Carrera, M.L., Gyakum, J.R.: Southeast Asian pressure surges and significant events of atmospheric mass loss from the northern hemisphere, and a case study analysis. J. Clim. **20**(18), 4678–4701(2007)
27. Chang, P., Wang, B., Li, T., Ji, L.: Interactions between the seasonal cycle and the Southern Oscillation—Frequency entrainment and chaos in a coupled ocean–atmosphere model. Geophys. Res. Lett. **21**(25), 2817–2820 (1994)
28. Chang, P., Ji, L., Wang, B., Li, T.: Interactions between the seasonal cycle and El Niño–Southern Oscillation in an intermediate coupled ocean–atmosphere model. J. Atmos. Sci. **52**(13), 2353–2372 (1995)
29. Chang, P., Wang, B., Lau, N-C.G. (eds.): The global monsoon system: Research and forecast. TRMP-70, WMO/TD-No. 1266. WMO, Geneva (2005)
30. Charney, J.G., DeVore, J.G.: Multiple flow equilibria in the atmosphere and blocking. J. Atmos. Sci. **36**(7), 1205–1216 (1979)
31. Fleming, E.L., Lim, G.-H., Wallace, J.M.: Differences between spring and autumn circulations in the Northern Hemisphere. J. Atmos. Sci. **44**(9), 1266–1286 (1987)
32. Gadgil, S.: The Indian monsoon 3. Physics of the monsoon. Resonance **12**(5), 4–20 (2007)
33. Gadgil, S., Sajani, S.: Monsoon precipitation in AMIP runs. Clim. Dynam. **14**(9), 659–689 (1998)
34. Gates, W.L.: AMIP: The atmospheric model intercomparison project. Bull. Am. Meteorol. Soc. **73**(12), 1962–1970 (1992)
35. Gates, W.L., Batten, E.S., Kahle, A.B., Nelson, A.B.: A documentation of the Mintz–Arakawa two-level atmospheric general circulation model. R-877-ARPA. Rand Corporation, Santa Monica (1971)

36. Gerstengarbe, F.-W. (ed.): 3. Deutsche Klimatagung, Potsdam, 11–14 April 1994, Tagungsband der Vorträge und Poster (Proceedings). PIK Report No. 1. Potsdam Institute for Climate Impact Research, Potsdam, Germany (1994)
37. Ghan, S.J., Lingaas, J.W., Schlesinger, M.E., Mobley, R.L., Gates, W.L.: A documentation of the OSU two-level atmospheric general circulation model. Report No. 35. Climatic Research Institute, Oregon State University, Corvallis (1982)
38. Ghil, M.: Climate sensitivity, energy balance models, and oscillatory climate models. J. Geophys. Res. **89**(D1), 1280–1284 (1984)
39. Ghil, M.: Dynamics, statistics and predictability of planetary flow regimes. In: Nicolis, C., Nicolis, G. (eds.) Irreversible Phenomena and Dynamical Systems Analysis in Geosciences, pp. 241–283. Reidel, Dordrecht (1987)
40. Ghil, M., Childress, S.: Topics in Geophysical Fluid Dynamics: Atmospheric Dynamics, Dynamo Theory, and Climate Dynamics. Springer, New York (1987)
41. Ghil, M., Mo, K.: Intraseasonal oscillations in the global atmosphere. Part I: Northern hemisphere and tropics. J. Atmos. Sci. **48**(5), 752–779 (1991)
42. Ghil, M., Kimoto, M., Neelin, J.D.: Nonlinear dynamics and predictability in the atmospheric sciences. Revs. Geophys. **29**(Suppl.), 46–55 (1991)
43. Goddard, L., Graham, N.E.: Precipitation forecast skill and the hydrological cycle. Preprint volume, 490–491. Second International Scientific Conference on the Global Energy and Water Cycle, Washington (1996)
44. Goswami, B.N.: South Asian summer monsoon: An overview. In: Chang, C.-P., Wang, B., Lau, N.-C. (eds.) The Global Monsoon System: Research and Forecast (IWM-III). TMRP-70, WMO/TD-No. 1266, pp. 47–71. WMO, Geneva (2004)
45. Graf, H.-F.: On El Niño/Southern Oscillation and northern hemispheric temperature. Gerlands Beitr. Geophys. **95**(1), 63–75 (1986)
46. Graf, H.-F., Zanchettin, D.: Central Pacific El Niño, the "subtropical bridge," and Eurasian climate. J. Geophys. Res. **117**, D01102 (2012). doi:10.1029/2011JD016493
47. Gutowski, W.J., Iacono, M.J., Liang, X.-Z., Wang, W.-C.: Simulating climate with two different numerical schemes. Tech. Rep. TR049, US Department of Energy (1990)
48. Held, I.M., Suarez, M.J.: Simple albedo feedback models of the icecaps. Tellus **26**(6), 613–628 (1974)
49. Hendon, H.H., Sperber, K.R., Waliser, D.E., Wheeler, M.C.: Modeling monsoon intraseasonal variability. Available from http://journals.ametsoc.org/toc/bams/92/9. Bull. Am. Meteorol. Soc., vol. 92, issue 9, p. 4, online only (2011)
50. Jin, F.-F., Ghil, M.: Intraseasonal oscillations in the extratropics: Hopf bifurcation and topographic instabilities. J. Atmos. Sci. **47**(24), 3007–3022 (1990)
51. Jin, F.-F., Neelin, J.D., Ghil, M.: El Niño on the devil's staircase: Annual subharmonic steps to chaos. Science **264**(5155), 70–72 (1994)
52. Johnson, D.R.: The forcing and maintenance of global monsoonal circulations: An isentropic analysis. Adv. Geophys. **31**, 43–316 (1989)
53. Krishnamurti, T.N., Bhalme, H.N.: Oscillations of a monsoon system. Part I. Observational aspects. J. Atmos. Sci. **33**(10), 1937–1954 (1976)
54. Krishnamurti, T.N., Sinha, N.C., Krishnamurti, R., Osterhof, D., Comeaux, J.: Angular momentum, LOD and monsoonal LF mode. J. Meteorol. Soc. Jpn. **70**(1), 131–165 (1992)
55. Lagarias, J.C.: Number theory and dynamical systems. Proc. Symp. Appl. Math. **46**, 35–72 (1992)
56. Langlois, W.E., Kwok, H.C.W.: Description of the Mintz–Arakawa numerical General Circulation Model. Tech. Rep. No. 3. Department of Meteorology, University of California, Los Angeles (1969)
57. Lau, K.M., Peng, L.: Origin of low frequency (intraseasonal) oscillations in the tropical atmosphere. Part III: Monsoon dynamics. J. Atmos. Sci. **47**(12), 1443–1462 (1990)
58. Lau, K.M., Yang, G.J., Shen, S.H.: Seasonal and intraseasonal climatology of summer monsoon rainfall over East Asia. Mon. Weather Rev. **116**(1), 18–37 (1988)

59. Le Roux, D.Y., Pouliot, B.: Analysis of numerically induced oscillations in two-dimensional finite-element shallow water equations. Part II: Free planetary waves. SIAM J. Sci. Comput. **30**(4), 1971–1991 (2008)
60. Lin, Z., Zeng, Q.: Simulation of East Asian summer monsoon by using an improved AGCM. Adv. Atmos. Sci. **14**(4), 513–526 (1997)
61. Lorenz, E.N.: Simplified dynamic equations applied to the rotating–basin experiments. J. Atmos. Sci. **19**(1), 39–51 (1962)
62. Lorenz, E.N.: Deterministic nonperiodic flow. J. Atmos. Sci. **20**(2), 130–141 (1963)
63. Lorenz, E.N.: The mechanics of vacillation. J. Atmos. Sci. **20**(5), 448–464 (1963)
64. Lorenz, E.N.: Low-order models of atmospheric circulation. J. Meteorol. Soc. Jpn. **60**(1), 255–267 (1982)
65. Lorenz, E.N.: Can chaos and intransitivity lead to interannual variability? Tellus **42A**(3), 378–389 (1990)
66. Madden, R.A., Julian, P.R.: Detection of a 40–50 day oscillation in the zonal wind in the tropical Pacific. J. Atmos. Sci. **28**(5), 702–708 (1971)
67. Magaña, V., Yanai, M.: Tropical–midlatitude interaction on the time scale of 30 to 60 days during the northern summer of 1979. J. Clim. **4**(2), 180–201 (1991)
68. Marcus, S.L., Ghil, M., Dickey, J.O., Eubanks, T.M.: Origin of the 30–60 day oscillation in lod and atmospheric angular momentum: New findings from the UCLA General Circulation Model. In: Boucher, C., Wilkins, G.A. (eds.) Earth Rotation and Coordinate Reference Frames, pp. 98–105. Springer, New York (1990)
69. Moiseev, N.N., Aleksandrov, V.V., Tarko, A.M.: Man and Biosphere (in Russian). Nauka, Moscow (1985)
70. Murakami, T., Chen, L.-X., Xie, A.: Relationship among seasonal cycles, low-frequency oscillations, and transient disturbances as revealed from outgoing longwave radiation data. Mon. Weather Rev. **114**(8), 1456–1465 (1986)
71. Nakazawa, T.: Tropical super clusters within intraseasonal variations over the western Pacific. J. Meteorol. Soc. Jpn. **66**, 823–839 (1988)
72. Normand, C.: Monsoon seasonal forecasting. Quart. J. Roy. Meteorol. Soc. **79**(342), 463–473 (1953)
73. Randall, D.A. (ed.): General Circulation Model Development. Academic, San Diego (2000)
74. Reznyanskij, Y.D., Trosnikov, I.V.: Parameterization of the oceanic mixing layer when modelling the zonal atmospheric circulation (in Russian). Trudy Gidrometcent. SSSR **229**, 18–31 (1980)
75. Sellers, W.D.: A climate model based on the energy balance of the earth–atmosphere system. J. Appl. Meteorol. **8**(3), 392–400 (1969)
76. Stenchikov, G.L., Carl, P.: Climatic consequences of nuclear war: Sensitivity against large–scale inhomogeneities in the initial atmospheric pollutions. GDR Acad. Sci. and Phys. Soc., Berlin (1985)
77. Suarez, M.J., Schopf, P.S.: A delayed action oscillator for ENSO. J. Atmos. Sci. **45**(21), 3283–3287 (1988)
78. Tao, S., Chen, L.: A review of recent research on the East Asian summer monsoon in China. In: Chang, C.-P., Krishnamurti, T.N. (eds.) Monsoon Meteorology, pp. 60–92. Oxford University Press, New York (1987)
79. Tapley, B.D., Bettadpur, S., Watkins, M., Reigbar, C.: The gravity recovery and climate experiment: Mission overview and early results. Geophys. Res. Lett. **31**, L09607 (2004). doi:10.1029/2004GL019920
80. Thompson, S.L., Aleksandrov, V.V., Stenchikov, G.L., Schneider, S.H., Covey, C., Chervin, R.M.: Global climatic consequences of nuclear war: Simulations with three dimensional models. AMBIO **13**(4), 236–243 (1984)
81. Tomita, T., Yasunari, T.: On the two types of ENSO. J. Meteorol. Soc. Jpn. **71**(2), 273–284 (1993)
82. Tschentscher, I., Carl, P.: Climate and climate change dynamics on the attractor sets of a tropospheric GCM. Report No. 6. Climate Dynamics Group, Berlin, Germany (1996)

83. Tschentscher, I., Carl, P.: Dynamische Selbstorganisation in einem Atmosphäre–Land–System. Ann. Meteorol. **34**, 19–20 (1997)
84. Tschentscher, I., Worbs, K.D., Carl, P.: Frequency drift and retreat variability of a GCM's monsoon oscillator. WCRP-84, WMO/TD-No. 619(II), pp. 781–788. WMO, Geneva (1994)
85. Turner, A., Sperber, K.R., Slingo, J.M., Meehl, G.A., Mechoso, C.R., Kimoto, M., Giannini, A: Modelling monsoons: Understanding and predicting current and future behaviour. In: Chang, C.-P., Ding, Y., Lau, N.-C., Johnson, R.H., Wang, B., Yasunari, T. (eds.) The Global Monsoon System: Research and Forecast, 2nd edition (IWM-IV), pp. 421–454. World Scientific, Singapore (2011). LLNL-PROC-407065, Lawrence Livermore National Laboratory (2008)
86. Tziperman, E., Stone, L., Cane, M.A., Jarosh, H.: El Niño chaos: overlapping of resonances between the seasonal cycle and the Pacific ocean–atmosphere oscillator. Science **264**(5155), 72–74 (1994)
87. Tziperman, E., Cane, M.A., Zebiak, S.E.: Irregularity and locking to the seasonal cycle in an ENSO prediction model as explained by the quasi-periodicity route to chaos. J. Atmos. Sci. **52**(3), 293–306 (1995)
88. Valdes, P.: Built for stability. Nat. Geosci. **4**(July), 414–416 (2011)
89. Wang, B., Ding, Q., Liu, J.: Concept of global monsoon. In: Chang, C.-P., Ding, Y., Lau, N.-C., Johnson, R.H., Wang, B., Yasunari, T. (eds.) The Global Monsoon System: Research and Forecast, 2nd edition (IWM-IV), pp. 3–14. World Scientific, Singapore (2011)
90. Webster, P.J.: The elementary monsoon. In: Fein, J.S., Stephens, P.L. (eds.) Monsoons, pp. 467–522. Wiley, New York (1987)
91. Webster, P.J., Chou, L.C.: Low-frequency transitions of a simple monsoon system. J. Atmos. Sci. **37**(2), 368–382 (1980)
92. Webster, P.J., Yang, S.: Monsoon and ENSO: Selectively interactive systems. Quart. J. Roy. Meteorol. Soc. **118**(507), 877–926 (1992)
93. Webster, P.J., Magaña, V.O., Palmer, T.N., Shukla, J., Tomas, R.A., Yanai, M., Yasunari, T.: Monsoon: Processes, predictability, and the prospects for prediction. J. Geophys. Res. **103**(C7), 14451–14510 (1998)
94. Werth, S., Güntner A., Schmidt, R., Petrovic, S.: Integration of GRACE mass variations into a global hydrological model. Earth Planet. Sci. Lett. **277**, 166–173 (2009). doi:10.1016/j.epsl2008.10.021
95. Yasunari, T.: Cloudiness fluctuations associated with the northern hemisphere summer monsoon. J. Meteorol. Soc. Jpn. **57**(3), 227–242 (1979)
96. Yasunari, T.: Structure of an Indian summer monsoon system with around 40-day period. J. Meteorol. Soc. Jpn. **59**(3), 336–354 (1981)
97. Yasunari, T.: Low-frequency interactions between the summer monsoon and the northern hemisphere westerlies. J. Meteorol. Soc. Jpn. **64**(5), 693–708 (1986)
98. Yeh, T.-C., Dao, S.-J., Li, M.-T.: The abrupt change of circulation over the northern hemisphere during June and October. In: Bolin, B. (ed.) The Atmosphere and the Sea in Motion, pp. 249–267. Oxford University Press, New York (1959)
99. Zhou, T., Wu, B., Scaife, A.A., Brönnimann, S., Cherchi, A., Fereday, D., Fischer, A.M., Folland, C.K., Jin, K.E., Kinter, J., Knight, J.R., Kucharski, F., Kusunoki, S., Lau, N.-C., Li, L., Nath, M.J., Nakaegawa, T., Navarra, A., Pegion, P., Rozanov, E., Schubert, S., Sporyshev, P., Voldoire, A., Wen, X., Yoon, J.H., Zeng, N.: The CLIVAR C20C project: Which components of the Asian–Australian monsoon circulation variations are forced and reproducible? Clim. Dynam. **33**(7–8), 1051–1068 (2009)

Part III
Fractal and Its Applications in Epileptic Seizure

Chapter 4
Fractal Dimension in Epileptic EEG Signal Analysis

R. Uthayakumar

Abstract Fractal Analysis is the well developed theory in the data analysis of non-linear time series. Especially Fractal Dimension is a powerful mathematical tool for modeling many physical and biological time signals with high complexity and irregularity. Fractal dimension is a suitable tool for analyzing the nonlinear behaviour and state of the many chaotic systems. Particularly in analysis of chaotic time series such as electroencephalograms (EEG), this feature has been used to identify and distinguish specific states of physiological function.

Epilepsy is the main fatal neurological disorder in our brain, which is analyzed by the biomedical signal called Electroencephalogram (EEG). The detection of Epileptic seizures in the EEG Signals is an important tool in the diagnosis of epilepsy. So we made an attempt to analyze the EEG in depth for knowing the mystery of human consciousness. EEG has more fluctuations recorded from the human brain due to the spontaneous electrical activity. Hence EEG Signals are represented as Fractal Time Series.

The algorithms of fractal dimension methods have weak ability to the estimation of complexity in the irregular graphs. Divider method is widely used to obtain the fractal dimension of curves embedded into a 2-dimensional space. The major problem is choosing initial and final step length of dividers. We propose a new algorithm based on the size measure relationship (SMR) method, quantifying the dimensional behaviour of irregular rectifiable graphs with minimum time complexity. The evidence for the suitability (equality with the nature of dimension) of the algorithm is illustrated graphically.

We would like to demonstrate the criterion for the selection of dividers (minimum and maximum value) in the calculation of fractal dimension of the irregular curves with minimum time complexity. For that we design a new method of computing

R. Uthayakumar (✉)
Department of Mathematics, The Gandhigram Rural Institute – Deemed University, Gandhigram 624 302, Dindigul, Tamil Nadu, India
e-mail: uthayagri@gmail.com

S. Banerjee and L. Rondoni (eds.), *Applications of Chaos and Nonlinear Dynamics in Science and Engineering 3*, Understanding Complex Systems,
DOI 10.1007/978-3-642-34017-8_4, © Springer-Verlag Berlin Heidelberg 2013

fractal dimension (FD) of biomedical waveforms. Compared to Higuchi's algorithm, advantages of this method include greater speed and the criterion to choose the maximum and minimum values for time intervals. Comparisons with the other waveform fractal dimension algorithms are also demonstrated. In order to discriminate the Healthy and the Epileptic EEGs, an improved method of Multifractal Measure such as Generalized Fractal Dimensions (GFD) is also proposed. Finally we conclude that there are significant differences between the Healthy and Epileptic Signals in the designed method than the GFD through graphical and statistical tools. The improved multifractal measure is very efficient technique to analyze the EEG Signals and to compute the state of illness of the Epileptic patients.

4.1 Introduction

In the last decade, time series analysis based on chaos theory and the theory of nonlinear dynamics, which are among the most interesting and growing research topics, has been applied to time series data with some degree of success. Moreover, a variety of experiments have shown that a recorded time series is driven by a deterministic dynamical system with a low dimensional chaotic attractor, which is defined as the phase space point or set of points representing the various possible steady-state conditions of a system. Characterization and quantification of the dynamics of nonlinear time series are also important steps towards understanding the nature of random behaviour and may enable us to predict the occurrences of some specific events which follow temporal dynamical patterns in the time series [59]. Recently, methods developed for the dynamical analysis of complex time series have been used with a relatively high degree of success in the investigation of signals produced by real biological systems. Fractals are closely related to chaotic systems. The fractal dimension of a waveform represents a powerful tool for transient detection.

4.1.1 Analysis of Electroencephalogram (EEG)

The electroencephalogram (EEG) is a bio-signal representing the electrical activity of the brain. Specifically, it mostly represents the electrical activity of the outer layer of the brain, the cortex. EEG is measured as a potential difference (voltage) between two electrodes placed on the scalp. It represents the time series that maps the voltage corresponding to neurological activity as a function of time. EEG is an observable property of large fields of real neurons [22]. Connecting EEG signals to computers has become widely used technique in mapping the brain. The status of EEG, despite for time being eclipsed by that of single neuron recording, has since become a classical area of investigating dynamics at the neurosystem level [12, 13].

In real systems like EEG the complexity is very difficult to prove or exclude [72]. The skepticism against finite dimension estimates is understandable. It is hard to believe that a complicated system as the brain, which is continually interacting with many other complex systems, should manifest as deterministic low-dimensional dynamics. Predominantly, it is a manifestation of a mixture of noise, some cyclic processes and random fractal signals. Each part of such a composition itself is frequently reported to fool the algorithms used to detect chaotic dynamics. Therefore, the fractal dimension estimates should be interpreted with extreme caution. Ascending number of experts believe that there is little evidence for deterministic chaos in brain dynamics. In particular, in analysis of electroencephalograms the fractal dimension feature has been used to identify and distinguish specific states of physiologic function [5, 53, 76].

4.1.2 Epileptic Seizure

Herein, we direct our fractal dimension applications to bioengineering problems, particularly epilepsy. Epilepsy is a disorder characterized by recurrent seizures. The seizures (ictal states) cause temporary disturbances of brain functions (e.g., motor control, responsiveness, recall), for periods ranging from seconds to minutes. Seizures may be followed by a post-ictal period of confusion or impaired sensorium that can last several hours. Seizures occur when a massive group of neurons in the cerebral cortex suddenly begin to discharge in a highly organized rhythmic pattern [48, 49]. This pattern usually begins and terminates spontaneously, without external triggers. The initiation and termination of epileptic seizures reflect intrinsic, but poorly understood properties of the epileptic brain. A seizure involves large portions of the cerebral cortex, thus, ten to hundreds of thousands of interacting neurons [59].

Therefore, it is likely that investigation into the epileptic brain as a system will elucidate important mechanisms underlying seizures. The macroscopic and microscopic features of the epileptogenic zone have been comprehended, the mechanism by which these fixed disturbances in local circuitry produce intermittent disturbances of brain function cannot be explained and understood. The development of the epileptic state can be considered as changes in network circuitry of neurons in the brain. When neuronal networks are activated, they produce a change in voltage potential, which can be captured by an EEG. The best way of understanding the epileptic brain is analyzing the EEG signals [59]. So we have considered EEG seizure signals for our study.

Most of the researchers from the field of Nonlinear Dynamics used Lyapunov exponents for their dynamical study [30, 35, 71]. They have shown that the rate of divergence is an important aspect of the system dynamics and is reflected in the value of Lyapunov exponents and dynamical phase. Due to the inconveniences of the methods from nonlinear dynamics we have used fractal dimension as a tool for the detection of critical cortical sites. It is sometimes desirable to identify a brief seizure, occasional spike, single evoked potential, or other transient in the EEG [75]. Transient detection in an EEG can be a difficult task, often requiring prior knowledge of the characteristics of the transient. Regarding the EEG as a fractal, we have shown that transient deterministic data in the EEG have a fractal dimension different from the quasirandom background [5]. From the obtained results presented in this chapter, analysis by fractal dimension is shown to be a promising method of transient detection, requiring no prior knowledge of the characteristics of the transient.

4.2 Fractal Dimension for Synthetic Waveforms

The dimension plays a vital role in the study of fractals and their applications. There is a need to study the dimensional properties of fractal sets and to extend this concepts to real world applications. Scientists compare complex patterns by summarizing them in standardized forms. One important comparative technique assigns patterns unique positions along a one-dimensional number line; a line is used because points only have a natural order in one-dimension and higher dimensions, there is no natural ordering for an arbitrary collection of points. To compare patterns by summarizing them as individual values on a number line, a set of pattern features must be transformed into a single number. There are no complete and general rules for picking these features: in any actual case, the specific features to be used for classifying a pattern depend on the ultimate goal of the analysis [40].

Waveforms are a common scientific pattern. A waveform is a collection of (x, y) point pairs, where the x values increase monotonically. Waveforms are planar curves proceeding resolutely forward, they do not go backwards and they do not cross over themselves. Chart recordings, oscilloscope records, time series, cumulative graphs of winnings in games of chance, and many linear (one-dimensional) waves, such as ballistic trajectories, are all examples of waveforms. Waveforms have been numerically categorized using a myriad of different schemes. Often, an ideal curve is chosen as a standard and the actual waveforms are then assigned individual numbers describing their closeness of fit to the ideal. For instance, data graphs are frequently compared to straight lines or to exponential curves, because these two classes of curves represent simple models of the underlying mechanisms by which such data may have been generated. When the standard curves are ideal and when, like straight lines or exponential curves, they can be characterized by a few elemental parameters, the ranking schemes are called "parametric" [40].

Non-ideal curves can also be used as standards; in such cases, the ranking schemes are usually "non-parametric", because the waveforms are compared to the whole standard curve and there need be no simple set of standard summary parameters. For a non-parametric ranking, a researcher might use the sign test: here, waveforms are compared as sets of (x, y) point pairs and the numerical ranking (the statistic) for any particular waveform is computed from the proportion of point pairs in which its y value is greater than the y value for the corresponding point pair in the standard curve.

Another approach to ranking waveforms is the direct quantification of particular pattern features. Waveforms are frequently thought of as periodic functions; thus, waveforms are commonly characterized in terms of such wave features as fundamental frequencies or mean amplitude. Periodic wave features are quantified through harmonic analysis, and the most widely used of these are the Fourier analysis, which approximate waveforms as a series of cosine and sine waves.

It would be useful also to have other general pattern features that can be quantified and that do not depend on the assumption that the waveforms have been generated by periodic phenomena. Here, we describe one such pattern feature—the fractal dimension of a waveform. The fractal dimension of a pattern is a measure of its complexity—its spatial extent, convolutedness, or space-filling propensity and it is intimately tied to the shape and the dimensionality of the pattern [40].

The dimension of fractal graphs was first studied by Besicovitch and Ursell in 1937. Burlaga and Klein [9] have presented a method to calculate the length of a curve and have obtained stable values of the fractal dimension of large-scale fluctuations of the interplanetary magnetic field. In 1986, Higuchi modified the method of Burlaga to calculate the fractal dimension and he showed that by using his method we can get a stable value of the fractal dimension. Also he applied his method to the time series of the natural phenomena which show the turbulent behavior. In 1988, Katz proposed a method to estimate the fractal dimension of waveforms. After the long time, Petrosian developed four different algorithms for computing fractal dimension of waveforms in 1995. Esteller et al. [20] compared all the above waveform fractal dimension algorithms based on their performance with synthetic waveforms and EEG signals. Qiong and Xiong [61] derived a method to calculate fractal dimension of digital voice signal waveforms. Kalauzi et al. [37] developed a fractal dimension method based on consecutive finite differences for analysis of fractal signals. The present study deals about the developments in the waveform fractal dimension methods. The synthetic waveform data used in this section has the property of a fractal curve over all time scales. The techniques developed in this section can easily give us stable indices and time scale corresponding to the characteristic frequency even for a small number of data.

4.2.1 *Fractal Dimension of Waveforms*

Fractal graphs have been used by Mathematicians over the years. Even among Mathematicians, these graphs had often been considered to have little interest. A variety of interesting fractals, both of theoretical and practical importance, occur as graphs of functions. Indeed, many phenomena display fractal features when plotted as functions of time. Examples include wind speed, levels of reservoirs, population data and prices on the stock market, atleast when recorded over fairly long time spans [21]. The special form of graphs give rise to the several definitions of dimension. Here we consider only rectifiable (have a finite length) graphs with irregular behaviour.

We define a graph Γ to the image of an interval $[a,b]$ under a continuous bijection $\psi : [a,b] \rightarrow \mathbb{R}^n$. Thus, we restrict attention to graphs that are non-self-intersecting. If Γ is a graph and $\delta > 0$, we define $M_\delta(\Gamma)$ to be the maximum number of points $x_0, x_1, \ldots, x_m$, on the graph Γ, in that order, such that $|x_k - x_{k-1}| = \delta$ for $k = 1, 2, \ldots, m$. Thus $(M_\delta(\Gamma) - 1)$ may be thought of as the 'length' of the graph Γ measured using a pair of dividers with points set at a distance δ apart. Under certain circumstances the graph

$$\text{graph } \Gamma = \{(t, \psi(t)) : a \leq t \leq b\} \tag{4.1}$$

regarded as a subset of the (t, x)-coordinate plane may be a fractal [21].

We consider functions $\psi : [a, b] \to \mathbb{R}$. If ψ has a continuous derivative, then it is not difficult to see that graph ψ has dimension 1 and indeed, is a regular 1-set.

The same is true if ψ is of bounded variation; that is, $\sum_{i=0}^{m-1} |\psi(t_i) - \psi(t_{i+1})| \leq$ constant for all dissections $0 = t_0 < t_1 < \ldots.. < t_m = 1$. However, it is possible for a continuous function to be sufficiently irregular to have a graph of dimension strictly greater than 1. In this section we concentrate on irregular graphs. The best known example is

$$\psi(t) = \sum_{k=0}^{\infty} \lambda^{Hk} sin(2\pi\lambda^k t) \tag{4.2}$$

where $0 < H < 1, \lambda > 1$ and $FD = 2 - H$.

This function, essentially Weierstrass's sine function, that is continuous and nowhere differentiable, has box dimension s, and is believed to have Hausdorff dimension s.

The another form of Weierstrass cosine function, given as follows:

$$\psi(t) = \sum_{k=0}^{\infty} \gamma^{kH} cos(2\pi\gamma^k t), \ 0 < H < 1 \tag{4.3}$$

here $\gamma > 1$ and $FD = 2 - H$ [20].

Given a function ψ and an interval $[t_1, t_2]$, we write R_ψ for the maximum range of ψ over an interval,

$$R_\psi[t_1, t_2] = \sup_{t_1 \leq t, u \leq t_2} |\psi(t) - \psi(u)| . \tag{4.4}$$

Theorem 1. *Let* $\psi : [0, 1] \to \mathbb{R}$ *be continuous. Suppose that* $0 < \delta < 1$ *and* m *is the least integer greater than or equal to* $1/\delta$*. Then, if* N_δ *is the number of squares of the* δ*-mesh that intersect graph* ψ,

$$\delta^{-1} \sum_{i=0}^{m-1} R_\psi [i_\delta, (i+1)\delta] \leq N_\delta \leq +2m + \delta^{-1} R_\psi [i_\delta, (i+1)\delta] . \tag{4.5}$$

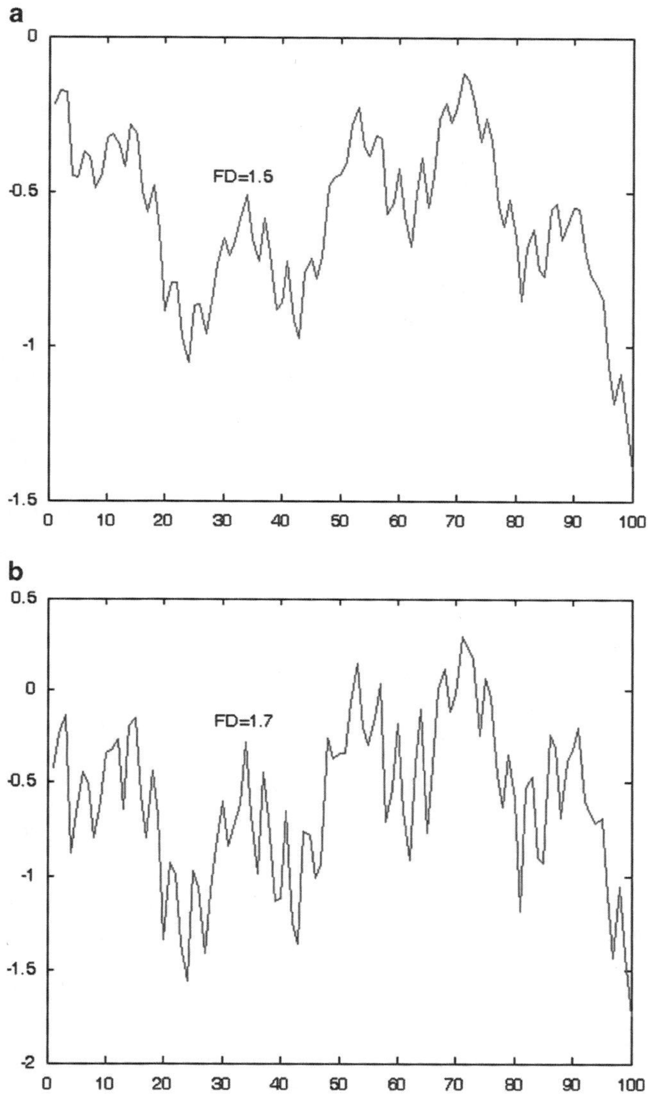

Fig. 4.1 Weierstrass sine function for fractal dimension equal to (**a**) 1.5, (**b**) 1.7

Fix $\lambda > 1$ and $1 < s < 2$. Define $\psi : [0,1] \to \mathbb{R}$ by

$$\psi(t) = \sum_{k=1}^{\infty} \lambda^{(s-2)k} sin(\lambda^k t) \quad (4.6)$$

Then, provided λ is large enough, dim_B graph $\psi = s$ [21]. Some waveforms are generated with known fractal dimension values using Weierstrass cosine and sine functions and are shown in Figs. 4.1–4.3.

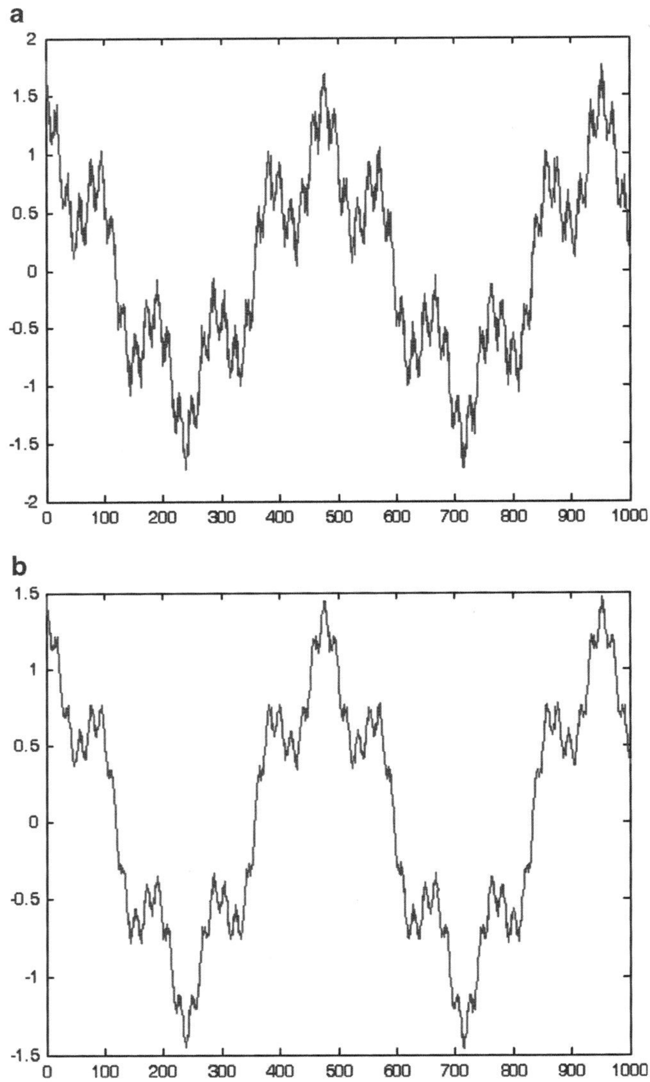

Fig. 4.2 Weierstrass cosine function for fractal dimension equal to (**a**) 1.5 and (**b**) 1.3

4.2.2 Existing Methods for the Computing Fractal Dimension

Many algorithms for estimating the *FD* of the waveforms have been proposed by various authors, but their computational requirements are expensive [37]. Among the algorithms, we consider three of the most prominent methods for computing the *FD* and the analysis of waveforms.

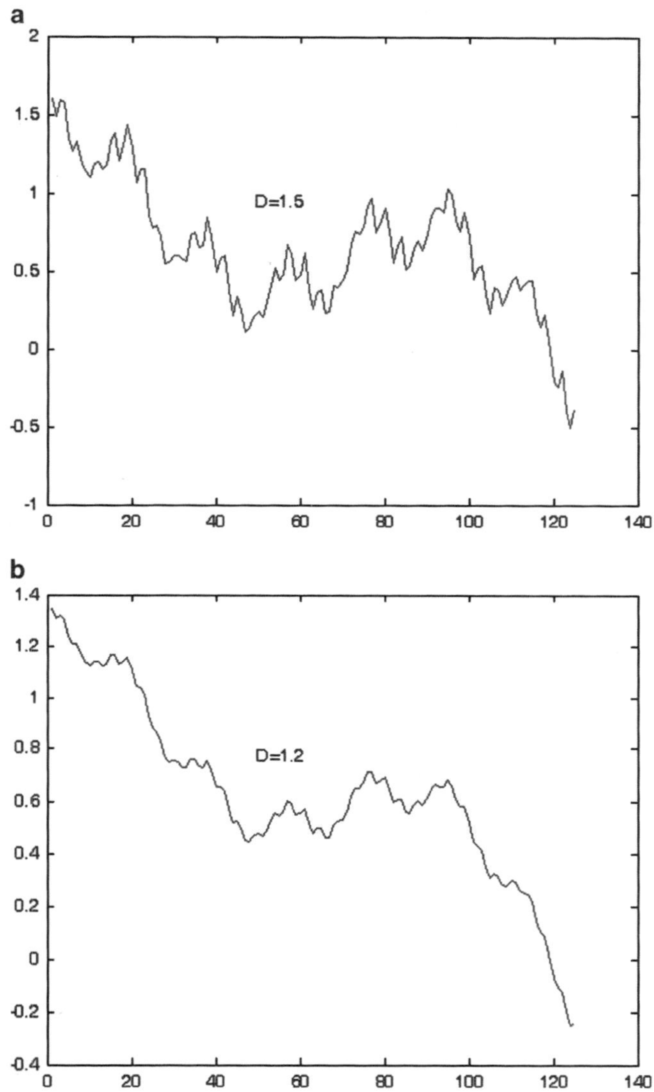

Fig. 4.3 Digitalized curves generated from Weierstrass cosine function for fractal dimension equal to (**a**) 1.5 and (**b**) 1.2

4.2.2.1 Higuchi's Method

This method was appreciated by the researchers in the field of waveform analysis. The procedure for the method is given below. Consider $x(1), x(2), \ldots., x(N)$, the time sequence to be analyzed. Construct k new time series x_m^k as

$$x_m^k = \left\{ x(m), x(m+k), x(m+2k), \ldots, x\left(m + \left\lfloor \frac{N-m}{k} \right\rfloor\right) \right\}, \tag{4.7}$$

for $m = 1, 2, \ldots..k$ where m indicates the initial time value, k indicates the discrete time interval between points (delay), and $\lfloor a \rfloor$ means integer part of a. For each of the curves or time series x_m^k constructed, the average length $L_m(k)$ is computed as

$$L_m(k) = \frac{\sum_{i=1}^{\lfloor \frac{N-m}{k} \rfloor} |x(m+ik) - x(m+(i-1)k)| \, (n-1)}{\lfloor \frac{N-m}{k} \rfloor k} \tag{4.8}$$

where N is the total length of the data sequence x and $\frac{(N-1)}{\lfloor \frac{N-m}{k} \rfloor k}$ is a normalization factor. An average length is computed for all time series having the same delay (or scale) k as the mean of the k lengths $L_m(k)$ for $m = 1, 2, \ldots, k$. This procedure is repeated for each k ranging from 1 to k_{max}, yielding a sum of average length $L(k)$ for each k as indicated in (4.9)

$$L(k) = \sum_{m=1}^{k} L_m(k). \tag{4.9}$$

The total average length for scale k, $L(k)$, is proportional to k^{-D}, where D is the *FD* by Highuchi's method. In the curve of $ln(L(k))$ versus $ln(1/k)$, the slope of the least squares linear best fit is the estimate of the fractal dimension [20]. The implementation of Higuchi's algorithm needs to determine the parameters k_{min} and k_{max} value.

4.2.2.2 Katz's Method

The method was based on a conjecture by Mandelbrot [46] who suggested that a fractal dimension of a river may be derived from the river's length and the distance between the river's spring and its end. Katz procedure implies that the fractal dimension of a waveform may be measured empirically by sampling N points evenly spaced on the abscissa [40].

The *FD* of a curve can be defined as

$$D = \frac{\log_{10}(L)}{\log_{10}(d)} \tag{4.10}$$

where L is the total length of the curve or sum of distances between successive points, and d is the diameter estimated as the distance between the first point of the sequence and the point of the sequence that provides the farthest distance. Mathematically, d can be expressed as

$$d = max(distance(1, i)). \tag{4.11}$$

Considering the distance between each point of the sequence and the first, point i is the one that maximizes the distance with respect to the first point. The *FD* compares the actual number of units that compose a curve with the minimum number of units required to reproduce a pattern of the same spatial extent. *FD*s computed in this fashion depend upon the measurement units used. If the units are different, then so are the *FD*s. Katz's approach solves this problem by creating a general unit or yardstick: the average step or average distance between successive points, $\underline{a}$. Normalizing distances in (4.10) by this average results in

$$D = \frac{\log_{10}(L/\underline{a})}{\log_{10}(d/\underline{a})}. \tag{4.12}$$

Defining n as the number of steps in the curve, then $n = \frac{L}{\underline{a}}$, and (4.12) can be written as

$$D = \frac{\log_{10}(n)}{\log_{10}(\frac{d}{L}) + \log_{10}(n)}. \tag{4.13}$$

Expression (4.13) summarizes Katz's approach to calculate the *FD* of a waveform [40]. The Katz's algorithm works with minimum window length and the signal contains minimum variance. The signal noise and the window length affect the dynamic range of Katz's algorithm.

4.2.2.3 Petrosian Method

Petrosian uses a quick estimate of the *FD*. However, this estimate is really the *FD* of a binary sequence as originally defined by Katz. Since waveforms are analog signals, a binary signal is derived from four different methods. Here we consider the fourth method. The differences between consecutive waveform values are given, the value of one or zero depending on whether their difference exceeds or not a standard deviation magnitude [60]. The *FD* of the binary sequence generated by this method is computed as

$$D = \frac{log_{10}n}{log_{10}n + log_{10}\left(\frac{n}{n+0.4\ N_{\Delta}}\right)} \tag{4.14}$$

where n is the length of the sequence (number of points), and N_{Δ} is the number of sign changes (number of dissimilar pairs) in the binary sequence generated. Petrosian's method performance depends on the type of binary sequence used. If a binary sequence based on slope-sign changes is utilized then this method becomes less suitable for analog signal analysis, its high sensitivity to noise and its poor reproducibility of dynamic range of synthetic *FD*.

4.2.3 *Novel Fractal Dimension Methods*

In this section we present our developed methods for computing *FD* and analysis of waveforms [56–58, 73, 74]. We have considered all the drawbacks of the existing algorithms.

4.2.3.1 Modified Divider Method

The application of divider method to determine the Fractal Dimension (FD) of the curves introduced by Shelberg [68], uses a chord length (step) and measures the number of chord lengths (length) needed to cover a fractal curve [21]. The technique is based on the principle of taking smaller and smaller rulers of step size to cover the curve and counting the number of rulers length required in each case [7]. It is a recursive process in which the step is decreased (typically halved) and the new length is calculated. Here, the input curves are taken to be of size N where N is a power of 2 because of the recursive nature of the method. A method of least squares fit to the bilogarithmic plot of length against step gives the slope s and the fractal dimension $D = -s$.

This part of the calculation provides and estimate the average gradient are illustrated in Fig. 4.3. If C is a curve and $r > 0$, we define $N_r(C)$ to be the maximum number of points $x_0, x_1, \ldots, x_m$, on the curve C, in that order, such that

$$|x_k - x_{k-1}| = r. \tag{4.15}$$

Thus the length of the curve C measured using a pair of dividers with points set at a distance r apart is

$$(N_r(C) - 1) \times r. \tag{4.16}$$

The dimension of C is determined by the power law obeyed by $N_r(C)$ as $r \to 0$. If

$$N_r(C) \sim c \times r^{-D} \tag{4.17}$$

for constants c and D, we might say that C has 'divider fractal dimension' D, with c regarded as the 'D-dimensional length' of C. Taking logarithms we get

$$logN_r(C) \simeq c - D \times log(r). \tag{4.18}$$

The fractal dimension of the curve is defined as

$$D = \frac{logN_r(C)}{-log(r)}. \tag{4.19}$$

The above formula is for computational or experimental purposes, *FD* can be estimated as minus the gradient of log-log graph plotted over a suitable range of r.

The problem arises when selecting the suitable range of the size r. Shelberg [68] describes an appropriate starting value as half of the average distance between the points.

That is

$$r = \frac{1}{m}\sum_{k=1}^{m} |x_k - x_{k-1}| \tag{4.20}$$

$$r_{min} = \frac{r}{2} \tag{4.21}$$

Most of the researchers had used divider method to measure the complexity of non fractal shapes, but did not describe the criteria to select the final step [10]. Solving this problem makes this algorithm less time consuming. For each r, we measure a set in a way that ignores irregularities of size less than r, and in the limit $r \to 0$ (where r is the step length), the length is $N_r(C)$, the validity necessarily hold in a finite size range (r_{min}, r_{max}) where $0 < r_{min} < r_{max}$. We describe the method for selecting the final step as half the original length of the curve, that is

$$r_{max} = \frac{(N_r(C) - 1) \times r}{2}. \tag{4.22}$$

4.2.3.2 Algorithm

The following algorithm will estimate the dimension of the digital curve.

- Consider the curve C.
- Calculate r_{min} using $r_{min} = \frac{\frac{1}{m}\sum_{k=1}^{m} |x_k - x_{k-1}|}{2}$.
- Calculate r_{max} using $r_{max} = \frac{(N_r(C)-1)\times r}{2}$.
- Calculate Length $(N_r(C) - 1) \times r$ for r_{min} to r_{max}.
- Plot $(N_r(C) - 1) \times r$ versus r, the slope of the plot is the required fractal dimension.

Deterministic fractals such as Von Koch Curve show repeated structure of all scales. Each subsequent magnification reveals ever more fine structure. The *FD* of Von Koch Curve ($FD = 1.262$) easily obtained by using similarity dimension [46].

For irregular curves (no self-similarity) similarity dimension cannot be applied [51]. The modified divider method proposed here is the effective tool to retrieve the fractal dimension of the irregular digitalized curves, the value of fractal dimension does not exceed the required $1 \leq D \leq 2$ interval.

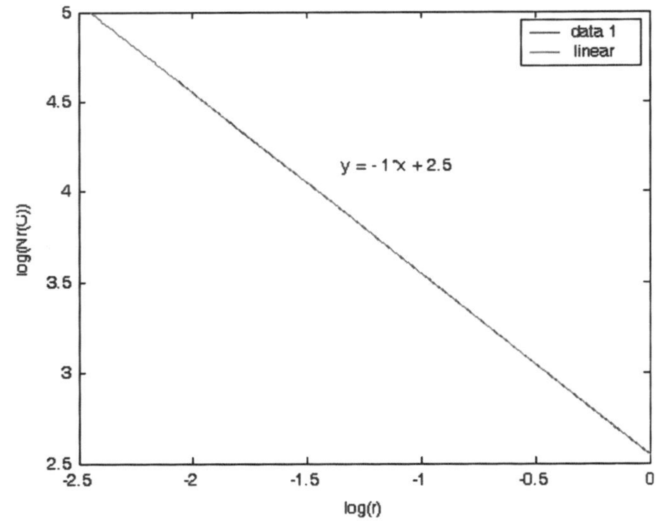

Fig. 4.4 *log* − *log* plot $N_r(C)$ versus r

4.2.3.3 Example and Results

The need to check the validity of the proposed method, we have used two simple synthetic curves Fig. 4.3a and b generated from Weierstrass cosine function of $m = 125$ points with known fractal dimensions 1.5 and 1.2. The algorithm described above was implemented in MATLAB and tested on synthetic curves with known fractal dimension. Note that the perfect reproduction of the known fractal dimension should yield a straight line of slope equal to one [70]. Figure 4.4 shows the known fractal dimension value of Fig. 4.3a is equal to the *FD* value estimated by modified divider method. The Weierstrass synthetic curve contains minimum irregularity. If digitalized curves contained chaotic behaviour then the preprocess normalization is needed to analyze the curves [51].

The values of r_{min} and r_{max} of the divider can be calculated from (4.20)–(4.22) is 0.0489 and 6.0175. The length calculated for the ranges of $r_{min} = 0.0490$ and $r_{max} = 6.0175$. Plot the values of $log(N_r(C))$ versus $log(r)$. The divider dimension is estimated from the slope value of the log-log plot. The slope value indicates the complexity presented in the curve. From Fig. 4.4 the slope value is one that the algorithm estimates *FD* of the curve (Fig. 4.3a) equal to the known fractal dimension 1.5. The MATLAB code for the corresponding procedure simply takes the following form. The curve is stored in the variable C,

- $r_{min} = mean(sum(diff(C)))$;
- $r_{max} = (sum(diff(C)))/2$;
- $F = polyfit(log(N_r(C)), log(r), 1)$;
- $FD = F(1)$;

In this way one may obtain the exact fractal dimension of the irregular digitalized curves.

4.2.3.4 Modified Higuchi's Method

Applications of *FD* in this setting include two types of approaches, those in the time domain and the ones in the phase space domain. The former approaches estimate the *FD* directly in the time domain or original waveform domain, where the waveform or original signal is considered a geometric figure [10]. Phase space approaches estimate the *FD* of an attractor in state-space domain [51, 70]. Calculating the *FD* of waveforms is useful for transient detection with the additional advantage of fast computation. Most of the algorithms do not contain the proper criteria to estimate the minimum and maximum values of the interval k. Higuchi describes an algorithm to compute *FD* of waveforms but he did not implement the criteria to choose the k_{min} and k_{max} values [29]. The above algorithms got computational difficulty to estimate the chaotic behaviour of the waveforms. Here the new algorithm which resolves all the problems of existing algorithms is presented.

Consider $x(1), x(2), \ldots .x(N)$ be a finite set of time series of observations taken at a regular interval. Construct k new time series x_m^k as

$$x_m^k = \left\{ x(m), x(m+k), x(m+2k), \ldots ., x\left(m + \left\lfloor \frac{N-m}{k} \right\rfloor \right) \right\}, \tag{4.23}$$

for $m = 1, 2, \ldots ..., k$ where m indicates the initial time value, k indicates the discrete time interval between points (delay), and $\lfloor a \rfloor$ means integer part of a. Here a description is given about the criterion for choosing k which consists of k_{min} and k_{max} respectively.

Shelberg [68] describes an appropriate starting value as half of the average distance between the points. That is

$$k = \frac{1}{m} \sum_{i=1}^{m} |x_i - x_{i-1}| \tag{4.24}$$

$$k_{min} = \frac{k}{2} \tag{4.25}$$

If C is a curve and $r > 0$, we define $N_r(C)$ to be the maximum number of points $x_0, x_1, \ldots ., x_m$, on the curve C, in that order, such that $|x_k - x_{k-1}| = r$. Thus the length of the curve C measured using a pair of dividers with points set at a distance r apart is $(N_r(C) - 1) \times r$. We describe the method for selecting the final step as half the original length of the curve $(N_k(C) - 1) \times k$, that is

$$k_{max} = \frac{(N_k(C) - 1) \times k}{2}, \text{ where } C \text{ is the waveform signal.} \quad (4.26)$$

For each k, we measure a set in a way that ignores irregularities of size less than k, and in the limit $k \rightarrow 0$, the length is $L_m(k)$, the validity necessarily hold in a finite size range (k_{min}, k_{max}) where $0 < k_{min} < k_{max}$. For each of the curves or time series x_m^k constructed, the average length $L_m(k)$ is computed as

$$L_m(k) = \frac{\sum_{i=1}^{\lfloor \frac{N-m}{k} \rfloor} |x(m+ik) - x(m+(i-1)k)| \, (n-1)}{\lfloor \frac{N-m}{k} \rfloor k} \quad (4.27)$$

where N is the total length of the data sequence x and

$$\frac{(N-1)}{\lfloor \frac{N-m}{k} \rfloor k}$$

is a normalization factor. An average length is computed for all time series having the same delay (or scale) k as the mean of the k lengths $L_m(k)$ for $m = 1, 2, \ldots., k$.

This procedure is repeated for each k ranging from 1 to k_{max}, yielding a sum of average lengths $L(k)$ for each k as indicated in (4.28)

$$L(k) = \sum_{m=1}^{k} L_m(k). \quad (4.28)$$

The total average length for scale k, $L(k)$, is proportional to k^{-D}, where D is the Fractal dimension. In the curve of $ln(L(k))$ versus $ln(1/k)$, the slope of the least squares linear best fit is the estimate of the fractal dimension.

4.2.3.5 Procedure for Computing Fractal Dimension

Here, a new algorithm for computing the fractal dimension of the waveforms based on the recursive length of the waveform is presented. The waveform generated from Weierstrass cosine function of different window lengths up to 2,000 points with known fractal dimension. The following algorithm will estimate the dimension of the waveform. The algorithm is written in MATLAB, corresponding procedure for calculating *FD*, simply takes the form:

1. Consider the time sequence $x(1), x(2), \ldots\ldots., x(N)$
2. Choose the range of k that is k_{min} and k_{max} using

$$k = \tfrac{1}{m} \sum_{i=1}^{m} |x_i - x_{i-1}|$$

$$k_{min} = \tfrac{k}{2}$$

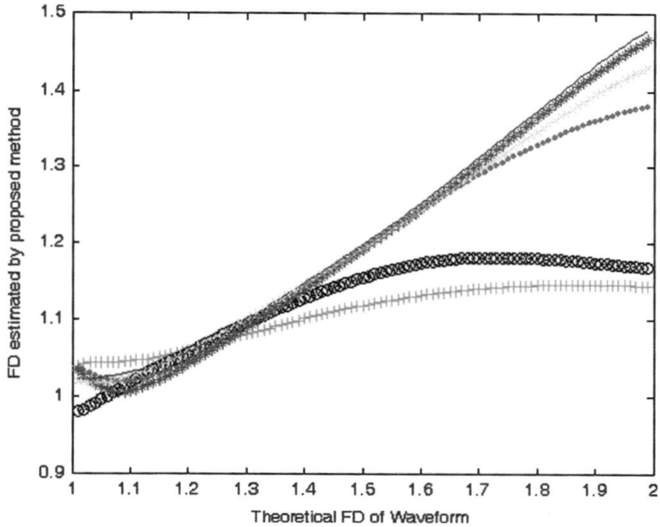

Fig. 4.5 Plot of theoretical *FD* and *FD* computed by proposed method, +++ N = 150, 000N = 250, - - -N = 750, ***N = 1,000, xxxN = 2,000

$$k_{max} = \frac{(N_k(C)-1)\times k}{2}.$$

3. Calculate the average length $L_m(k)$ using (4.27).
4. Compute the sum of average lengths $L(k)$ using (4.28).
5. Plot the values of $ln(k)$ versus $ln(1/k)$, the slope is the *FD* of the waveform.

The above procedure is repeated for different window lengths from 250 to 2,000 points. Figure 4.5 shows the reliability of the algorithm with different window lengths. A set of 100 sequences, each with different *FD*, was generated using Weierstrass function. We tested our algorithm with respect to reliability, efficiency (computational time), noise sensitivity, and record length. The *FD* computed by the proposed method is compared with other fractal dimension algorithms. Also the criteria for choosing k_{min} and k_{max} is given below. The computed *FD* with different window length is compared with theoretical *FD*. The comparison results are presented in following section.

4.2.3.6 Example and Results

Fractal dimension of synthetic signals ranges from 1.01 to 1.99. Figure 4.5 shows the *FD* values obtained by our proposed method with different window lengths. We can easily observe the reliability of the algorithm from Fig. 4.5.

The noise and ranges of window lengths do not affect the effectiveness of the algorithm. No window length effect is observed in the range of 250 to 2,000 points.

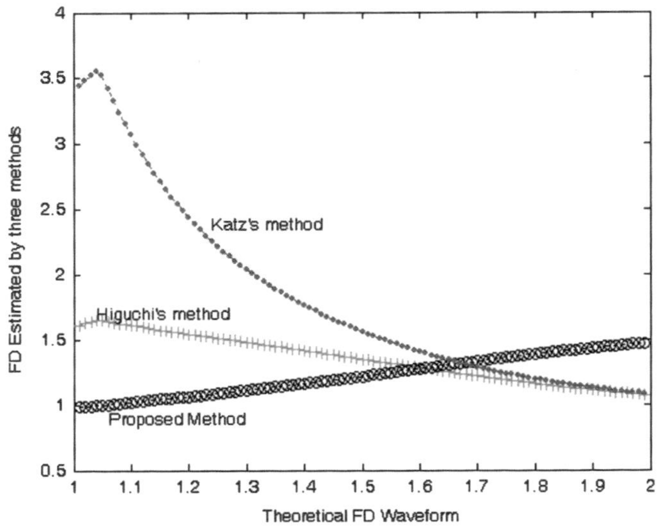

Fig. 4.6 The performance of three methods with synthetic waveform, N = 1,000 points

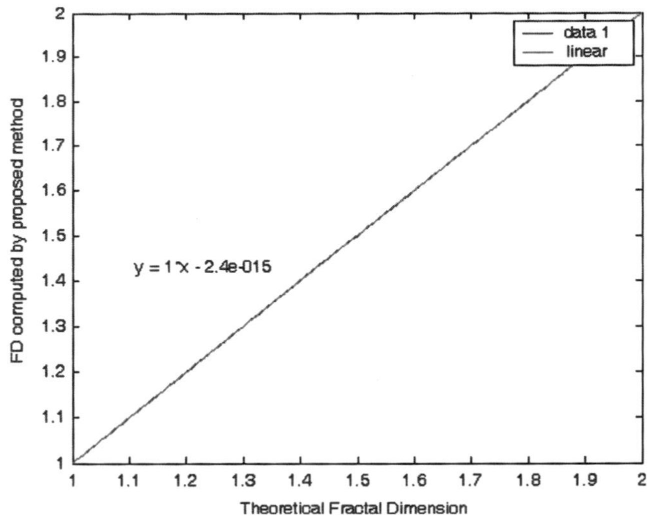

Fig. 4.7 Plot of theoretical *FD* versus *FD* computed by modified Higuchi's method

It works with synthetic signal data for 8,000 points and gives useful results. The comparison with other famous fractal dimension algorithms such as Katz's and Higuchi's are represented in Fig. 4.6. The *FD* computed by our method is similar with theoretical *FD*. That is the curve closest to the ideal straight line of slope one is shown in Fig. 4.7. In Higuchi's algorithm there is no criterion to choose the value of k_{min} and k_{max}. But the criterion described in our method gives effective result

Table 4.1 Comparison of run time of FD methods with modified Higuchi's method

Window length	k_{min}, k_{max}	Run-time (seconds) Modified Higuchi's method	Katz's method	Higuchi's method
250	1,6	0.0120	0.1720	0.0420
500	1,24	0.0260	0.3440	0.0680
1,000	1,49	0.0480	0.9530	0.950
2,000	1,98	0.1510	3.0160	0.1950
4,000	1,196	0.4230	12.9310	0.4250
8,000	1,393	1.4670	52.9210	1.4850

with window lengths from 250 to 2,000 points. Table 4.1 shows the computational burden between the algorithms and the values of k_{min} and k_{max} for different window length. Compared to other fractal dimension algorithms our method works with less time complexity. If the record length is increased to 8,000 points, then our algorithm performance improves and becomes very faster than Higuchi's and Katz's.

4.2.3.7 Size Measure Relationship Method (SMR)

Hausdorff dimension is computationally difficult process while we measure the fractal dimension of irregular graphs [7]. Also there is a problem in the box counting dimension that is choosing of initial and final size of the magnification factor, the algorithm takes more time complexity [10]. This causes the need of new algorithm which solves the above problems and it requires minimum time complexity. As with many other techniques of fractals, the computation of the fractal dimension can be undertaken in real place (processing the data directly) [46]. Here a new algorithm is formulated for computing the fractal dimension of the rectifiable irregular graphs based on the Size Measure Relationship (SMR) method. This method is based on recursive length of the irregular curves using different measuring scales. The *FD* of a irregular graph F can be defined in SMR method as

$$D = \frac{log_{10}(L)}{log_{10}(R)} \tag{4.29}$$

where L is the total length of the curve or sum of distances between successive points, it can be represented as

$$L = \sum_{i=1}^{N-n} |F_i^n| \tag{4.30}$$

where N denotes the total number of points to be taken. Since we are dealing with sampled graphs, the software implementation of these steps should be based on consecutive calculation of finite differences. For the graph F_i, with N samples and with normalized amplitudes, first order consecutive difference is calculated as

$$F_i^n = F_{i+1}^{n-1} - F_i^{n-1}, i = 1, 2 \ldots, n. \tag{4.31}$$

The R is the range estimated as the distance between maximum value point and the minimum value point. Mathematically R can be expressed as

$$R = distance(max(F_i^{nor}) - min(F_i^{nor})). \tag{4.32}$$

Sample differences depend on graphs amplitude, a preprocessing step comprising graphs amplitude normalization [20], is necessary for graph sample F_i, $i = 1, 2, \ldots, N$ and N is the total number of points in the graph sample. The graph amplitudes normalized by using

$$F_i^{nor} = \frac{F_i'}{\frac{1}{N}\sum_{j=1}^{N} |F_j'|} \tag{4.33}$$

where

$$F_i' = F_i - \frac{1}{N}\sum_{j=1}^{N} F_j, \; i = 1, 2 \ldots, N.$$

After normalization of amplitudes we get F_i^{nor}, it is used to calculate the Fractal Dimension of the graph.

4.2.3.8 Algorithm

The following algorithm will estimate the dimension of the irregular graphs. The program for the SMR algorithm is written in MATLAB and the graph F_i, $i = 1, 2, \ldots, N$, is stored in vector F, corresponding procedure for calculating FD, simply takes the form:

- Normalizing the irregular graph using

$$s = abs(mean(diff(F)));$$
$$F_i' = F - s;$$
$$F_i^{nor} = F_i'/s;$$

- Computing the total length of the graph using

$$L = sum(abs(diff(F_i^{nor}))).$$

- Computing the range in F_i^{nor} using

$$R = distance(max(F_i^{nor}) - min(F_i^{nor})).$$

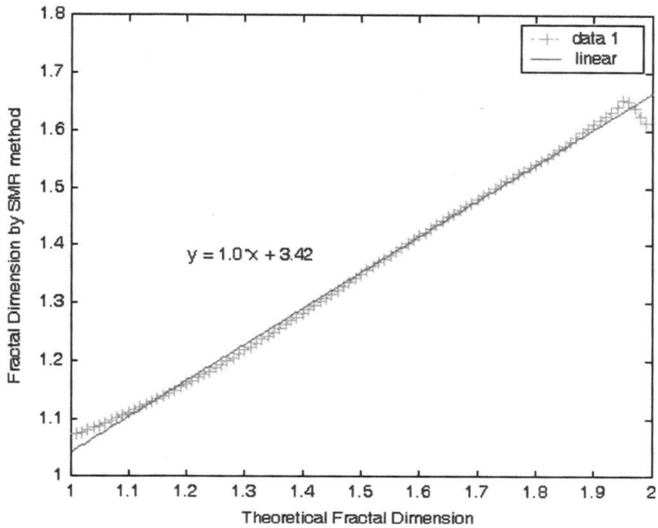

Fig. 4.8 The plot of theoretical fractal dimension versus fractal dimension by SMR method

- Applying formula $D = \frac{log_{10}(L)}{log_{10}(R)}$.

In this way one may obtain the exact fractal dimension of the rectifiable irregular graphs. The purpose of the above algorithm is to estimate the complexity present in the irregular graphs.

4.2.3.9 Example and Results

The SMR algorithm is tested with the rectifiable irregular synthetic graphs generated from Weierstrass sine function of known fractal dimensions 1.5 and 1.7 as shown in Fig. 4.1a and b.

The *FD* values for these curves can be calculated using the procedure for SMR algorithm implemented in MATLAB. Here we take the rectifiable graphs with the length of 100 points. The graphs with normal amplitudes that is dissections are equal contains less irregular [51, 70]. But our algorithm works both regular and irregular rectifiable graphs. In Fig. 4.2a and b, the length of the curve L and the size R can be calculated from (4.30) and (4.32) as follows $L = 10,572$, $R = 167.66$, $D = 1.5$ and $t = 0.0305$ s for Fig. 4.2a and $L = 30,357$, $R = 267.24$, $D = 1.7$ and $t = 0.0310$ s for Fig. 4.2b. Figure 4.8 shows that the perfect reproduction of calculated fractal dimension values should yield a straight line of slope equal to one with the known fractal dimension values. This is the evidence for the suitability of the algorithm with the calculation of fractal dimension of irregular graphs. Comparing with other computational methods such as Box counting, Prism Counting and Perimeter-area

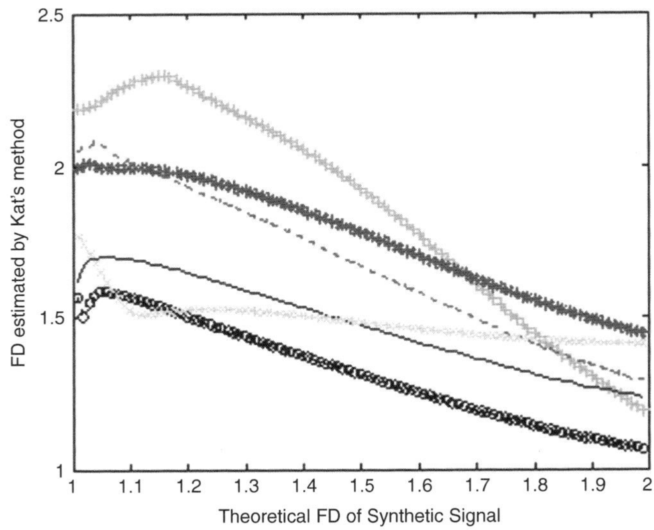

Fig. 4.9 Theoretical FD of synthetic signal versus Higuchi's method, for different number of points (N): +++ N = 150, OOON = 250, - - -N = 750, ***N = 1,000, xxxN = 2,000

relationship method our algorithm requires minimum time complexity, minimum number of steps and estimate the exact fractal dimension of irregular graphs by evidence.

4.2.4 Comparison of Fractal Dimension Methods

Among the developed methods, the SMR method is suitable for synthetic waveform analysis. So we restrict our attention to SMR method and comparison with other existing methods. The SMR approach of estimating the fractal dimension of a signal yields good results with various window lengths.

The reliability of the algorithm was tested with synthetic signal ranging from 1.001 to 1.009 using Weierstrass functions with known *FD*. Figures 4.9–4.12 show the *FD* values obtained by the Higuchi's, Katz's, Petrosian's and SMR methods against the known *FD* of the synthetic data. Note that perfect reproduction of the known *FD* should yield a straight line of the slope equal to one. Higuchi's algorithm and the SMR method provides the most accurate estimate of the *FD*.

Katz's method is less linear. Its calculated *FD* are exponentially related to the known *FD*. The *FD* estimates with Higuchi's method improve as window length increases. The curve that is closest to the ideal straight line of slope was obtained for a window size of 250 points.

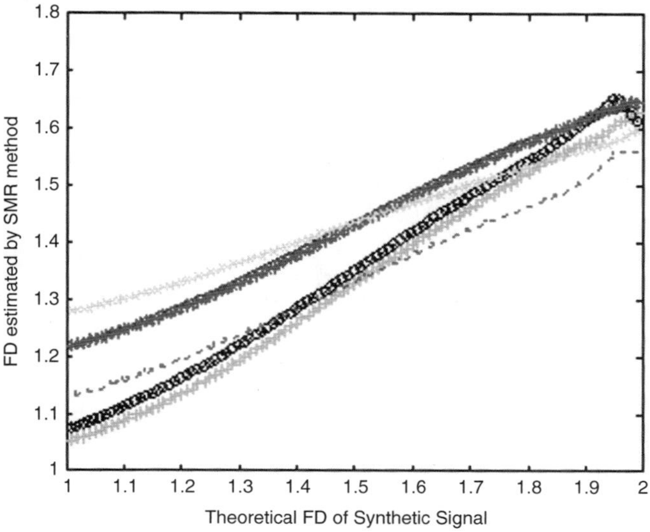

Fig. 4.10 Theoretical FD of synthetic signal versus Katz's method, for different number of points (N): +++ N = 150, OOON = 250, - - -N = 750, ***N = 1,000, xxxN = 2,000

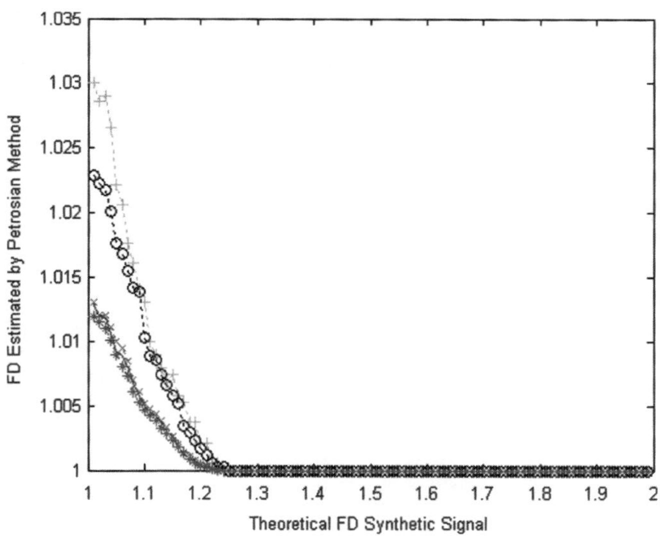

Fig. 4.11 Theoretical FD of synthetic signal versus Petrosian's method, for different number of points (N): +++ N = 150, OOON = 250, —N = 750, N = 1,000 points

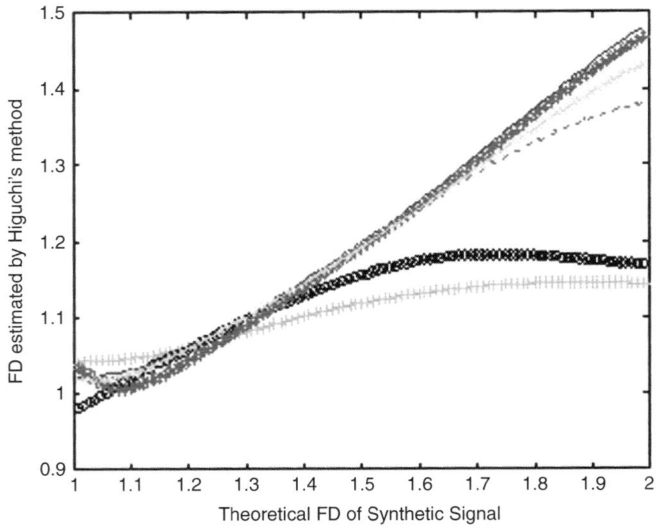

Fig. 4.12 Theoretical FD of synthetic signal by SMR method, for different number of points (N): +++ N = 150, 000N = 250, - - -N = 750, ***N = 1,000, xxxN = 2,000

In Katz's method the window length affects the dynamic range of the estimated FD yielding a dynamic range between 1 and 1.2 for window length greater than 750 window lengths lower than 250 points. Petrosian's algorithm works with synthetic signals, and gives less accurate results. Its results are not useful in estimating the chaotic nature of the signal. Petrosian's method gives good results with EEG data, but it does not discriminate the complex patterns in the EEG signals.

In the SMR method the window length does not affect the dynamic range of the estimated FD yielding a dynamic range between 1 and 2. It works with synthetic signal data for 8,000 points and gives useful results. The advantages of SMR method are greater speed and eliminating the need to choose the k_{min} and k_{max} values. It is much easier than the other methods to calculate the FD. The run time for SMR method is less than the other methods. The window length does not affect the SMR method and it works with various window lengths. The SMR method gives the distinguishability in results between different time points. A comparison of computational burden between the methods for different window lengths is presented in Table 4.2.

4.2.5 Concluding Remarks

The purpose of this text is to introduce effective methods for measuring the fractal dimension and calculating a more precise and stable characteristic time scale than those presented before.

Table 4.2 Comparison of run time of the four FD methods with synthetic waveforms

	Run-time (seconds)			
Window length	SMR method	Higuchi's method	Katz's method	Petrosian's Method
250	0.0320	0.1720	0.0420	0.0470
500	0.0470	0.3440	0.0680	0.0470
1,000	0.0780	0.9530	0.950	0.0780
2,000	0.1410	3.0160	0.1950	0.1720
4,000	0.3280	12.9310	0.4250	0.3750
8,000	0.8570	52.9210	1.0120	1.000

By applying modified divider method, we can determine the divider fractal dimension of irregular digitalized curves. Due to the modification in r_{min} and r_{max} this method becomes effective and less time consuming.

The modified Higuchi's algorithm is required for specific applications. The results show that the algorithm is the most consistent method for computing *FD* and discrimination of patterns in the waveforms. The reliability of the method of choosing k_{min} and k_{max} is proved from the results. Due to k_{min} and k_{max} this method becomes effective and less time consuming.

Size Measure Relationship method is suitable for computation of Fractal dimension and estimation of complexity presented in the waveforms with minimum time complexity. Therefore, we strongly suggest the application of the above methods to compute *FD* is best for waveform signals.

Fractal dimensions are an addition to the roster of techniques for quantifying waveforms. The fractal dimension *FD*, of a waveform is a measure of its convolutedness: perfectly straight waveforms have a minimal dimensionality ($FD = 1.0$), whereas highly spiked waveforms have a maximal dimensionality (approaching $FD = 1.5$).

As with any single pattern-recognition criterion, a fractal characterization cannot completely duplicate the performance of human pattern recognition. To mimic humans, fractal characterizations should be used in conjunction with different techniques, such as harmonic analysis. On the other hand, even when used individually, objective automatable pattern-recognition schemes can be especially useful in two arenas: first, for rigorously surveying large numbers of patterns, and second, for reproducibly comparing complex patterns. The fractal dimension emphasizes complexity of shape, and it is in the realm of very complex patterns that fractal analyses should prove most helpful.

One automatable technique for numerically classifying and comparing waveforms assesses their fractal dimensionality. Such a fractal characterization may be especially useful for analyzing and comparing very complex biological waveforms such as electroencephalograms (EEGs).

4.3 Fractal Dimension in Epileptic EEG Signals

The novel methods for computing fractal dimension of synthetic waveforms and the importance of fractal dimension in the analysis of waveforms are presented. The method of discriminating the patterns in the waveforms through fractal dimension is explored. We got the idea from the workability of the algorithms, they are suitable for analysis of human brain signals. Working towards real signal analysis is the hot topic in the signal processing field. During the last two decades, we have become fascinated with the realization that some nonlinear dynamical systems have very complex behaviours. Electrical signals of Electroencephalogram (EEG) from the brains' of seizure patients also exhibits complexity in the behaviour. EEG monitoring systems have become important clinical tools for evaluation and treatment of seizure. Improving seizure prediction through approximating the critical cortical site is the hot topic in epilepsy research.

Detection of critical cortical site (recurrent seizure portion) is used to give treatment for the patient. The major physiological brain states that have distinct EEG and behavior correlates are concerned with the degree of vigilance. Some totally different states could present apparently similar EEG patterns. The sites affected in the brain area which are related to other areas may also be affected [66]. The variation of activity in the brain differs from area to area. There is no similarity in the random collection of cortical areas. The EEG signal is the gateway to measure the complexity of the nonlinear behavior of the brain.

Human brain is considered as a dynamical system. Estimating the complexity in the signal gives the state of the dynamical system [70]. We found out the dimension for various signals obtained from different positions of the electrodes, which are needed to specify the state of a dynamical system [21]. One extreme in the spectrum of deterministic behaviours for nonlinear systems is chaos, in which the system shows no evidence of settling down to any stable state and shows extreme (exponential) sensitivity to small perturbations [23].

The classification of EEG signals gives useful results about the behaviour of the human brain. For this we need a tool from nonlinear dynamics. Hazarika [27] used wavelet transform to classify the EEG signals. Maiwald [45] compared three seizure prediction methods by means of the seizure prediction characteristic. The techniques from nonlinear dynamics such as Lyapunov exponents and Correlation dimension can describe the complexity of an EEG data. But both work with small samples and they also need large amount of computations. A hallmark of chaotic systems is that the dimension seems fractional or fractal, that is 1.26 rather than 1 or 2. A global value that is relatively simple to compute is the fractal dimension (FD); it refers to a noninteger or fractional dimension of a geometrical object [70]. The FD can give an indication of the dimensionality and complexity of the system. Since actual living biological systems are not stable and the system complexity varies with time, one can distinguish between different states of the system by the FD; it can also determine whether a particular system is more complex than other systems [21,70].

The algorithms developed by Higuchi [29], Katz [40] and Petrosian [60], for estimating the FD of the waveform signals, but their computational requirements are expensive. These algorithms contain more number of steps with lot of comparisons, so the time complexity is very high [25,26,29,40]. The Higuchi and Katz algorithms work only when the waveform signal contains less variance, so the algorithms do not give better results for EEG signals. The main contribution of this section is presenting an efficient algorithm for calculating FD of EEG signal with less time and space complexity. Along with other nonlinear methods, the fractal dimension has been used in the analysis of biomedical signals exhibiting non-stationary and transient characteristics such as the EEG. The short-term analysis of EEG signals and their dimension analysis will provide the measure of the rate of complexity. We compared the results of our algorithm with the results of Higuchi, Katz and Petrosian algorithms. From our results the complexity decrease prior to the seizure is confirmed with fractal dimension values. It discriminates the ictal (seizure period) and pre-ictal (before seizure) portions in the brain effectively. The clustering technique (K-means algorithm) is used to locate the critical cortical site in the brain.

4.3.1 Background

In the last decade, several quantitative system approaches incorporating statistical techniques nonlinear methods based on chaos theory have been successfully used to study epilepsy because the aperiodic and unstable behaviour of the epileptic brain is suitable to nonlinear techniques that allow precise tracking of the temporal evolution. Some previous studies shown that seizures are deterministic rather than random. Consequently, studies of the spatiotemporal dynamics in long-term intracranial EEG's, from patients with temporal lobe epilepsy, demonstrated the predictability of epileptic seizures; that is seizures develop minutes to hours before clinical onset. The period of seizure's development is called a preictal transition period, which is characterized by gradual dynamical changes in EEG signals of critical electrode sites approximately $1/2$ to 1 h duration before the ictal onset [30–34, 64, 65]. During a preictal transition period, gradual dynamical changes can be exposed by a progressive convergence (entrainment) of fractal dynamical measures at specific anatomical areas and cortical sites, in the neocortex and hippocampus [23].

The existence of the pre-ictal transition period has recently been confirmed and further defined by other investigators [3,14,19,43,44,62], the characterization of this spatiotemporal transition is still far from complete. For instance, even in the same patient, different set of cortical sites may exhibit preictal transition from one seizure to the next. In addition, this convergence of the normal sites with the epileptogenic focus (critical cortical sites) is reset after each seizure [35]. Therefore, complete or partial postictal resetting of preictal transition of the epileptic brain, affects the route to the subsequent seizure, contributing to the apparently non-stationary nature of the entrainment process. In those studies, however, the critical site selections are not trivial but extremely important since most groups of brain sites are irrelevant to the occurrences of the seizures and only certain groups of sites have dynamical convergence in preictal transition [59].

Since the brain is a nonstationary system, algorithms used to estimate measures of the brain dynamics should be capable of automatically identifying and appropriately weighing existing transients in the data. Complex dynamics is an important aspect of the chaotic system and it is reflected in the value of fractal dimension.

4.3.2 Methods

4.3.2.1 Electroencephalogram Signal

The Electroencephalogram (EEG) is a unique and valuable measure of the brain's electrical function. It is a graphic display of a difference in voltages from two sites of brain function recorded over time. Electroencephalography involves the study of recording these electrical signals that are generated by the brain. Extracranial EEG provides a broad survey of the electrocerebral activity throughout both hemispheres of the brain. Intracranial EEG provides focused EEG recording directly from the brain through surgically implanted electrodes that are targeted at specific regions of the brain. Information about a diffuse or focal cerebral dysfunction, the presence of interictal epileptiform discharges (IEDs), or patterns of special significance may be revealed.

For the successful interpretation of an abnormal EEG, one must first understand the criteria necessary to define normal patterns. While a normal EEG does not exclude a clinical diagnosis (i.e., epilepsy), an abnormal finding on EEG may be supportive of a diagnosis (i.e., in epilepsy), be indicative of cerebral dysfunction (i.e., focal or generalized slowing), or have nothing to do with the reason that the study was performed (i.e., in headache). It is the clinical application of the EEG findings that imparts the utility of EEG [6].

Scalp EEG recordings display the difference in electrical potentials between two different sites on the head overlying cerebral cortex that is closest to the recording electrode. From the patient scalp, electrodes conduct electrical potentials to an electrode box. Electrode placement has been standardized by an international 10–20 system that uses anatomical landmarks on the skull. These sites are then subdivided by intervals of 10–20 % and to designate the site where an electrode will be placed. A minimum of 21 electrodes are recommended for clinical study, although digital EEG now has the capability for a greater number. During infant EEG recordings, fewer electrodes are used depending upon age and head size. A newer modified combinatorial electrode system uses electrode placement with more closely spaced electrodes in a 10-10 system [22].

4.3.2.2 Spontaneous Ongoing Signal

In general, one or more tape-recorded channels of suitably filtered and amplified EEG signal, extending over the period of interest, are digitized at a sampling frequency m to form the database. The T second long and $L(=T \times m)$ sample long digitized signal,

$$F(n), n = 1, 2.., (M-1) \times D + N, \tag{4.34}$$

is split into Δt second long overlapping segments or epochs which are arranged as the columns or pattern vectors of an $N \times M$ matrix F,

$$F = \begin{bmatrix} F(1) & F(D+1) & \cdots & F((M-1) \times D + 1) \\ F(2) & F(D+2) & \cdots & F((M-1) \times D + 2) \\ \vdots & \vdots & \vdots & \vdots \\ F(N) & F(N+2) & \cdots & F((M-1) \times D + N) \end{bmatrix} \tag{4.35}$$

where $N(=\Delta t \times m)$ is the dimension of each pattern vector, $M = (\lfloor (L + D - N)/D \rfloor)$ is the number of patterns, and D is the delay between patterns.

In our calculation, we choose $N = 2,500$ points, the length of the signal is estimated by consecutive differences of samples. Out of the EEG data observed from the seizure patient with 32 electrode channels, we have considered the most seizure exhibited channels of 14 only [35].

4.3.2.3 Recording Procedure and EEG Data

The fractal dimension methods were applied using intracranial EEG data from four patients suffering from epilepsy of temporal and extratemporal origin. EEG recordings from bilaterally, surgically implanted microelectrodes in the hippocampus, temporal and frontal lobe cortices of epileptic patients with temporal lobe epileptic seizures (focal, complex, with or without secondary generalization) were analyzed (see Fig. 4.13 for our typical electrode montage). The EEG signals were recorded using amplifiers with an input range of ±0.6 mV, and a frequency range of 0.5 ∽ 70 Hz. Prior to storage, signals were sampled at 250 Hz using an analog-to-digital converter with 10-bit quantization. The multielectrode EEG signals (26–30 common reference channels) were obtained from long-term (3.6–7 days), continuous recordings from four patients and 2–5 seizures were examined for every patient [35].

4.3.2.4 Computing the Fractal Dimension of EEG Signals

We utilized an estimate of the Fractal dimension (FD) as the dynamical measure of the electroencephalogram [8]. Estimation of FD of synthetic signals described

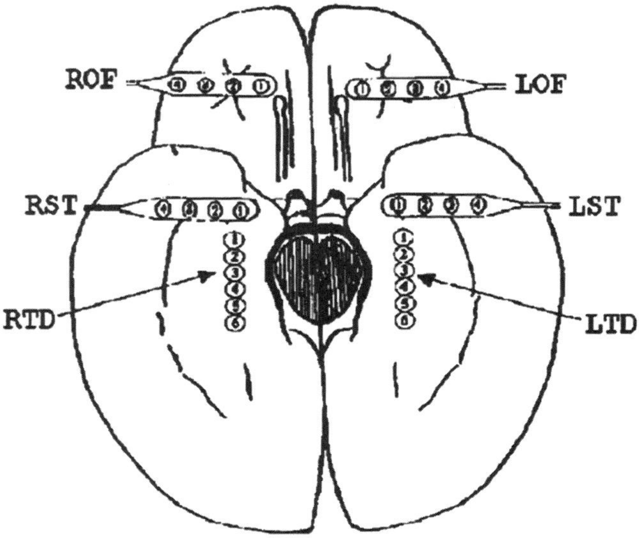

Fig. 4.13 Diagram of the depth and subdural electrode placement

Table 4.3 Comparison of run time of FD methods with EEG signals

	Run-time (seconds)			
Window length	SMR method	Higuchi's method	Katz's method	Petrosian's method
250	0.0660	0.1830	0.0460	0.5420
500	0.1370	0.5460	0.1780	0.5530
1,000	0.3750	0.9620	0.1250	0.6010
2,000	0.1010	3.1160	0.2950	0.6210
4,000	1.3250	13.9610	1.6260	0.6330
8,000	1.3390	52.9990	1.8920	1.5210

in the previous section. Now the same procedure (SMR) is applied to EEG signals. Estimation of FD was calculated by dividing the EEG into non-overlapping segments of 10.24 s each. The fractal dimension is calculated by our developed method. Also we have compared our results with other fractal dimension algorithms such as Higuchi's algorithm and Katz algorithm. The procedure for the algorithms are discussed in the previous section. The time consumption of fractal dimension algorithms works with EEG signals are presented in Table 4.3. The EEG signals collected from 14 electrodes and its fractal dimension values are presented in Table 4.6.

4.3.2.5 Properties of EEG Signals with Fractal Dimension

The fractal dimension profiles show two patterns pertaining to all the types of epileptic seizures. They are as follows:

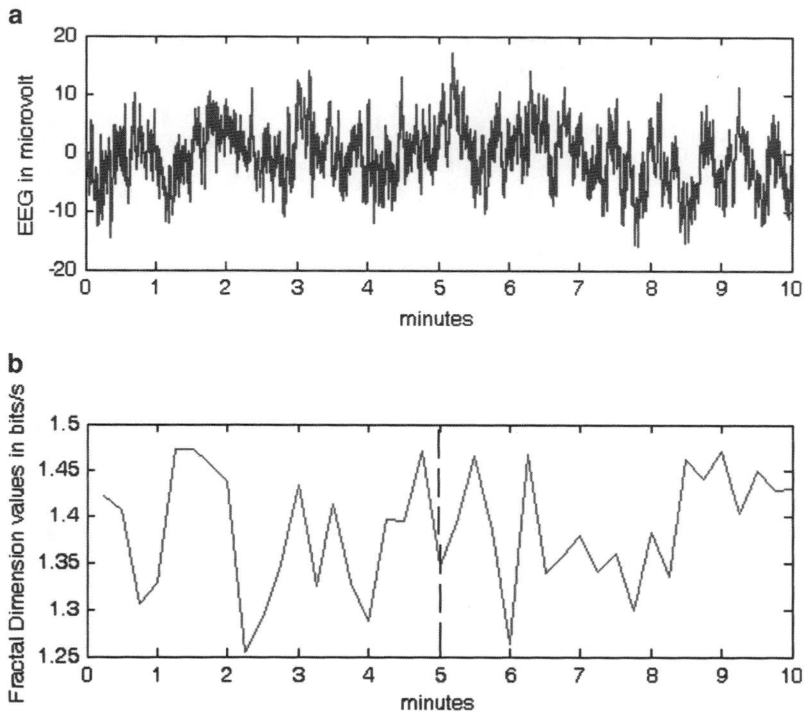

Fig. 4.14 EEG signal from the channel c3 and its fractal dimension signatures

1. A drop in the fractal dimension mean value with respect to the pre-ictal state, corroborating the hypothesis that the brain dynamics more ordered. It is interesting that this feature (a drop in value) is very different from the observed amplitude of the EEG, which normally increases during seizures. This drop can be used to detect critical brain sites by a range of fractal dimension values.
2. Entrainment between the fractal dimension of different channels before seizure. In signal processing terms, this means that the distance between the fractal dimension values will decrease when an ictal event is pending.

Figure 4.14 shows a typical seizure from the channel c3 where these two characteristics are clearly visible. However, it should be noted that the fractal dimension does not always show such obvious characteristics for all seizures and all electrodes, either because of inter-seizure dynamical differences, or because of seizure specific parameters to estimate fractal dimension. The set of parameters to estimate fractal dimension was held fixed throughout our experiments and works very well for complex partial seizures evolving into secondarily generalized seizures.

Figures 4.14–4.17 show examples of fractal dimension profiles for different seizure types from different channels. Figure 4.18b shows the performance of

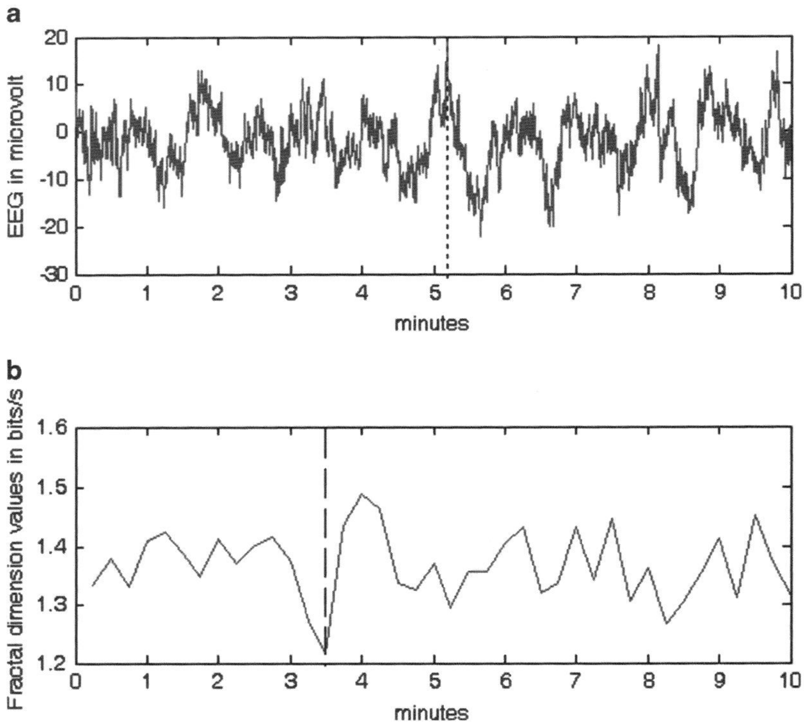

Fig. 4.15 EEG signal from the channel c4 and its fractal dimension signatures

three fractal dimension methods with EEG seizure signal displayed in Fig. 4.18a. Figures 4.19–4.22 present the EEG signal of one epileptic patient analyzed by Higuchi, Katz, Petrosian and SMR algorithms. Figures 4.23–4.25 illustrate the box plot for the fractal dimension values of 14 electrodes computed by SMR algorithm displayed in Table 4.6.

4.3.2.6 Selection of Critical Cortical Sites

One of the most important tasks to detect the chaotic dynamics is to identify the most possible group of cortical sites which will participate in the pre-ictal transition of an impending seizure. Cortical sites participating in the preictal transition vary somewhat from seizure to seizure. Here we identify the most critical group of cortical sites based on the chaotic dynamics estimated by fractal dimension values in the 10-min time window within the seizure period. More specifically, the fractal dimension and K-means algorithms are assisting to select the group of cortical sites which are most entrained prior to the seizure.

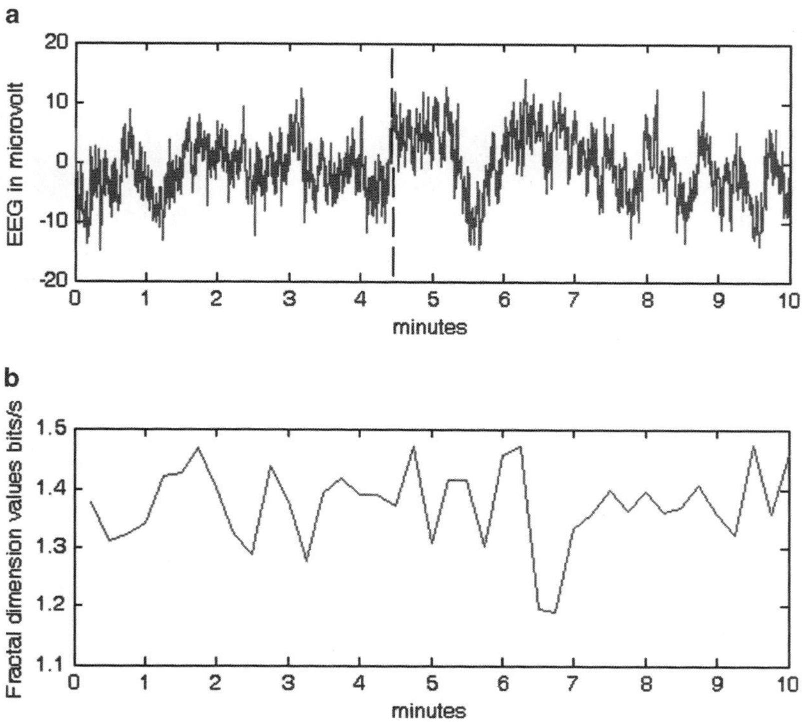

Fig. 4.16 EEG signal from the channel p3 and its fractal dimension signatures

4.3.2.7 Clustering of Electrode Sites

Cluster analysis has been applied in many fields. Here cluster analysis has been employed to find groups of electrode sites with similar and minimal distance (or nearest neighbour) of its fractal dimensions. There are large numbers of clustering algorithms used by the researchers [20]. Among them, we have used the effective clustering algorithm known as 'K-means algorithm' for grouping the electrode sites [36]. Besides the data, input to the algorithm consists of K, the number of clusters to be constructed. The K-means algorithm consists of the following steps.

1. Begin with K clusters, each consisting of one of the first k samples.
2. For each sample, find the centroid nearest to it. Put the sample in the cluster identified with this nearest centroid.
3. If no samples changing clusters, stop.
4. Recompute the centroids of altered clusters and go to step 2.

Once it is obtained the clusters can be used to classify patterns into distinct classes. Here the signals from each electrode are sampled at 250 Hz. The signals observed form electrode strips are placed over the left orbitofrontal (LOF), right orbitofrontal (ROF), left subtemporal (LST) and right subtemporal cortex(RST).

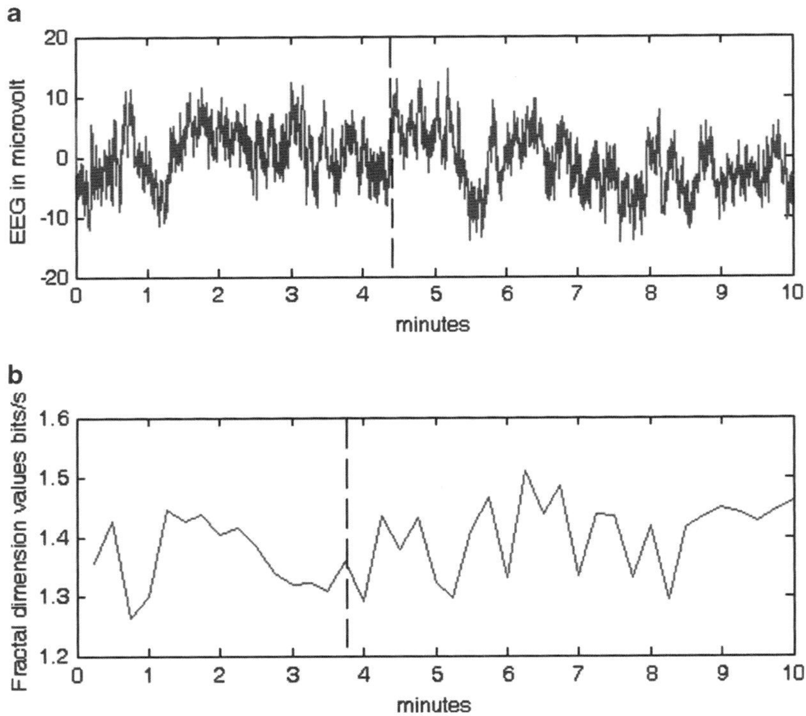

Fig. 4.17 EEG signal from the channel p4 and its fractal dimension signatures

Depth electrodes are placed in the left temporal depth (LTD) and right temporal depth (RTD) to record hippocampus EEG activity [69]. We have selected positions of 14 electrodes among the above locations.

The EEG data in the format of 14 columns (electrode positions) and 10 rows (FD values of segments signals from electrode positions) are used for clustering. The K-means algorithm makes the clusters depending on the dimension values. Similar behavior electrodes are in the same group. The grouped electrodes are displayed in Table 4.4 with their respective fractal dimension algorithms. Also the calculation of variation among the electrode groups are presented in Table 4.5.

4.3.3 Results and Discussion

In this section, results obtained from the application of developed fractal dimension method and the comparison with other three fractal dimension methods to estimate the complexity in the epileptic seizure EEG signal are presented. Also the application of fractal dimension methods to detect the critical cortical site in the human brain and their comparisons are displayed. Critical cortical sites are electrode sites

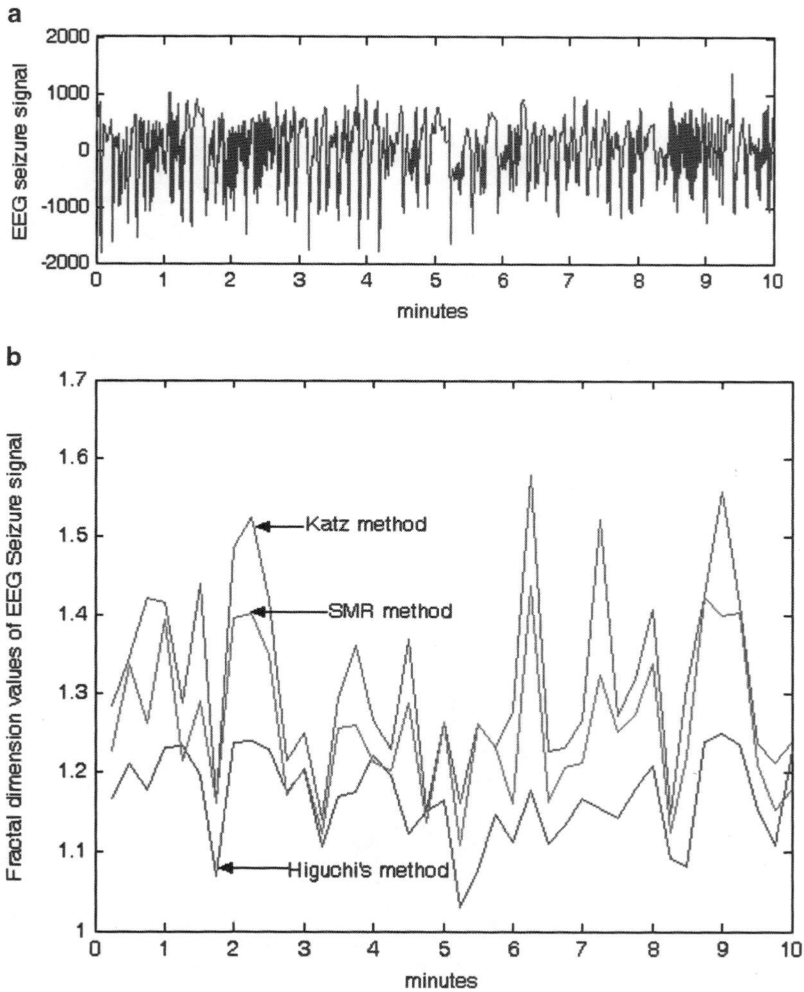

Fig. 4.18 (**a**) Seizure EEG signal (**b**) Performance of three fractal dimension methods with seizure EEG signal

that are most entrained during the current seizure and disentrained after the seizure onset should be more likely to be entrained prior to the next seizure than the other entrained sites. As a result, it is possible to predict a seizure if one can identify critical cortical sites in advance. As the critical cortical site detection characteristic depends on the three different fractal dimension methods, it is necessary to fix atleast one of them to display the result.

The assessment and comparison of the three fractal dimension methods is presented by the critical cortical site detection characteristics depending on, first the fractal dimension values of the seizure period, second the clustering based

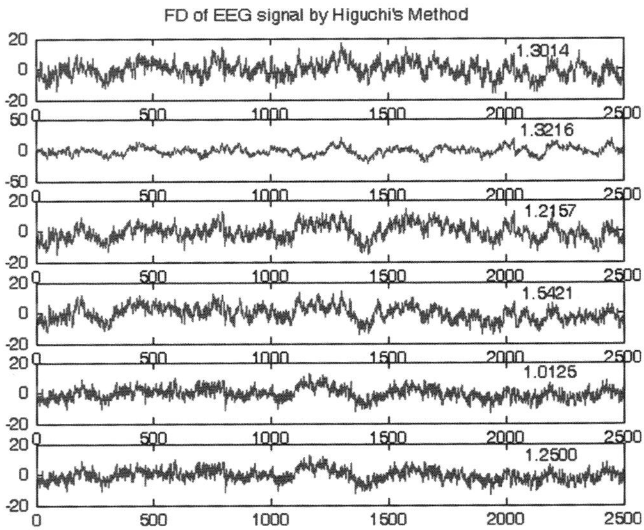

Fig. 4.19 FD of EEG signals for Higuchi's method

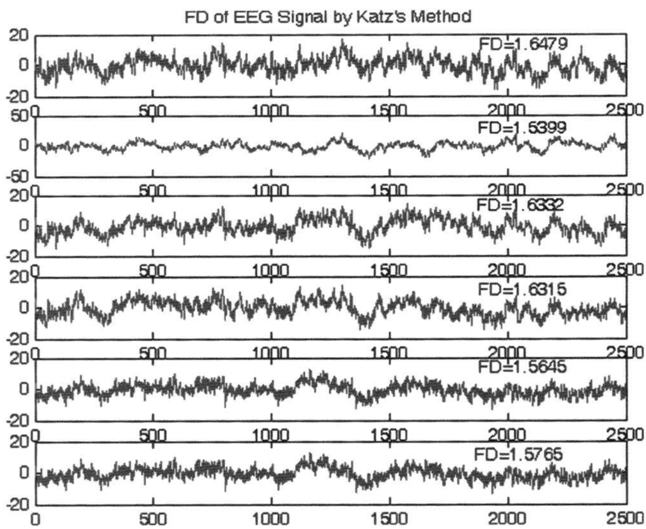

Fig. 4.20 FD of EEG signals for Katz's method

on fractal dimension values, and third the variations among the fractal dimension values. For estimating the critical cortical site, we use fractal dimension methods except the Petrosian method. Among the algorithms, the Petrosian method not discriminate the complex patterns in the EEG seizure signal. From the Fig. 4.14 we observed that, the EEG seizure signal from the channel c3 (Fig. 4.14a) is effectively analyzed by the SMR method. The seizure periods are clearly displayed

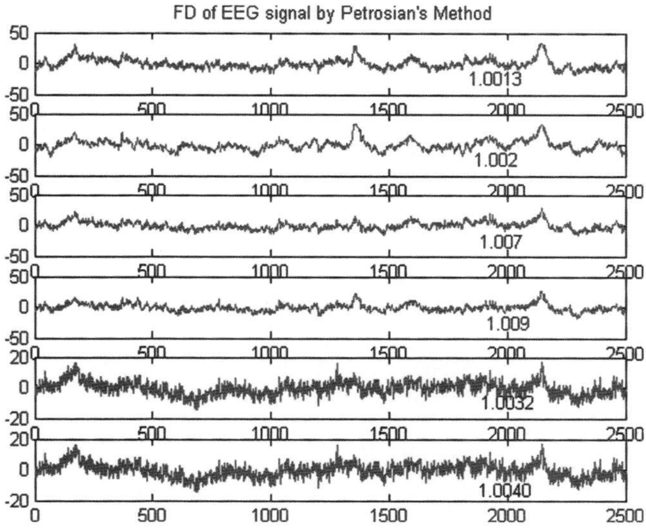

Fig. 4.21 FD of EEG signals for Petrosian's method

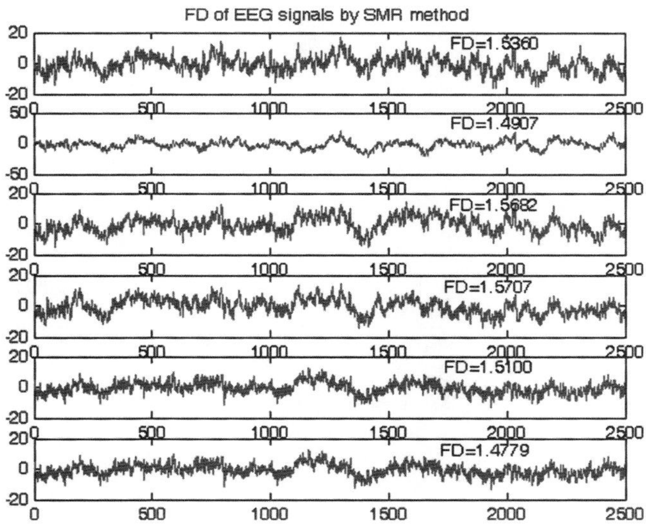

Fig. 4.22 FD of EEG signals for SMR method

with fractal dimension signatures in bits per seconds (Fig. 4.14b). Also we have analyzed another three EEG seizure signals with their fractal dimension signatures are displayed in Figs. 4.15–4.17. The performance of each method with EEG seizure signal give good results and it is observed from Figs. 4.19–4.22. But the SMR method gives effective results than the other methods. The performance of the three methods with their fractal dimension values is illustrated in Fig. 4.18b. From

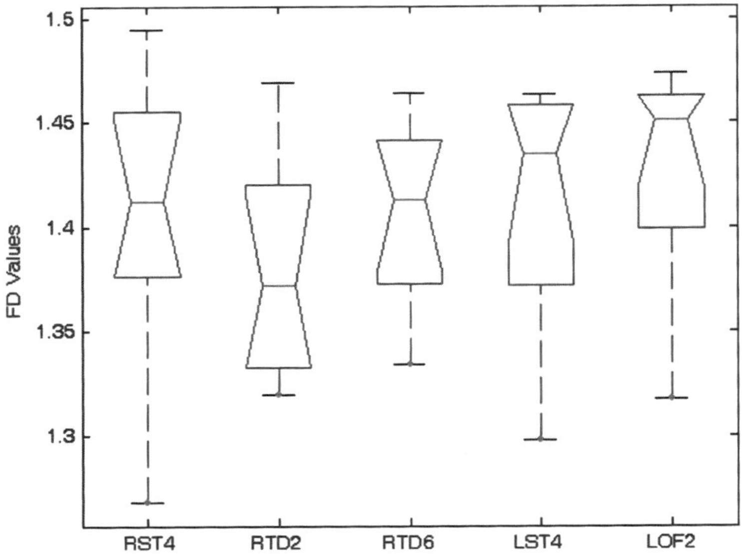

Fig. 4.23 Box plot of the fractal dimension values of the electrode channels RST4, RTD2, RTD6, LST4 and LOF2

Table 4.4 Selection of electrode sites

Fractal dimension algorithms	Electrode groups
SMR method	LST1, LST2, LTD1, LTD2, LTD3, LST4
Higuchi's method	RST1, RST4, LTD3, LST4, LTD2, RST2
Katz's method	LTD2, LTD3, RST4, RTD2, RTD6, RST1

Table 4.5 Calculation of variation

Fractal dimension algorithms	Coefficient of variation
SMR method	2.1430
Higuchi's method	2.4820
Katz's method	2.3510

Fig. 4.18a we have observed that the seizure onset and seizure ending can only be determined with 10 s precision, and the dynamic transition itself takes several samples and it varies from seizure to seizure. The same observation can be made for the transition to postictal state. These features imply that the fractal based detectors will provide exact warnings in real time applications.

From Fig. 4.18b the exact fractal dimension signatures for the EEG seizure signal is observed in middle wave. That is the middle wave is the fractal dimension signatures computed from SMR method. The FD results obtained with experimental EEG data reveals that our method is the most accurate of the three. Katz's method yielded the most consistent results regarding discrimination between states of the

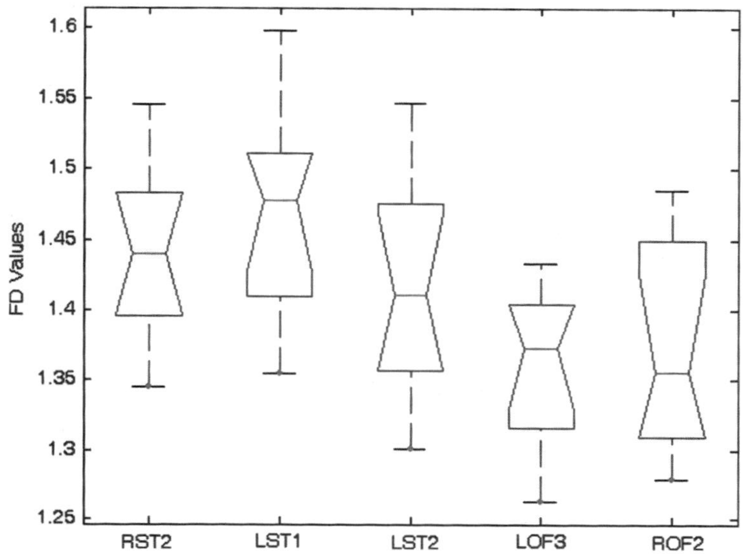

Fig. 4.24 Box plot of the fractal dimension values of the electrode channels RST2, LST1, LST2, LOF3 and ROF2

Table 4.6 FD values of EEG data segments

	EEG data segments (N = 2,500 points, sampled at 250 Hz)									
Electrodes	1	2	3	4	5	6	7	8	9	10
RST4	1.4252	1.4565	1.3764	1.4946	1.4000	1.3837	1.4555	1.3180	1.4538	1.2680
RTD2	1.4692	1.3787	1.4339	1.3290	1.3678	1.3331	1.4200	1.3762	1.3603	1.3200
RTD6	1.4639	1.4413	1.3698	1.4094	1.4042	1.3344	1.4574	1.3726	1.4281	1.4167
LST4	1.4614	1.4583	1.4222	1.4495	1.2975	1.3670	1.3763	1.4482	1.4630	1.3723
LOF2	1.4460	1.4732	1.4267	1.3994	1.3172	1.3658	1.4552	1.4567	1.4628	1.4667
RST2	1.4303	1.3899	1.3955	1.4025	1.3449	1.4498	1.4837	1.4731	1.5458	1.4835
LST1	1.4565	1.5113	1.5143	1.5983	1.4280	1.4080	1.3548	1.4985	1.4099	1.5041
LST2	1.4097	1.4899	1.4754	1.4544	1.5463	1.3141	1.3768	1.4117	1.3014	1.3570
LOF3	1.3208	1.3990	1.4025	1.4331	1.4049	1.2633	1.3481	1.3157	1.3145	1.4048
ROF2	1.4199	1.4850	1.3100	1.4501	1.4568	1.2791	1.3398	1.3036	1.3101	1.3723
LTD1	1.4311	1.4171	1.3405	1.3647	1.4822	1.3425	1.4288	1.4573	1.2875	1.3786
LTD2	1.4346	1.5420	1.4716	1.4143	1.5460	1.3970	1.5021	1.4626	1.3754	1.3779
LTD3	1.3746	1.5881	1.4503	1.4146	1.5432	1.3738	1.4306	1.3628	1.4348	1.4985
RST1	1.2459	1.5080	1.4754	1.4775	1.5075	1.1592	1.4497	1.5557	1.1638	1.5368

brain, but the dynamic ranges of the results are affected by window length and noise. The FD estimated with 14 channels for each method when the synthetic signal is contaminated with white noise, yielding a signal to noise ratio (SNR) of 10 db [20]. The run time and steps for the proposed method are less than that of other methods; window length does not affect and it works with various window lengths. The new method gives the distinguishability in results between different

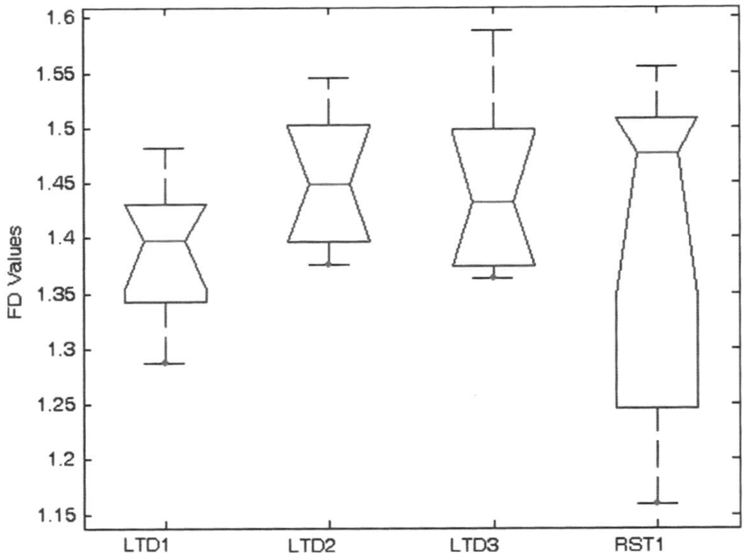

Fig. 4.25 Box plot of the fractal dimension values of the electrode channels LTD1, LTD2, LTD3 and RST1

time points. The SMR algorithm needs less time complexity compared to Higuchi and Katz algorithms. The new algorithm separates the ictal (seizure period) and pre-ictal (before seizure period) portions of the EEG signal with respect to FD values, which are presented in Table 4.6. This can be observed from the box plots for the fractal dimension values with respective channels (Figs. 4.23–4.25).

The ictal portions show greater dimension values and pre-ictal portions show lesser dimension values. The portion which has fractal dimension between 1 and 1.5 is the pre-ictal portions and which has the fractal dimension between 1.5 and 2 is known as ictal portion. The fractal dimension values discriminate the seizure and non-seizure portions. The EEG data observed from 14 electrode positions are clustered into four groups. The four groups obtained from our method and the fourth cluster contains the minimum variance electrode positions, which is presented in Table 4.4. Figures 4.19–4.22 show the behavior of electrodes group performed by Higuchi, Katz, Petrosian and SMR algorithms. The electrodes sites in the fourth cluster are LST1, LST2, LTD1,LTD2 and LTD3.

Computing the coefficient of variation (a relative measure of the differences of the values of various groups of electrodes from a measure of central tendency of electrode groups, while comparing the coefficient of variation of two or more series of data, the series having lesser coefficient of variation is less variable, more stable, more uniform and more consistent) among the electrode groups of three methods, the minimum variation appears in the proposed method which is presented in Table 4.5. The six electrodes express the related brain activity (recurrent seizure portion) and the surface of the above electrode portions are the critical brain sites estimated by our algorithm. For records from 150 to 2,500 points length, run time for our method is very small when compared with Higuchi's and Katz's method. If the record length is increased to 8,000 points, then our algorithm becomes more faster than Higuchi's and Katz's. All the three methods can be run in real time. If the window length increases up to 8,000 points then our algorithm performance improves and becomes faster than Higuchi's and Katz's method.

The results show that the detected critical cortical sites exhibit the pre-ictal transition using SMR algorithm. When we compare this results with preictal transition from any entrained sites, we obtain our results, which shows better performance. The located critical cortical sites detecting preictal transition better than the randomly selected entrained cortical sites.

4.3.4 Concluding Remarks

The results of this study in epilepsy confirm our aim that the critical cortical sites during the current seizure and reset after seizure onset is more likely to be entrained again during the next seizure than other randomly selected cortical sites. These results indicate that it may be possible to develop automated seizure warning devices for diagnostic and therapeutic purposes.

Investigations of brain dynamics by analysis of EEG signals and computing the fractal dimension of the brain activity are used to locate the critical brain sites of recurrent seizure portions. The developed method is easier than the methods from nonlinear dynamics. The treatment of epilepsy starts from the localization of brain sites. We have shown the assessment and comparison of three fractal dimension methods. From the discussions, we conclude that the algorithm of size measure relationship method effectively compute the fractal dimension and estimate of complexity presented in the EEG signal with minimum time complexity.

Therefore, we strongly suggest the application of SMR method to compute FD is best for EEG seizure signals. Further research will lead to the prediction of the seizure before their onset using the theory of nonlinear dynamical system and fractal analysis.

4.4 Improved Generalized Fractal Dimensions in Epileptic EEG Signals

In the recent years, various signal processing techniques are used to process the EEG signals. Fourier Transforms and Wavelet Transforms are widely used in the analysis of EEG data. These traditional linear analysis, both in time and frequency domains, has been used for Epileptic Seizure detection but it has its limits [55]. But our

Human Brain is a highly complex and a non-linear system. So the non-linear data analysis methods have been effectively applied to the studies of brain functions and pathological changes in Epileptic EEG Signals [2, 38, 39, 52, 54, 55]. The non-linear measures used, including Correlation Dimension, Largest Lyapunov Exponent and Approximate Entropy quantify the degree of complexity and irregularity in a Fractal Time Series [39, 52].

In the literature when Fractal theory has been applied to the EEG data analysis, the dimensional complexity has mainly been used to analyze the change in the chaotic nature in different physiopathological conditions [1]. Among all the non-linear techniques as mentioned the correlation dimension measurement is more accessible in dealing with experimental data such as the EEG records [50].

The absolute value of the estimated correlation dimension does not represent complexity of the signal since it is only one scalar value from the system of fractal dimension spectrum. This scalar value can represent dimension only in simple monofractal systems where all the fractal dimensions are same. This Dimensional quantity is insufficient to characterize the nonuniformity or inhomogeneity of the Signal. Generally, Chaotic Attractors are inhomogeneous. Such an inhomogeneous set is called a Multifractal and is characterized by Generalized Fractal Dimensions (GFD) or Rényi Fractal Dimensions. EEG time series is essentially multi scale fractal i.e., Multifractal. Multifractal signals are intrinsically more complex and inhomogeneous than monofractals. Therefore, quantifying the complexity of the EEG signal requires estimation of the fractal dimension spectrum where the complexity means higher variability in general fractal dimension spectrum [24, 42, 47].

The non-linear measure, correlation dimension belongs to an infinite family of Generalized Fractal Dimensions (GFD) [28]. The usage of the whole family of fractal dimensions should be very useful in comparison with using only some of the dimensions. Unlike the Fourier Spectra, the Fractal Spectra consists of a family of Fractal Dimensions that characterize the Fractal Time Series from both the amplitude and the frequency point of view [15–18, 41]. So Generalized Fractal Spectra is very efficient technique to quantify the chaotic nature of the EEG Signals and employs in the classification of Healthy and Epileptic EEG Signals [15, 16, 18].

Thus the Generalized Fractal measure is very efficient technique to quantify the chaotic nature of the EEG Signals. The Multifractal techniques in Signal Processing and Non-linear Data Analysis are devoted in the development of Epileptic research by using EEG Signals. So we could improve the method of Generalized Fractal Dimensions for the Discrimination between the Healthy and Epileptic EEGs. Besides that, we compare our proposed GFD Method with the GFD to distinguish the Healthy and Epileptic EEGs through graphically using the Fractal Spectra and the graph plotted with the absolute values of the Entropy against the log values of the corresponding scaling factor, and statistically using the one-way ANOVA Test with Box Plot. Hence we are assuring that the roll of the Improved GFD will be very efficient than the GFD.

4.4.1 Mathematical Analysis

4.4.1.1 Rényi Entropy

Rényi Entropy [63, 67] played a significant role in the Information theory. Rényi Entropy, a generalization of Shannon entropy, is one of the family of functionals for quantifying the diversity, uncertainty or randomness of a given system. It was introduced by Alfréd Rényi [63]. Rényi Entropy is also known as generalized entropy of a given probability distribution.

The *Rényi Entropy* of order $q(\neq 1)$, where q is a real number, of the given probability distribution is defined as

$$S_q = \frac{1}{1-q} log_2 \left(\sum_{i=1}^{N} p_i^q \right) \tag{4.36}$$

where $p_i \in [0, 1]$ are the probabilities of the random variable which takes the values $x_1, x_2, \ldots, x_N$.

If the probabilities are all the same then all the Rényi Entropies of the distribution are equal, with $S_q = log_2 N$. Otherwise the entropies are decreasing as a function of q.

4.4.1.2 Multifractal Analysis

The Rényi Entropies are important in Non-linear Analysis and Statistics as indices of uncertainty or randomness. They also lead to a spectrum of indices of Fractal Dimension (Rényi Fractal Dimensions or Generalized Fractal Dimensions). Grassberger [24] and Hentschel et al. [28] systematically developed the multifractal theory, which is based upon Generalized Fractal Dimensions (GFD). In this section, we describe the GFD [24, 28, 63, 67] and the Improved form of the usual GFD Method [18].

4.4.1.3 Generalized Fractal Dimensions

Now we define a probability distribution of a given Fractal Time Series by the following construction.

The total range of the Signal Time Series is divided into N intervals (bins) such that

$$N = \frac{V_{max} - V_{min}}{r}$$

where V_{max} and V_{min} are the maximum and the minimum values of the signal received in the experiments, respectively; and r is the uncertainty factor, that may be depend on the measuring device used to record the EEG signal.

Now the probability that the signal passes through the ith interval of length r is given by

$$p_i = \lim_{N \to \infty} \frac{N_i}{N}, \qquad i = 1, 2, \ldots, N$$

where N_i is the number of times the signal passes through the ith interval of length r.

Then, the *Rényi Fractal Dimensions or Generalized Fractal Dimensions (GFD)* of order $q \in (-\infty, \infty)$ for the known probability distribution, denoted by D_q, can be defined as

$$D_q = \lim_{r \to 0} \frac{1}{q-1} \frac{log_2 \sum_{i=1}^{N} p_i^q}{log_2 r}. \tag{4.37}$$

Here D_q is defined in terms of generalized Rényi Entropy. Note that $D_q = D_0$, for all q for a Self-similar signal with probabilities $p_i = 1/N$, for all i. Also observe that $D_q = D_0 = 0$, for all q for a constant signal because all probabilities except one equal to zero, whereas the exceptional probability value is one. For all q, we have $D_q > 0$. It can be shown that if $q_1 < q_2$, $D_{q_1} \geq D_{q_2}$ such that D_q is a monotone decreasing function of q.

Some Special Cases

- If $q = 0$, then

$$D_0 = -\frac{log_2 N}{log_2 r}$$

  which is nothing but the *Fractal Dimension*.
- As $q \longrightarrow 1$, D_q converges to D_1, which is given by

$$D_1 = \lim_{r \to 0} \frac{\sum_{i=1}^{N} p_i log_2 p_i}{log_2 r}.$$

  This is called as *Information Dimension.*
- If $q = 2$, then D_q is called the *Correlation Dimension.*
- There are two limit cases when $q = -\infty$ and $q = \infty$, which is given as

$$D_{-\infty} = \lim_{r \to 0} \frac{log_2 (p_{min})}{log_2 r}$$

$$D_{\infty} = \lim_{r \to 0} \frac{log_2 (p_{max})}{log_2 r}$$

where

$$p_{min} = \min\{p_1, p_2, \ldots, p_N\}$$

$$p_{max} = \max\{p_1, p_2, \ldots, p_N\}$$

Range of Generalized Fractal Dimensions

The two limit cases, $D_{-\infty}$ and D_{∞}, define the Range of Generalized Fractal Dimensions of a given Fractal Time Series as

$$R_{GFD} = D_{-\infty} - D_{\infty} \tag{4.38}$$

4.4.1.4 Improved Generalized Fractal Dimensions

In order to improve our GFD Method, we define a probability distribution of a given Fractal Time Series as follows.

The total range of the Signal Time Series is divided into $N_V \times N_t$ bins (boxes) such that

$$N_V = \frac{V_{max} - V_{min}}{r} \quad \text{and} \quad N_t = \frac{t_{max} - t_{min}}{r}$$

where V_{max} and V_{min} are the maximum and the minimum values of the signal received in the experiments and t_{max} and t_{min} are the maximum and minimum time of the experiments, respectively; and r is the uncertainty factor, that may be depend on the measuring device used to record the EEG signal.

Now the probability that the signal passes through the ijth bin (box) of size r is given by

$$p_{I_{ij}} = \lim_{N_V, N_t \to \infty} \frac{N_{ij}}{N_V \times N_t}, \tag{4.39}$$

$$i = 1, 2, \ldots, N_V \quad \text{and} \quad j = 1, 2, \ldots, N_t$$

where N_{ij} is the number of times the signal passes through the ijth bin of size r.

Then, the *Improved Rényi Fractal Dimensions or Improved Generalized Fractal Dimensions (GFD)* of order $q \in (-\infty, \infty)$ for the known probability distribution, denoted by ID_q, can be defined as

$$ID_q = \lim_{r \to 0} \frac{1}{q-1} \frac{log_2 \sum_{i=1}^{N_V} \sum_{j=1}^{N_t} p_{I_{ij}}^q}{log_2 r}. \tag{4.40}$$

Here ID_q is also defined in terms of generalized Rényi Entropy with the probability given in (4.39). Equation (4.40) is called the Improved form of the Generalized Fractal Dimensions as in (4.37).

Limiting Cases

The two limit cases in the Improved GFD Method, when $q = -\infty$ and $q = \infty$, are

$$ID_{-\infty} = \lim_{r \to 0} \frac{log_2\left(p_{I_{min}}\right)}{log_2 r}$$

$$ID_{\infty} = \lim_{r \to 0} \frac{log_2\left(p_{I_{max}}\right)}{log_2 r}$$

where

$$p_{I_{min}} = \min_{1 \le i \le N_V \ \& \ 1 \le j \le N_t} \{p_{I_{ij}}\}$$

$$p_{I_{max}} = \max_{1 \le i \le N_V \ \& \ 1 \le j \le N_t} \{p_{I_{ij}}\}$$

Range of Improved Generalized Fractal Dimensions

The two limit cases, $ID_{-\infty}$ and ID_{∞}, define the Range of Improved Generalized Fractal Dimensions of a given Fractal Time Series as

$$R_{IGFD} = ID_{-\infty} - ID_{\infty} \tag{4.41}$$

4.4.1.5 Method to Analyze the EEG Time Signals

For a given probability distribution of a given Fractal Time Series *(EEG Signal)*, the GFD function (the Improved GFD function), D_q (ID_q), is called as *Fractal Spectrum (Improved Fractal Spectrum)*. This Fractal Spectrum describes information about both the amplitudes and frequencies of the given Signal. So we consider that it is a significant tool to characterize the EEG Signals.

We can calculate the GFD (D_q) and IGFD (ID_q) for the given EEG Time Signals by using (4.37) and (4.40), respectively, through the graph plotted the absolute values of the Rényi Entropy against the log values of the corresponding scaling factor. From these measures we compare the Healthy and Epileptic EEG Time signals.

In order to discriminate the Healthy and Epileptic EEGs, set

$$REN = \frac{1}{q-1} log_2 \sum_{i=1}^{N} p_i^q \qquad \text{from the Equation (4.37)} \tag{4.42}$$

and

$$IREN = \frac{1}{q-1} log_2 \sum_{i=1}^{N_V} \sum_{j=1}^{N_t} p_{I_{ij}}^q \qquad \text{from the Equation (4.40)} \tag{4.43}$$

Then plot the graph of *REN* and *IREN* versus corresponding $log(r)$ for the given EEG signals and compare these graphs to analyze and discriminate the Healthy and Epileptic EEGs.

One-way ANOVA (Analysis of Variance) Test is one of the statistical tools to analyze the mean and variance of the given data set. ANOVA uses variances to decide whether the means are different. If the observed differences are high, then it is considered to be statistically significant. The p-value can be obtained using analysis of variance between groups (ANOVA) test. In this test, if the p-value is near zero, this casts doubt on the null hypothesis and suggests that at least one sample mean is significantly different than the other sample means.

Box Plot technique shows the differences between the data sets significantly. The Box Plot produces a box and whisker plot for each set of data. The box has lines at the lower quartile, median, and upper quartile values. The whiskers are lines extending from each end of the box to show the extent of the rest of the data. Outliers are data with values beyond the ends of the whiskers. If there is no data outside the whisker, a dot is placed at the bottom whisker. We can compare the Box Plots of GFD methods among the Healthy and Epileptic EEG data.

4.4.2 Experimental EEG Data

The EEG clinical data consists of two different sets each containing 100 single-channel EEG segments with 4,096 samples of 23.6-s duration, which were obtained from the EEG Database available with the Clinic of the Department of Epileptology in University of Bonn Medical Centre, Bonn, Germany [4, 11]. The EEG data was recorded with International 10–20 system and digitized at a sampling rate (sampling frequency) of 173.61 Hz. Note that the time series have the spectral bandwidth of the acquisition system, which is 0.5–85 Hz. We considered only 20 representative single-channel EEG segments out of 100 segments from each of the two data sets, which were more appropriate to this study.

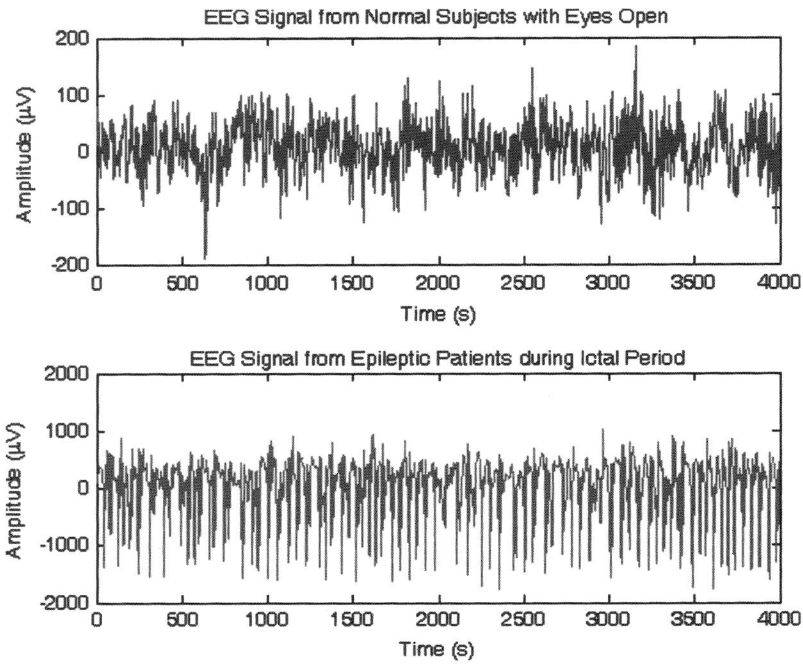

Fig. 4.26 Sample EEG epochs from both healthy subject (eyes open) and epileptic patient (ictal period)

The first Data sets consists of segments taken from surface EEG recordings that were carried out on five Healthy (Normal) persons with eyes open (Healthy EEG). The data in the second set was recorded during seizure activity (Ictal periods) using depth electrodes placed within the Epileptogenic zone of the Epileptic Patients' Brain (Ictal EEG). A sample of EEG epochs from each of the two data sets are plotted in Fig. 4.26.

4.4.3 Results and Discussions

The computations in this section are performed through the MATLAB Software.

The probability distribution of 20 representative Clinical EEG segments from each of the two data sets taken from the Healthy subjects and the Epileptic patients during Ictal period, were obtained and the corresponding Generalized Fractal Dimensions and Improved Generalized Fractal Dimensions for q varies from 2 to 100 were computed. For instance, Generalized Fractal Spectra and Improved Generalized Fractal Spectra of a sample EEG Segments from each of the two data sets are depicted in Fig. 4.27.

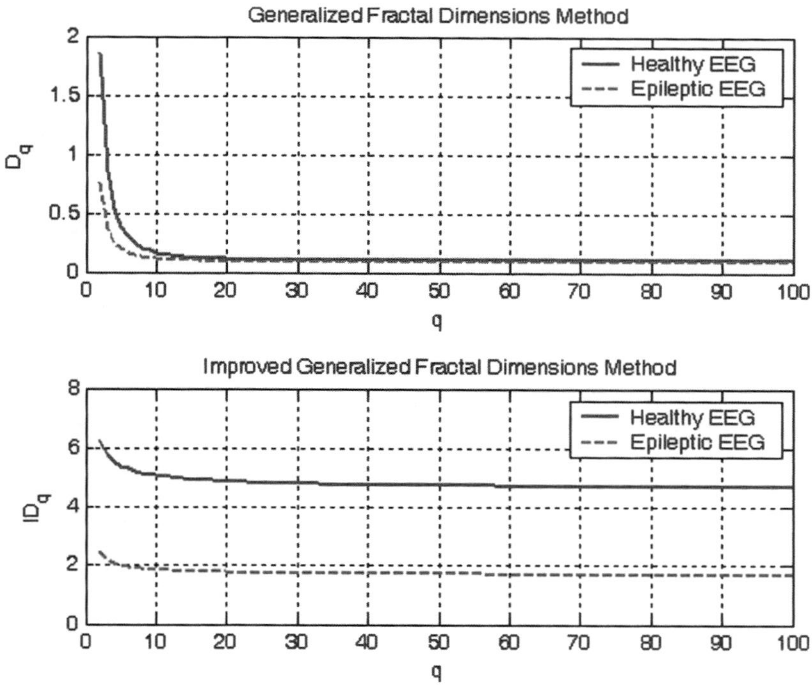

Fig. 4.27 Generalized fractal spectra and improved generalized fractal spectra for healthy and epileptic EEGs

For $q = 0.5$, all the values of *REN*, as in (4.42), and *IREN*, as in (4.43) with the corresponding scaling factors r for all the representative EEG signals from both the Healthy subjects and the Epileptic patients during Ictal period were calculated. Then Plotted the values of *REN* and *IREN* against the corresponding values of $log(r)$ for both Normal and Ictal EEGs. The plotted graphs for the GFD method (*REN*) and Improved GFD method (*IREN*) for a sample EEG segment are depicted in the Fig. 4.28a and b, respectively.

In order to test the mean differences among Normal and Ictal EEG Segments statistically, a repeated measures one-way analysis of variance (ANOVA) is performed on the Fractal Spectra for GFD and IGFD Methods, using a standard tool of Statistical analysis (MATLAB Statistical Tool Box), as shown in Table 4.7. The Box Plots of Fractal Spectra for GFD and IGFD Methods among Normal and Ictal EEG Signals are achieved as shown in Fig. 4.29.

Figure 4.27 shows that, as q increases, there is a specific difference among the values of curves plotted for Normal and Ictal EEGs in the Improved GFD Method than the GFD Method for a sample EEG segment. For all the representative EEG segments we observed that, as q increases, the values of D_q for Normal and Ictal EEGs would coincide in the GFD Method. But in our designed Improved GFD Method, there is a significant differences among the values of ID_q for Normal and Ictal EEGs even for high values of q.

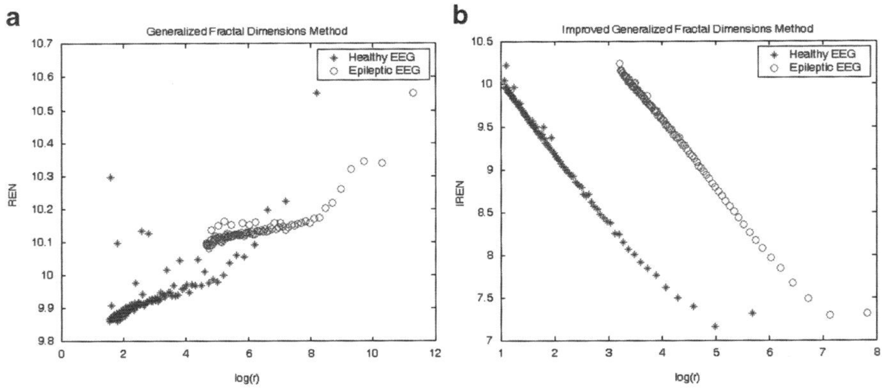

Fig. 4.28 *REN* (**a**) and *IREN* (**b**) versus $log(r)$ for both healthy and epileptic EEGs for $q = 0.5$

Table 4.7 One-way ANOVA Tables for Fractal Spectra of Healthy and Epileptic EEGs

	ANOVA Table				
Source	SS	df	MS	F	Prob > F
(a) GFD Method					
Columns	0.08207	1	0.08207	3.68	0.0564
Error	4.36795	196	0.02229		
Total	4.45002	197			
(b) Improved GFD Method					
Columns	471.949	1	471.949	14691.55	0
Error	6.296	196	0.032		
Total	478.246	197			

In the Fig. 4.28a, b, *REN* and *IREN* against the corresponding values of $log(r)$ for both Normal and Ictal EEGs were plotted. We observed that, by using GFD Method, the values of *REN* for Normal EEG would be mingled with the vales of *REN* for Ictal EEG against the values of $log(r)$ in the Fig. 4.28a and the linearity of values in the GFD Method is very low. Also we noticed that, in the Method of Improved GFD, the values of *IREN* versus $log(r)$ for Normal EEG is specifically different from the values of *IREN* versus $log(r)$ for Ictal EEG as in Fig. 4.28b and the proposed method have high degree of linearity than GFD Method. Since the linearity of such graphs in Fig. 4.28 decides the dimensionality of the EEG Signals, so our designed method gives the accurate Dimensional values than the GFD Method in the discrimination of Healthy and Epileptic EEGs.

ANOVA Test also supports our designed method statistically, than the GFD Method. The p-value in the Table 4.7a is greater than the p-values in the Table 4.7b, which is zero. Hence the values corresponding to Normal and Ictal EEGs are

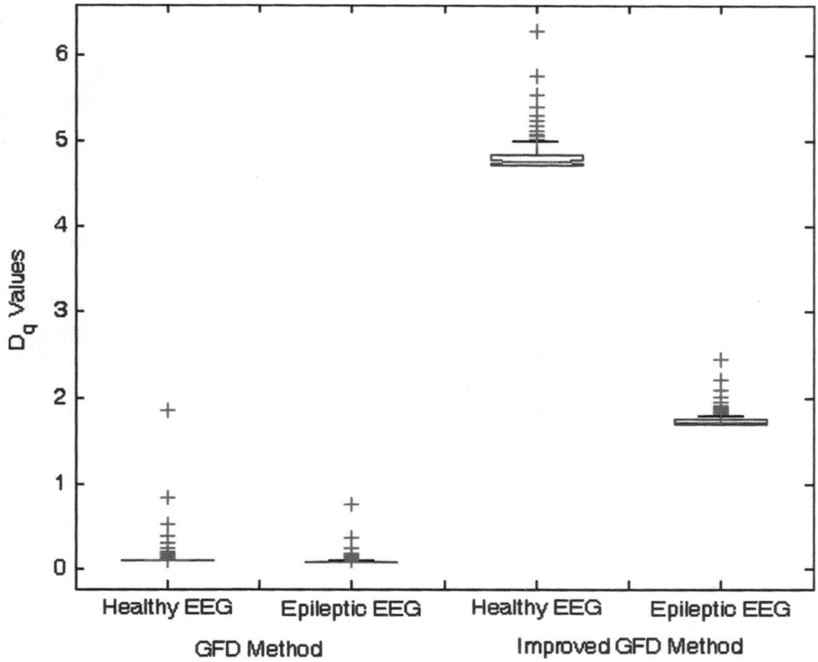

Fig. 4.29 Notched box plots for fractal spectra of healthy and epileptic EEGs

differed significantly in the designed method than the GFD Method. It is also observed from the Box-Whiskers Plots of Fig. 4.29 that there is a significant variability in the Fractal spectra of Improved GFD Method among the Normal and Ictal EEGs when compared with the GFD Method.

Hence, the Figs. 4.27–4.29 and the Table 4.7 are the evidence that the Improved Method of GFD plays an efficient role than the GFD Method in the analysis of EEG and in the discrimination of Normal and Epileptic EEGs.

4.4.4 Concluding Remarks

In this study, we have proposed an improved form of GFD, developed from the concept of GFD for discrimination between the Seizure free (Normal) and Epileptic (Ictal) EEGs. We have calculated the Fractal Spectra (D_q and ID_q) and *REN* and *IREN* for the Normal and Ictal EEGs using both the GFD Method and the Improved GFD Method. Finally, we compared the Fractal Spectra and the values of *REN* and *IREN* against $log(r)$ for Normal and Ictal EEGs using the both Methods to distinguish the Normal and Ictal EEGs through the graphical techniques. The statistical tool namely ANOVA Test also illustrated that there is significant differences in the values corresponding to the Normal and Ictal EEGs in our proposed method than the GFD Method. This scheme demonstrated that the Improved form of GFD is well proficient mathematical tool in the analysis of Epileptic EEG than the GFD Method. Hence we concluded that the Multifractal Analysis using Improved form of GFD plays a crucial role in the area of Epileptic research using EEG signals to analyze, detect or predict the state of illness of the Epileptic patients. We hope that not only in the EEG Signals but also, the Improved from of GFD Method; will be useful to analyze all the critical time signals in the area of Non-linear Analysis.

Acknowledgements The research work has been supported by University Grants Commission *(UGC - MRP & SAP)*, Government of India, New Delhi, India.

References

1. Accardo, A., Affinito, M., Carrozzi, M., Bouquet, F.: Use of the fractal dimension for the analysis of electroencephalographic time series. Biol. Cybern. **77**, 339–350 (1997)
2. Acharya, U., Faust, O., Kannathal, N., et al.: Non-linear analysis of EEG signals at various sleep stages. Comp. Meth. Prog. Biomed. **80**, 37–45 (2005)
3. Adeli, H., Zhou, Z., Dadmehr, N.: Analysis of EEG records in an epileptic patient using wavelet transform. J. Neurosci. Meth. **123**(1), 69–87 (2003)
4. Andrzejak, R., Lehnertz, K., Mormann, F., et al.: Indications of nonlinear deterministic and finite-dimensional structures in time series of brain electrical activity: Dependence on recording region and brain state. Phys. Rev. E **64**, 061907-1–061907-8 (2001)
5. Arle, J., Simon, R.: An application of fractal dimension to the detection of transients in the electroencephalogram. Electroencephalogr. Clin. Neurophysiol. **75**, 296–305 (1990)
6. Ayala, M., Cabrerizo, M., Tito, M., Barreto, A., Adjouadi, M.: A spreadsheet application for processing long-term EEG recordings. Comp. Biol. Med. **39**, 844–851 (2009)
7. Barnsley, M.: Fractals Everywhere, 2 edn. Academic, USA (1993)
8. Bullmore, E., Brammer, M., Bourlon, P., Alarcon, G., Polkey, C., Elwes, R., Binnie, C.: Fractal analysis of electroencephalographic signals intracerebrally recorded during 35 epileptic seizures: Evaluation of a new method for synoptic visualisation of ictal events. Electroencephalogr. Clin. Neurophysiol. **91**, 337–345 (1994)
9. Burlaga, L., Klein, L.: Fractal structure of the interplanetary magnetic field. J. Geophys. Res. **91**, 347 (1986)
10. Carlin, M.: Measuring the complexity of non-fractal shapes by a fractal method. Pattern Recognit. Lett. **21**(11), 1013–1017 (2000)
11. Clinic: Department of Epileptology University of Bonn Medical Centre, B.G.: http://www.epileptologie-bonn.de/. Accessed January 2010
12. Crevecoeur, G., Hallez, H., Hese, P.V., Asseler, Y., Dupre, L., de Walle, R.V.: EEG source analysis using space mapping techniques. J. Comput. Appl. Math. **215**, 339–347 (2008)
13. Cuffin, B.: A method for localizing EEG sources in realistic head models. IEEE Trans. Biomed. Eng. **42**, 68–71 (1995)
14. Diambra, L., Figueiredo, J., Malta, C.: Epileptic activity recognition in EEG recording. J. Phys. A: Math. Gen. **15**, L611–L615 (1999)
15. Easwaramoorthy, D., Uthayakumar, R.: Analysis of biomedical EEG signals using wavelet transforms and multifractal analysis. In: Proceedings of the 1st IEEE International Conference on Communication Control and Computing Technologies, pp. 544–549. IEEE Xplore Digital Library, IEEE, USA (2010)

16. Easwaramoorthy, D., Uthayakumar, R.: Analysis of EEG signals using advanced generalized fractal dimensions. In: Proceedings of the Second International Conference on Computing, Communication and Networking Technologies, pp. 1–6. IEEE Xplore Digital Library, IEEE, USA (2010)
17. Easwaramoorthy, D., Uthayakumar, R.: Estimating the complexity of biomedical signals by multifractal analysis. In: Proceedings of the IEEE Students' Technology Symposium, pp. 6–11. IEEE Xplore Digital Library, IEEE, USA (2010)
18. Easwaramoorthy, D., Uthayakumar, R.: Improved generalized fractal dimensions in the discrimination between healthy and epileptic EEG signals. J. Comput. Sci. **2**(1), 31–38 (2011)
19. Elger, C., Lehnertz, K.: Seizure prediction by non-linear time series analysis of brain electrical activity. Eur. J. Neurosci. **10**, 786–789 (1998)
20. Esteller, R., Vachtsevanos, G., Echauz, J., Litt, B.: A comparison of waveform fractal dimension algorithms. IEEE Trans. Circ. Syst. **48**(2), 177–183 (2001)
21. Falconer, K.: Fractal Geometry: Mathematical Foundations and Applications, 2 edn. Wiley, England (2003)
22. Gevins, A., Remonds, A.: Handbook of Electroencephalography and Clinical Neurophysiology, vol. 1. Elsevier, University of Chicago (1987)
23. Goldberger, A., West, B.: Fractals in physiology and medicine. Yale J. Biol. Med. **60**, 421–435 (1987)
24. Grassberger, P.: Generalized dimensions of strange attractors. Phys. Lett. A **97**, 227–320 (1983)
25. Grassberger, P., Proccacia, I.: Characterization of strange attractors. Phys. Rev. Lett. **50**(5), 346–349 (1983)
26. Grassberger, P., Proccacia, I.: Measuring the strangeness of strange attractors. Phys. D **9D**, 189–208 (1983)
27. Hazarika, N., Chen, J., Tsoi, A., Sergejew, A.: Classification of EEG signals using the wavelet transform. Signal Process. **59**(1), 61–72 (1997)
28. Hentschel, H., Procaccia, I.: The infinite number of generalized dimensions of fractals and strange attractors. Physica **8D**, 435–444 (1983)
29. Higuchi, T.: Approach to an irregular time series on the basis of the fractal theory. Phys. D **31**, 277–283 (1988)
30. Iasemidis, L.: Epileptic seizure prediction and control. IEEE Trans. Biomed. Eng. **50**, 549–558 (2003)
31. Iasemidis, L., Sackellares, J.: The evolution with time of the spatial distribution of the largest lyapunov exponent on the human epileptic cortex. In: Measuring Chaos in the Human Brain. pp. 49–82. World Scientific Publishing, Singapore (1991)
32. Iasemidis, L., Zaveri, H., Sackellares, J., Williams, W.: Phase space topography of the electrocarticogram and the lyapunov exponent in partial seizures. Brain Topogr. **2**, 187–201 (1990)
33. Iasemidis, L., Shiau, D., Sackellares, J., Pardalos, P.: Transition to epileptic seizures: Optimization. DIMACS Series in Discrete Mathematics and Theoretical Computer Science, pp. 55–74. American Mathematical Society, USA (1999)
34. Iasemidis, L., Pardalos, P., Sackellares, J., Shiau, D.: Quadratic binary programming and dynamical system approach to determine the predictability of epileptic seizures. J. Combin. Optim. **5**, 9–26 (2001)
35. Iasemidis, L., Shiau, D., Pardalos, J.S.P., Prasad, A.: Dynamical resetting of the human brain at epileptic seizures: Application of nonlinear dynamics and global optimization techniques. IEEE Trans. Biomed. Eng. **51**(3), 493–506 (2004)
36. Jain, A., Dubes, R.: Algorithm for Clustering Data. Prentice-Hall, Englewood Cliffs (1988)
37. Kalauzi, A., Spasic, S.: Consecutive differences as a method of signal fractal analysis. Fractals **13**(4), 283–292 (2005)
38. Kannathal, N., Acharya, U., Lim, C., et al.: Characterization of EEG - A comparative study. Comp. Meth. Progr. Biomed. **80**, 17–23 (2005)
39. Kannathal, N., Lim, C., Acharya, U., et al.: Entropies for detection of epilepsy in EEG. Comp. Meth. Progr. Biomed. **80**, 187–194 (2005)

40. Katz, M.: Fractals and the analysis of waveforms. Comp. Biol. Med. **18**(3), 145–156 (1988)
41. Kulish, V., Sourin, A., Sourina, O.: Human electroencephalograms seen as fractal time series: Mathematical analysis and visualization. Comp. Biol. Med. **36**, 291–302 (2006)
42. Lakshmanan, M., Rajasekar, S.: Nonlinear Dynamics: Integrability, Chaos and Patterns. Springer, Berlin (2003)
43. Lehnertz, K., Elger, C.: Can epileptic seizures be predicted? evidence from nonlinear time series analysis of brain electrical activity. Phys. Rev. Lett. **80**, 5019–5022 (1998)
44. Litt, B., Esteller, R., Echauz, J., Maryann, D., Shor, R., Henry, T., Pennell, P., Epstein, C., Bakay, R., Dichter, M., Vachtservanos, G.: Epileptic seizures may begin hours in advance of clinical onset: A report of five patients. Neuron **30**, 51–64 (2001)
45. Maiwald, T., Winterhalder, M., Aschenbrenner-Scheibe, R., Voss, H., Schulze-Bonhage, A., Timmer, J.: Comparison of three nonlinear seizure prediction methods by means of the seizure prediction characteristics. Phys. D **194**, 357–368 (2004)
46. Mandelbrot, B.: The Fractal Geometry of Nature. W.H. Freeman and Company, New York (1983)
47. Mandelbrot, B.: Negative fractal dimensions and multifractals. Phys. A **163**, 306–315 (1990)
48. Meyer-Lindenberg, A.: The evolution of complexity in human brain development: An EEG study. Electroencephalogr. Clin. Neurophysiol. **99**, 405–411 (1996)
49. Munck, J., Goncalves, S., Mammoliti, R., Heethaar, R., Silva, F.: Interactions between different EEG frequency bands and their effect on alpha-fMRI correlations. Neuroimage **47**, 69–76 (2009)
50. Nan, X., Jinghua, X.: The fractal dimension of EEG as a physical measure of conscious human brain activities. Bull. Math. Biol. **50**(5), 559–565 (1988)
51. Normant, F., Tricot, C.: Method for evaluating the fractal dimension of curves using convex hulls. Phys. Rev. A **43**, 6518–6525 (1991)
52. Ocak, H.: Automatic detection of epileptic seizures in EEG using discrete wavelet transform and approximate entropy. Expert Syst. Appl. **36**, 2027–2036 (2009)
53. Ouyang, G., Li, X., Guan, X.: Application of wavelet based similarity analysis to epileptic seizure prediction. Comp. Biol. Med. **37**, 430–437 (2007)
54. Pachori, R.: Discrimination between ictal and seizure-free EEG signals using empirical mode decomposition. Res. Lett. Signal Process. Article ID 293056 (2008)
55. Paivinen, N., Lammi, S., Pitkanen, A., et al.: Epileptic seizure detection: A nonlinear view point. Comp. Meth. Progr. Biomed. **79**, 151–159 (2005)
56. Paramanathan, P., Uthayakumar, R.: An algorithm for computing the fractal dimension of waveforms. Appl. Math. Comput. **195**(2), 598–603 (2008)
57. Paramanathan, P., Uthayakumar, R.: Application of fractal theory in analysis of human electroencephalogram signals. Comp. Biol. Med. **38**(3), 372–378 (2008)
58. Paramanathan, P., Uthayakumar, R.: Size measure relationship method for fractal analysis of signals. Fractals **16**, 235–241 (2008)
59. Pardalos, P., Sackellares, J., Carney, P., Iasemidis, L.: Quantitative Neuroscience: Models, Algorithms, Diagnostics and Therapeutic Applications. Springer, Berlin (2004)
60. Petrosian, A.: Kolmogorov complexity of finite sequences and recognition of different preictal EEG patterns. In: Proceedings of IEEE Symposium on Computer Based Medical Systems, pp. 212–217. IEEE Xplore Digital Library, IEEE, USA (1995)
61. Qiong, X., Xiong, W.: Fractal dimension of voice signal waveforms. Wuhan Univ. J. Nat. Sci. **7**(4), 399–402 (2002)
62. Quyen, M., Martinerie, J., Baulac, M., Varela, F.: Anticipating epileptic seizures in real time by non-linear analysis of similarity between EEG recordings. Neuro Rep. **10**, 2149–2155 (1999)
63. Renyi, A.: On a new axiomatic theory of probability. Acta Math. Hung. **6**, 285–335 (1955)
64. Sackellares, J., Iasemidis, L., Shiau, D.: Detection of the preictal transition in scalp EEG. Epilepsia **40**, 176 (1999)
65. Sackellares, J., Iasemidis, L., Gilmore, R., Roper, S.: Epilepsy - When Chaos Fails, Chaos in the Brain? World Scientific, Singapore (2002)

66. Sebastian, M., Navascues, M., Valdizan, J.: Surface laplacian and fractal brain mapping. J. Comput. Appl. Math. **189**(1), 132–141 (2004)
67. Shannon, C.: The Mathematical Theory of Communication. University of Illinois Press, Champaign (1998)
68. Shelberg, M.: The development of a curve and surface algorithm to measure fractal dimensions. Master's thesis, Ohio State University (1982)
69. Spasic, S., Kalauzi, A., Culic, M., Grbic, G., Martac, L.: Fractal analysis of rat brain activity injury. Med. Biol. Eng. Comput. **43**(4), 345–348 (2005)
70. Tricot, C.: Curves and Fractal Dimension. Springer, New York (1995)
71. Ubeyli, E.: Analysis of eeg signals using lyapunov exponents. Neural Netw. World **16**(3), 257–273 (2006)
72. Ubeyli, E.: Statistics over features: EEG signals analysis. Comp. Biol. Med. **39**, 733–741 (2009)
73. Uthayakumar, R., Paramanathan, P.: An algorithm for computing fractal dimension of rectifiable irregular graphs. Appl. Math. Comput. **190**(1), 305–308 (2007)
74. Uthayakumar, R., Paramanathan, P.: Fractal dimension of irregular digitalized curves by divider method. Appl. Math. Comput. **189**(1), 68–71 (2007)
75. Wu, L., Gotman, J.: Segmentation and classification of EEG during epileptic seizures. Electroencephalogr. Clin. Neurophysiol. **106**, 344–356 (1998)
76. Zhang, J., Yang, X., Luo, L., Shao, J., Zhang, C., Ma, J., Wang, G., Liu, Y., Peng, C., Fang, J.: Assessing severity of obstructive sleep apnea by fractal dimension sequence analysis of sleep EEG. Phys. A **388**, 4407–4414 (2009)

Part IV
Chaos Synchronization: Communications and Symbolic Analysis

Chapter 5
Inferring Global Synchrony from Local Symbolic Dynamics

Sarika Jalan, Fatihcan M. Atay, and Jürgen Jost

Abstract Symbolic dynamics based on specific partitions prevents the occurrence of certain symbolic sequences that are characteristics of the dynamical function. Such partitions lead to a maximal difference in the permutation entropy of a chaotic and the corresponding random system. The symbolic dynamics defined by such partition has several practical applications, one of which is the detection of global synchrony in coupled systems. The synchronized state is detected by observing the complete absence or at least low frequency of particular symbol sequences. The method uses short time series and is hence computationally fast. Also, because it compares the symbol sequence of one single unit in the network with some model behavior, it does not depend on the size of the network and is robust against external noise.

5.1 Introduction

Synchronization is a prototype of emerging system level behavior of interacting dynamical units. This behavior, in large ensembles of coupled systems, is studied in many different fields [1–3], describing synchronous behavior in various natural and artificial settings such as the Belousov-Zhabotinsky reaction [4], neuronal activities in different cortical regions of the cat brain [5], brain signals during epileptic seizures [6, 7], technical systems such as power grids to achieve secure communication [8], climate behavior such as solar forcing of Indian monsoon, etc. Depending

S. Jalan (✉)
Indian Institute of Technology Indore, IET-DAVV Campus Khandwa Road, Indore 452017, Madhya Pradesh, India
e-mail: sarika@iiti.ac.in

F.M. Atay · J. Jost
Max Planck Institute for Mathematics in the Sciences, Inselstrasse, 04103 Leipzig, Germany
e-mail: fatay@mis.mpg.de; jjost@mis.mpg.de

S. Banerjee and L. Rondoni (eds.), *Applications of Chaos and Nonlinear Dynamics in Science and Engineering 3*, Understanding Complex Systems,
DOI 10.1007/978-3-642-34017-8_5, © Springer-Verlag Berlin Heidelberg 2013

on the field, synchronization can be a desired or undesirable behavior. Detection of synchronization in extended systems from local measurements has important applications. For example, during anesthesia, it is found that the dynamical activity passes reversibly through a sequence of different synchronized cardio-respiratory states as the anesthesia level changes, and thus the synchronization state may be used to characterize the depth of anesthesia [9]. Certain pathologies in the neural system, such as epileptic seizures, manifest themselves by synchronized brain signals [6], and there is some evidence that they can be predicted by changing levels of synchronization [7]. It is thus of interest to be able to determine synchronization in a brain area from local measurements, such as EEG recordings. Most existing methods used to detect synchronization are based on cross-correlation function analysis. For multivariate time series, this relies on the proper embedding of the phase space. Mutual false nearest neighbor or mixed state embedding methods are also used to detect synchronization. All these methods have their practical limitations and are difficult to implement when the data are noisy and of limited length [10, 11]. Here, we describe a new method to detect synchronization based on symbolic dynamics. Symbolic dynamics is a fundamental tool for describing a complicated time evolution of chaotic dynamical systems [12, 13]. Instead of representing a trajectory by an infinite sequences of numbers, one watches the alternation of symbols. In doing so one 'loses' a great amount of information but some invariant, robust properties of the dynamics may be kept. A good symbolic dynamics representation crucially depends on the partition of phase space [14, 15]. Recently, a method to control spatio-temporal chaos in coupled map lattices using symbolic dynamics is proposed [16].

Symbolic dynamics based on specific partitions prevents the occurrence of certain symbolic sequences characteristic of the dynamical function. Such a partition leads to the maximal difference in the permutation entropy [17] of a chaotic and the corresponding random system [18]. The symbolic dynamics defined by such partition has several practical applications [18–20]. One such application is the detection of global synchronization in coupled chaotic systems. The synchronized state is detected by simply observing the complete absence or at least low frequency of particular symbol sequences. The method uses short time series and is hence computationally fast. Also, because it compares the symbol sequence of one single unit in the network with some model behavior, it does not depend on the size of the network and is robust against external noise.

5.2 Definition of Symbolic Dynamics

Consider a dynamical system described by

$$\dot{\mathbf{x}}(t) = \mathbf{F}(\mathbf{x}(t)) \tag{5.1}$$

where $\mathbf{x} = (x^{(1)}, x^{(2)}, \ldots, x^{(k)}) \in \mathbb{R}^k$, $\mathbf{F}(\mathbf{x}) = (F^{(1)}(\mathbf{x}), F^{(2)}(\mathbf{x}), \ldots, F^{(k)}(\mathbf{x}))$ is a k-dimensional vector function of $\mathbf{x}$, and $\dot{\mathbf{x}}$ denotes the time derivative $d\mathbf{x}/dt$. Let $S \subset \mathbb{R}^k$ be an invariant set for (5.1), and $\{S_i : i = 1, \ldots, m\}$ be a partition of S, i.e., a collection of mutually disjoint and nonempty subsets satisfying $\cup_{i=1}^m S_i = S$. If the state of the system is considered at discrete instances of time that are uniformly separated by Δt, then the sequence of symbols $\{\ldots, s_{t-\Delta t}, s_t, s_{t+\Delta t}, \ldots\}$, where $s_t = i$ if $\mathbf{x}(t) \in S_i$, defines a symbolic dynamics corresponding to (5.1). Of course, symbols from any m-letter alphabet can be equally well used instead of the integers i.

5.3 Choice of the Partition

For applications, a judicious choice of partition is crucial. Suppose the scalar component $x^{(n)}$, $1 \leq n \leq k$, is available for measurement. For a given threshold value $x^* \in \mathbb{R}$, define the sets

$$S_1 = \{\mathbf{x} \in S : x^{(n)} < x^*\}$$
$$S_2 = \{\mathbf{x} \in S : x^{(n)} \geq x^*\}$$

The value of $x^{(n)}$ can be chosen to make the sets S_1, S_2 nonempty, in which case they form a non-trivial partition of S. For this special partition, two-symbol dynamics generated by the following equations are used:

$$s_t = \begin{cases} \alpha \text{ if } x^{(n)}(t) < x^* \\ \beta \text{ if } x^{(n)}(t) \geq x^*. \end{cases} \tag{5.2}$$

The symbolic dynamics depends only on the measurements $x^{(n)}$, yielding a sequence of symbols determined by whether a measured value exceeds the threshold x^* or not. Essentially any choice of the threshold x^* will yield a non-generating partition. We say that the set S_i *avoids* S_j if

$$x(t) \in S_i \Rightarrow x(t + \Delta t) \notin S_j \tag{5.3}$$

It is useful to search for partitions containing avoiding sets. The significance is that if S_i avoids S_j, then the symbolic dynamics cannot contain the symbol sequence ij, and the transition $i \to j$ is *forbidden*. The essence of this symbolic dynamics is based on choosing a partition where (5.3) holds for some sets in the partition. The probability $P(i, j)$ of the occurrence of the subsequence ij in the symbolic dynamics can be estimated from time series data by the ratio $n(s_t = i, s_{t+\Delta t} = j)/n(s_t = i)$, where n is the count of the number of occurrences.

For practical calculations using short time series, the choice of x^* becomes important. The choice should make certain transition probabilities very small.

Clearly, increasing the threshold decreases the probability of occurrence of the repeated sequence $\beta\beta$. However, it also decreases the probability of observing the single symbol β, making it difficult to work with short time series. Hence, the choice of the threshold is a compromise between these two effects. The natural density defined by the data provides a choice for the threshold. In the following section we will present the method by taking examples of two different maps. The method can simply be extended for the continuous systems, where Δt has to be chosen such that for the appropriate x^*, we get certain transition probabilities zero. Note that this can always be done if Δt is chosen sufficiently small; however, this is not the most useful procedure as too many transition probabilities become zero simultaneously. Moreover, measurement conditions may already impose a lower bound on Δt.

5.4 Coupled Dynamics on Networks and Global Synchrony

Consider the following coupled oscillator model,

$$\dot{\mathbf{x}}_i(t) = \mathbf{F}(\mathbf{x}_i(t)) + \varepsilon \sum_{j=1}^{N} C_{ij}[g(\mathbf{x}_i(t)) - g(\mathbf{x}_j(t))] \tag{5.4}$$

where $\mathbf{x}_i(t)$ is the state of the ith node at time t, $i = 1, \ldots, N$, C_{ij} are the elements of the adjacency matrix $\mathbf{C}$ with value 1 or 0 depending upon whether i and j are connected or not, $\varepsilon \in [0, 1]$ is the coupling strength. The system can exhibit a wide range of behavior depending upon the local dynamics (5.1) and coupling structure $\mathbf{C}$, so corresponding symbol sequences observed from a node can vary widely. However, at the globally synchronized state $x_i(t) = x_j(t)$ for all i, j and t, and all nodes evolve according to the rule (5.1). It follows that when the network is synchronized, the symbolic sequence measured from a node will be subject to the same constraints as that generated by (5.1). Thus, an easy way of detecting synchronization of the network is simply by choosing a random node and calculating the transition probabilities. Synchronization is indicated if these transition probabilities match with the transition probabilities of the isolated dynamics.

A discrete-time version of (5.4) is [21]

$$\mathbf{x}_i(t+1) = \mathbf{f}(\mathbf{x}_i(t)) + \frac{\varepsilon}{k_i} \sum_{j=1}^{N} C_{ij} \left[f(\mathbf{x}_j(t)) - f(\mathbf{x}_i(t)) \right] \tag{5.5}$$

where, for simplicity and because this occurs in many other applications, the same function f for the local and the interaction dynamics are used, and $k_i = \sum_j C_{ij}$ denotes the degree of node i. For the discrete dynamical system (5.5) the criteria for forbidden set (5.3) can be written as

$$\mathbf{f}(S_i) \cap S_j = \emptyset. \tag{5.6}$$

In the following, the logistic and tent maps and the Lorenz oscillator are used as the examples for numerical studies.

5.5 Detection of Synchronization

For detection of global synchrony the transition probabilities of any randomly selected node with the transition probabilities of the isolated function are compared. Often a partition that has one or more forbidden sets for the dynamics of an isolated unit can be found, which is the case for logistic and tent maps. Therefore, in that case, the situation is rather simple as one has to look for the presence or absence of symbol subsequences which are forbidden in the dynamics of the map. We first demonstrate the results for the tent map and after that turn to the Lorenz system as the local dynamical function. For the latter case, one does not have strictly forbidden transitions for the dynamics of the individual map, but the method nevertheless applies rather well.

5.5.1 Synchronization Measures

The variance of the variables over the network is given by $\sigma^2 = \left\langle \frac{1}{N-1} \sum_i [x_i(t) - \bar{x}(t)]^2 \right\rangle_t$, where $\bar{x}(t) = \frac{1}{N} \sum_i x_i(t)$ denotes an average over the nodes of the network and $\langle \dots \rangle_t$ denotes an average over time. This quantity acts as a measure indicating the global synchrony: σ^2 drops to zero when the whole network is synchronized. Symbolic measure for the synchronization is defined as follows. First the transition probability $P(i, j)$ is estimated by the ratio,

$$P(i, j) = n(s_t = i, s_{t+1} = j)/n(s_t = i), \tag{5.7}$$

where n is a count of the number of times of occurrence. The deviation of $P(i, j)$ of any node is defined as

$$\delta^2_{i,j} = \left\langle \frac{1}{m-1} \sum_{k=1}^{m} [P_k(i, j) - \overline{P(i, j)}]^2] \right\rangle, \tag{5.8}$$

where $\overline{P(i, j)} = \frac{1}{m} \sum_k P^0_k(i, j)$ is calculated at $\varepsilon = 0$, and $k = 1, \dots\dots$ are m different sets of random initial conditions drawn from the invariant measure of dynamics at the node. Note that δ^2 is calculated by using only single node dynamics, whereas calculation of σ^2 involves all the dynamics of all the nodes. Furthermore we show that symbolic measure based on the transition probabilities calculated for a single node is sufficient to detect global synchrony.

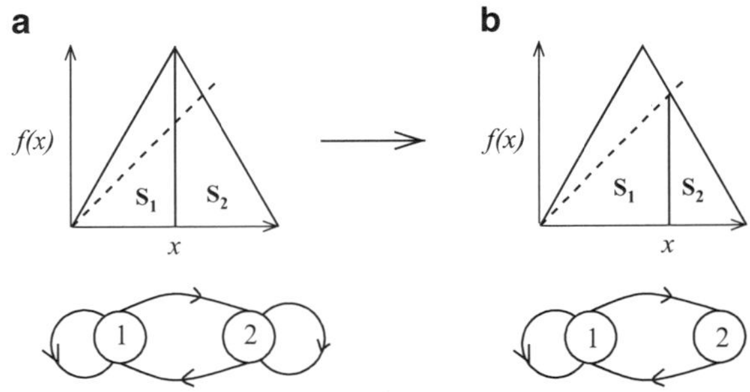

Fig. 5.1 The figure shows two different partition for the tent map (5.9). Subfigure (**a**) shows the generating partition, and the corresponding transition possibilities. (**b**) shows the partition that gives forbidden transitions (transition $2 \to 2$ is absent)

5.5.2 *Tent Map*

The tent map is given by

$$f(x) = 1 - 2|x - \tfrac{1}{2}|. \tag{5.9}$$

Its stationary density on [0,1] is the uniform density, $p(x) \equiv 1$. If we take the generating partition [14], which is simply a partition of [0, 1] with dividing point $l_p = 1/2$ [13], then the above tent map and a random sequence from a uniform distribution give equal Kolmogorov-Sinai (KS) entropy [22]. For this partition the iteration takes a trajectory to the left or right set of the partition with equal probabilities. As we shift the partition from $l_p = 1/2$, the transition probabilities remain no more the same and for $l_p = 2/3$, i.e., $S_1 = [0, l_p], S_2 = (l_p, 1]$, the transition $2 \to 2$ does not occur (see Fig. 5.1). Note that for this partition the difference between the permutation entropy of the tent map and the corresponding random system is maximal [18]. As the partition point is moved further to the right in the range $a/(a+1) \leq l_p < 1$, the transition remains forbidden, but the difference between entropies decreases, reaching zero at $l_p = 1$. The implication is that a longer time series would now be needed as the observed occurrence of the symbol 2 becomes less frequent. Hence, the optimal partition would be the one for which the self-avoiding set S_2 is largest. Note that the choice of l_p in the range $(1/2, a/(a+1))$ gives a non-generating partition, but has no two-symbol forbidden sequences. Although some longer sequences may be forbidden, their detection requires longer time series and more computational effort.

For numerical calculations a slightly modified definition of the symbolic dynamics [18, 23] can be used. To two consecutive measurements $x_t x_{t+1}$ we assign the symbol α if $x_{t+1} \geq x_t$ and the symbol β otherwise.[1] This is equivalent to the symbol sequences defined by the sets S_1, S_2 because of the specific partition point x^*; thus

[1]Such a comparison between successive values of the variables has been introduced in [17] for the calculation of permutation entropy, and used in [23] for studying phase synchronization in coupled map networks. The first study [17] has shown that permutation entropy based on the comparison of n successive values gives a good estimate of the KS entropy, and it matches the KS entropy for $n \to \infty$.

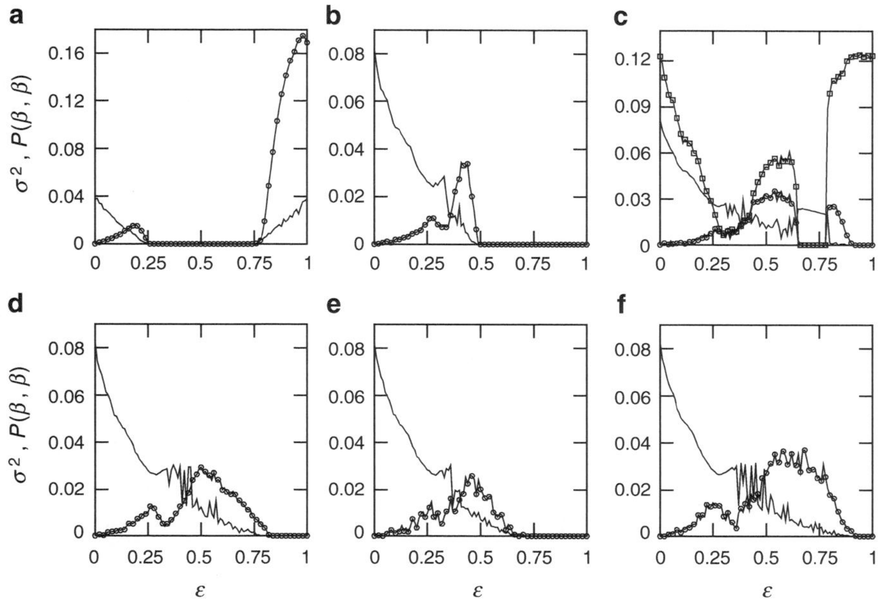

Fig. 5.2 Detection of synchronization in various networks of coupled maps. Figures are plotted for coupled tent maps (5.9). The networks consist of (**a**) two coupled maps, (**b**) a globally coupled network of size 50, (**c**) a small-world network of size 100 and average degree 30, (**d**) a scale-free network of size 100 and average degree 15, (**e**) a scale-free network of size 200 and average degree 10, and (**f**) a random network of size 200 and average degree 8. The horizontal axis is the coupling strength, and the vertical axis gives the synchronization measure σ^2 (–) for the whole network, as well as the transition probability $P(\beta,\beta)$ (∘) calculated using a scalar time series from a randomly selected node. In all cases, the synchronization region ($\sigma^2 = 0$) coincides with the region where $P(\beta,\beta) = 0$ and the other transition probabilities are nonzero, which is the situation for the uncoupled map. Note that in subfigure (**c**) there is an interval of ε roughly between 0.65 and 0.8 such that $P(\beta,\beta) = 0$, but there is no synchronization as $P(\alpha,\alpha)$ (shown by □) is also zero here, unlike the case for the isolated tent map

the transition $\beta \to \beta$ does not occur for the single tent map. The advantage of using this definition of α, β is that one only needs to check increases and decreases in the measured signal, which adds more robustness in case the fixed point is not precisely known. We evolve (5.5) starting from random initial conditions and estimate the transition probabilities using time series of length $\tau = 1{,}000$ from a randomly selected node. Note that the length of the time series is much shorter than would be required by standard time-series methods which use embedding to reconstruct the phase space for large networks [24–26]. For the method considered here, however, the length of the time series is independent of the network size. We estimate the transition probability $P(i,j)$ (i,j being α,β) as given by (5.7). Synchronization is signaled when the variance of variables over the network σ^2 drops to zero. Figure 5.2 summarizes the results. It is seen in all cases that the region for synchronization exactly coincides with the range for which $P(\beta,\beta)$ is zero

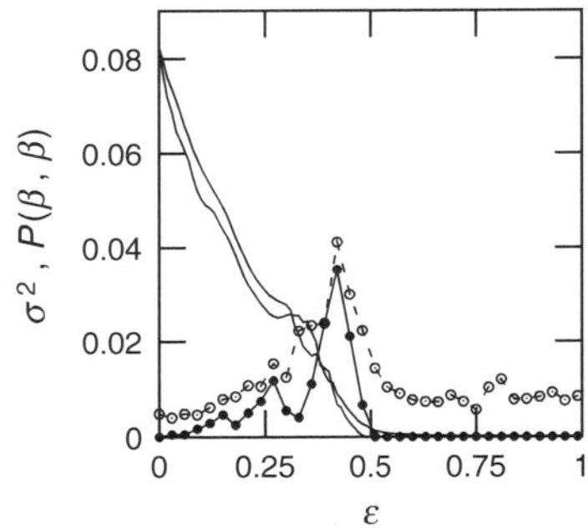

Fig. 5.3 Figure is plotted for globally connected networks with $N = 50$ and shows $P(\beta, \beta)$ for Gaussian noise with strength 2 % (•) and 5 % (∘)

(and other transition probabilities are nonzero; see subfigure (c)). Hence, regardless of network topology and size, both synchronized and unsynchronized behavior of the network can be accurately detected over the whole range of coupling strengths using only measurements from an arbitrarily selected node. The method has some robustness against external noise. Figure 5.3 plots $P(\beta, \beta)$ when the measurements are taken in a noisy environment.

For higher-dimensional systems, for example the Lorenz attractor, finding optimal partitions corresponding to the maximal permutation entropy difference may be more difficult, but the method also works for other partitions, as global synchrony is detected by comparing all the transition probabilities measured from a time series of an arbitrary node with those of the isolated function.

5.5.3 Lorenz Oscillator

This section shows the applicability of the method to coupled Lorenz oscillators. The Lorenz oscillator

$$\begin{aligned} \dot{x} &= 10(y - x) \\ \dot{y} &= 28x - y - xz \\ \dot{z} &= -(8/3)z + xy \end{aligned} \tag{5.10}$$

has a chaotic attractor [27]. We use only the x variable to detect synchronization. This, in fact, again exhibits a principal feature of the approach discussed here, namely that we only need to evaluate partial information about the dynamics. Symbolic dynamics is defined as earlier, with $\Delta t = 0.2$ (Sect. 5.2). We define a partition $S_1 = (a, x^*]$, $S_2 = (x^*, b)$, where (a, b) is the x-range for the Lorenz attractor. The threshold value x^* is obtained as earlier by checking the values of transition probabilities ($x(t) \in S_2$ and $x(t + \Delta t) \in S_2$) with varying x^* (Fig. 5.4). We evolve (5.4) starting from random initial conditions with (5.10) as local dynamics, and calculate the transition probabilities by (5.7), from the x time series of length $\tau = 1{,}000$. Figure 5.5 is plotted for two coupled Lorenz oscillators.

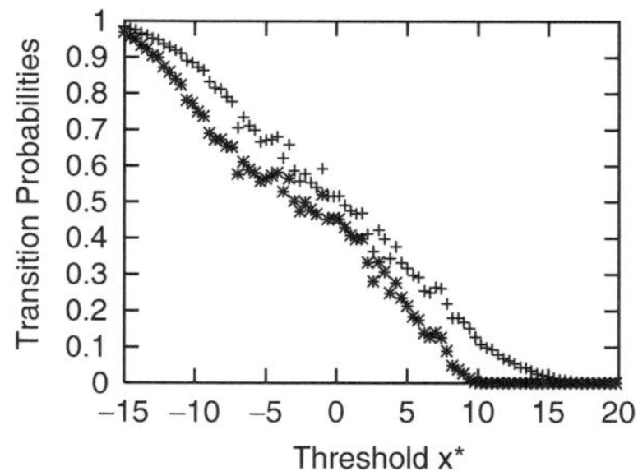

Fig. 5.4 Figure plots $P(\beta)(+)$ and $P(\beta,\beta)(*)$ for the Lorenz oscillator (5.10). The threshold for the partition is the point, $x^* \sim 10$, where $P(\beta,\beta)$ drops to zero (with $P(\beta)$ being non zero)

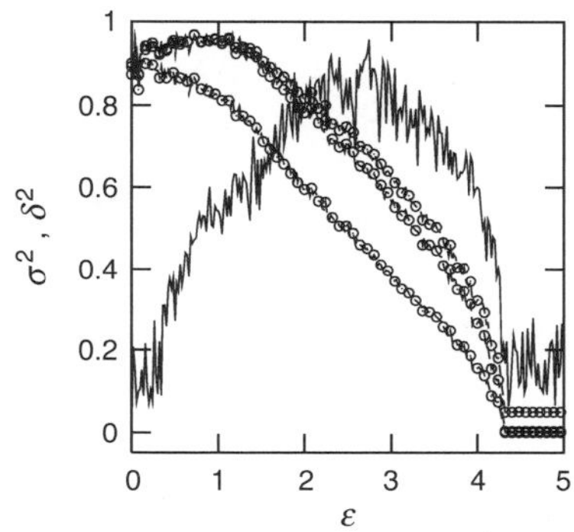

Fig. 5.5 The transition probability measure for coupled Lorenz oscillators. δ^2(——) (calculated by a scalar from a variable randomly selected node) and all the three σ^2's corresponding to the x, y and z variables of the Lorenz oscillator are plotted as a function of the coupling strength

Global synchrony ($\dot{\mathbf{x}}_i(t) = \dot{\mathbf{x}}_j(t); \forall i, j, t$) is detected by comparing the transition probabilities measured from the x time series of any randomly selected node with those of the x time series of the uncoupled dynamics (5.10). σ^2 and δ^2 are calculated as given in Sect. 5.5.1. For $\varepsilon > 4.5$, the deviation of transition probabilities from those of the uncoupled case ($\varepsilon = 0$) is very small indicating the synchronization (σ^2 is also zero for this region).

5.6 Conclusion

We have discussed a simple and effective method based on symbolic dynamics for the detection of synchronization in diffusively coupled networks. The method works by taking measurements from as few as a single node, and can utilize rather short sequences of measurements, and hence is computationally fast.

The special symbolic sequences discussed here are not drawn from the Markovian (generating) partitions which is the usual practice for symbolic dynamics. Rather, the symbolic sequences presented here are generated by non-generating partitions. The partitions that lead to the maximal difference between chaotic dynamics and the corresponding random dynamics are the best ones for our purpose, and they prevent occurrence of certain symbol sequences related to the characteristics of the dynamics; so synchronization can be detected by checking for these forbidden symbolic sequences. For higher-dimensional maps, finding the optimal partition (leading to maximal difference between the chaotic and corresponding random system) can be difficult, but the method works also for other partitions and global synchrony is detected by comparing all transition probabilities measured from a time series of an arbitrary node with those of the isolated function. Synchrony can be detected so long as the synchronized dynamics is identical to the isolated dynamics, and the fact that certain transition probabilities are exactly zero makes this procedure especially robust. The method is independent of the size and the connection architecture of the underlying network, and also robust against external noise.

References

1. Winfree, A.T.: Oscillating systems. On emerging coherence. Science **298**(5602), 2336–2337 (2002)
2. Nadis, S.: Nature (London) **421**, 780 (2003)
3. Pikovsky, A., Rosenblum, M., Kurths, J.: Synchronization: A Universal Concept in Nonlinear Dynamics. Cambridge University Press, Cambridge (2001)
4. Vanag, V.K., Epstein, I.R. Science, **294**, 835 (2001)
5. Roelfsema, P.R., Engel, A.K., Konig, P., Singer, W.: Nature (London) **385**, 157 (1997)
6. Engel, J. Jr., Pedley, T.A. (eds.): Epilepsy: A Comprehensive Textbook. Lippincott-Raven, Philadelphia (1997)
7. Mormann, F., Kreuz, T., Andrzejak, R.G., David, P., Lehnertz, K., Elger, C.E.: Epilepsy Res. **53**, 173 (2003)
8. Kye, W.H., Choi, M., Kim, C.-M., Park, Y.-J.: Phys. Rev. E **71**, 045202(R) (2005)
9. Stefanovska, A., Haken, H., McClintock, P.V.E., Hozic, M., Bajrovic, F., Ribaric, S. Phys. Rev. Lett. **85**, 4831 (2000)
10. Kantz, H., Schreiber, T.: Nonlinear Time Series Analysis. Cambridge University Press, London (1997)
11. Boccaletti, S., Kurths, J., Osipov, G., Valladare, D.L. Zhou, C.S.: Phys. Rep. **366**, 1 (2002)
12. Lind, D., Marcus, B.: Symbolic Dynamics and Coding. Cambridge University Press, London (1995)
13. Hao, B.-L., Zheng, W.-M.: Applied Symbolic Dynamics and Chaos. World Scientific, Singapore (1998)
14. Rudolph, D.J. Fundamentals of Measurable Dynamics, Ergodic theory on Lebesgue Spaces. Clarendon Press, Oxford (1990)
15. Bollt, E.M., Stanford, T., Lai, Y.-C., Życzkowski, K.: Phys. D **154**, 259 (2001)
16. Pethel, S.D., Corron, N.J., Bollt, E.: Phys. Rev. Lett. **99**, 214101 (2007)
17. Bandt, C., Pompe, B.: Phys. Rev. Lett. **88**, 174102 (2002)
18. Atay, F.M., Jalan, S., Jost, J.: Complexity **15**(1), 29 (2009)
19. Jalan, S., Jost, J., Atay, F.M.: Chaos **16**, 033124 (2006)
20. Atay, F.M., Jalan, S., Jost, J.: Phys. Lett. A **375**, 130 (2010)
21. Kaneko, K.: Phys. D **34**, 1 (1998); Phys. Rev. Lett. **65**, 1391 (1990); Phys. D **41**, 137 (1990); Phys. D **124**, 322 (1998)
22. Grassberger, P., Procaccia, I.: Phys. Rev. A **28**, 2591 (1983)
23. Jalan, S., Amritkar, R.: Phys. Rev. Lett. **90** 014101 (2003)
24. Takens, F.: Lecture Notes in Mathematics **818**, 366 (1981)
25. Farmer, J.D., Sidorowich, J.J.: Phys. Rev. Lett. **59**, 845 (1987)
26. Kennel, M.B., Brown, R., Abarbanel, H.D.I.: Phys. Rev. A **45**, 3403 (1992)
27. Lichtenberg, A.J., Lieberman, M.A.: Regular and Chaotic Dynamics. Springer, Berlin (1983)

Chapter 6
Evaluation of the Number of Keys in a Chaotic Cryptographic Method

A.A. Dmitriev, A.S. Dmitriev, Y.V. Andreyev, E.V. Efremova, I.P. Antoniades, A.N. Miliou, and A.N. Anagnostopoulos

Abstract Data stream coder based on chaotic synchronous response is considered. An estimate of the number of keys available in this scheme is obtained by cascading the basic building block of the system and thus repeating the encoding procedures. Efficiency of the discussed algorithm (in the sense of computational expenses) is evaluated and compared to known cryptographic algorithms. As it is shown, the efficiency increases in the case of smaller number of repetitions and greater number of encoding function parameters.

6.1 Introduction

Rapidly developing information networks require secure and confidential channels in order to transmit various kinds of data. There are many methods for encoding data that provide the necessary security level for networking tasks. Yet, the problem of designing efficient and versatile processes for secure communications is still open. Cryptographic methods based on dynamical chaos attract the attention of researchers for more than a decade [1–4, 9, 11–16, 18, 20, 23–26]. Chaos-based

A.A. Dmitriev · A.S. Dmitriev · Y.V. Andreyev · E.V. Efremova
Information and Communication Technologies based on Dynamic Chaos, Kotelnikov Institute of Radio Engineering and Electronics of RAS, Moscow, Russia
e-mail: chaos@mail.cplire.ru; dmitr@chaos.msk.edu

I.P. Antoniades · A.N. Miliou (✉)
Department of Informatics, Aristotle University of Thessaloniki, Greece
e-mail: antoniades33@yahoo.gr; amiliou@csd.auth.gr

A.N. Anagnostopoulos
Department of Physics, Aristotle University of Thessaloniki, Greece
e-mail: anagnost@physics.auth.gr

S. Banerjee and L. Rondoni (eds.), *Applications of Chaos and Nonlinear Dynamics in Science and Engineering 3*, Understanding Complex Systems,
DOI 10.1007/978-3-642-34017-8_6,

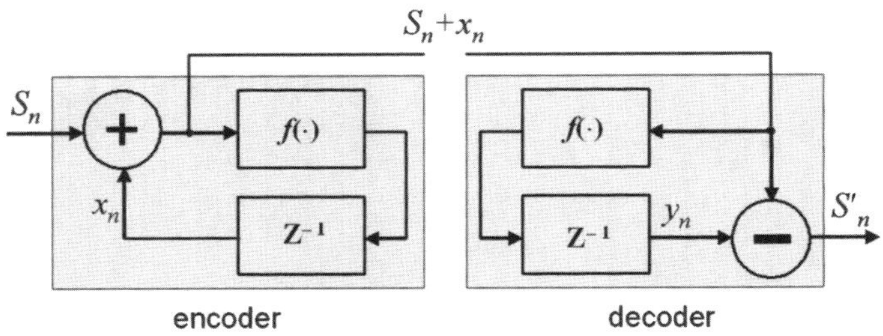

Fig. 6.1 Stream coder system based on chaotic synchronous response

systems have a number of nontrivial features such as random behavior, ergodicity, and sensitivity to initial conditions and system parameters demonstrating high diffusion and confusion properties that are highly desirable for cryptosystems.

During the last 15 years numerous methods for chaotic cryptography have been proposed in the literature. These methods can be divided into three main groups, judging by the features of dynamical chaos employed in each case. The first one comprises methods in which chaotic dynamical systems are used as random number generators for well-known cryptographic schemes [1, 12, 13, 15, 18, 25, 26]. The second group uses the property of chaotic systems to mix (shuffle) the phase space ("stretching and folding") for designing coders (mostly block-wise). The algorithms are based on permutations of text blocks, bits, picture fragments, etc. [2,3,11,14,15, 24]. The third group of chaos applications to cryptography [4,9,16,20,23] employs the phenomenon of chaotic synchronization [19] (specifically, what is known as chaotic synchronous response [22]). In these systems the data stream is added to the signal and therefore pervades the chaotic dynamical system (Fig. 6.1). Information is decoded using an identical dynamical system with its feedback loop disconnected.

In this work, we consider a chaotic cryptographic scheme based on chaotic synchronization and explore it from the viewpoint of the number of available cryptographic keys (or the potential key length). The aim is to study the system with a simple encoding function, estimate the number of keys in such a system, generalize the results, and draw some conclusions for the choice of encoding function.

The book chapter is organized as follows: in Sect. 6.2, the architecture of a chaotic cryptographic scheme based on chaotic synchronization is introduced. We use a simple piecewise-linear function to encode/decode a test image, and derive an approach to evaluate the number of cryptographic keys in such system. In Sect. 6.3, we consider an advanced cryptographic system by cascading a number of the basic building blocks and a multi-segment piecewise-linear encoding function. Section 6.4 presents a derivation of a general estimate of the number of keys in such scheme while in the following Sect. 6.5 the estimation of the length of the key is introduced. General conclusions along with the performance analysis are drawn in Sect. 6.6.

6.2 Encoding System Based on Nonlinear Mixing of Information with Chaotic Signal

The scheme for transmitting information using nonlinear mixing of information and chaotic signals [22] is considered for many years as a prototype for the systems of secure communications (e.g., [6, 7, 10]) and it is used as the basic element in a number of encoding proposals [4, 9, 16, 20, 23]. The main advantages of this scheme with respect to other communication schemes based on dynamic chaos are the self-synchronization of the transmitter and the receiver, the possibility of exact information retrieval, and the operation with analog signal.

The system considered for encoding/decoding information, is depicted in Fig. 6.1. This is a stream coder employing a discrete-time nonlinear dynamical system. The coder is composed of a nonlinear map, i.e. a nonlinear function $f(x)$ with unit delay Z^{-1} and an adder. In the absence of information stream ($S_n = 0$), the map generates a chaotic signal x_n. When the data stream S_n is fed to the input, it is added to the chaotic signal x_n and participates in the dynamic process of the chaotic generator. Modulo summation can be used in the adder (transmitter and receiver) since it has been shown to increase the confidentiality of the transmitted information signal along with the robustness of the receiver performance in comparison to the circuit with ordinary summation [8].

The output signal of the encoder is $(S_n + x_n)$. The decoder is built of the same elements as the encoder. Data is retrieved by means of subtracting the signal that has passed through disconnected feedback loop, from the input signal.

More analytically, the encoding-decoding scheme works as follows: Starting from any initial condition x_0 at time step $t = 0$, the transmitter transmits x_0 (without adding any information signal to it) to the receiver. At the receiver x_0 is passed to the chaotic map $f(x)$ and produces $x_1' = f(x_0)$ at the output. In the next step the transmitter produces $x_1 = f(x_0)$, then adds to x_1 (with modulo 1 summation) the information signal S_1 and transmits $x_1 + S_1$ to the receiver. Subsequently, at the receiver's subtractor, x_1' is subtracted from $x_1 + S_1$. Since the system parameters of chaotic map f, at receiver and transmitter are identical, then $x_1 \equiv x_1'$ and thus S_1 is recovered. The process is repeated for all time steps n, and thus the entire information signal S_n can be reproduced. Notice that the process does not depend at all on the choice of initial condition x_0.

To facilitate the study of chaotic encoding schemes we use as examples encoding images, which is a more complex task than working with texts, since images have some special features such as bulk capacity, high redundancy and strong correlations between their different parts. Moreover, images are also interesting, because any flaws of the method are actually "visible".

In order to estimate the number of possible keys in the discussed scheme, we consider the simple 1D piecewise linear tent map and encode a test image with it. The map is described by the following function

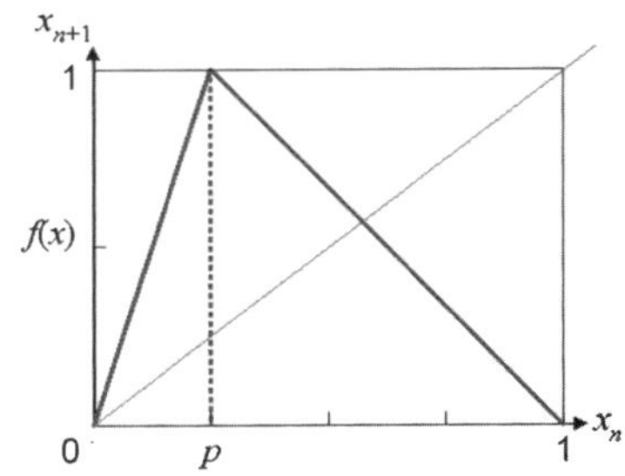

Fig. 6.2 1D piecewise linear tent map

Fig. 6.3 Test image

$$x_{n+1} = f(x_n) = \begin{cases} \frac{x_n}{p}, & \text{if } x_n < p \\ \frac{(1-x_n)}{(1-p)}, & \text{if } x_n \geq p \end{cases} \tag{6.1}$$

where the parameter $p \in (0, 1)$ (Fig. 6.2). For all values of p the map is chaotic.

The test image is shown in Fig. 6.3. It is an $m \times n = 512 \times 512$ picture with 256 gray levels. Let us denote the intensity of a pixel (i, j) by s_{ij}, with $s_{ij} \in [0, 255]$. To obtain the data stream, the image is scanned row by row top to bottom, and is normalized in the range [0, 1] using the following expression

$$S_k = \frac{(s_{ij} + 0.5)}{L} \tag{6.2}$$

where $k = n(i-1) + j$, $n = 512$ is the number of pixels in a row and $L = 256$ is the number of gray levels.

The encoded image is shown in Fig. 6.4 while the parameter $p = 0.7$ and the initial value of the map is $x_0 = 0.1$. However, the initial value of x_0 is not necessary in order to decode the image, because as it was mentioned earlier the proposed scheme possesses the property of self-synchronization. Consequently, the value of the parameter p is adequate and is used as a key. From the viewpoint of cryptography, with the same key (i.e. equal values of the parameter p) in the encoder and decoder, the original image could be completely retrieved at the decoder whereas for different values of p, the retrieval is unattainable.

Fig. 6.4 Encoded image ($p = 0.7$)

Fig. 6.5 Decoded image ($p = 0.7$)

Simulations show that in the case of equal values of p in the nonlinear functions of the encoder and decoder, the original image is recovered without any flaws (Fig. 6.5). For a slightly different parameter value in the decoder the recovered image is distorted, but it is still easily recognized (Fig. 6.6). For larger discrepancies of the parameter, visual recognition becomes impossible (Fig. 6.7). This means that the parameter space (the open unit interval in our case) may be divided into clusters, such that a pair of parameter values from one cluster, one used for encoding and the other for decoding, provides recognizable image recovery, whereas when a pair of parameter values is taken from different clusters the recovered image is distorted compared to the original one.

Obviously, the number of different keys in the system is equal to the number of the clusters of the parameter p. The size of the clusters can be determined by the degree of correlation of the original and decoded image. The correlation of the two images is calculated by the expression

$$C = \frac{\sum_i^n \sum_j^m (x_{ij} - \bar{x})(y_{ij} - \bar{y})}{\sqrt{\sum_i^n \sum_j^m (x_{ij} - \bar{x})^2 \sum_i^n \sum_j^m (y_{ij} - \bar{y})^2}} \tag{6.3}$$

Fig. 6.6 Decoded image ($p = 0.6$)

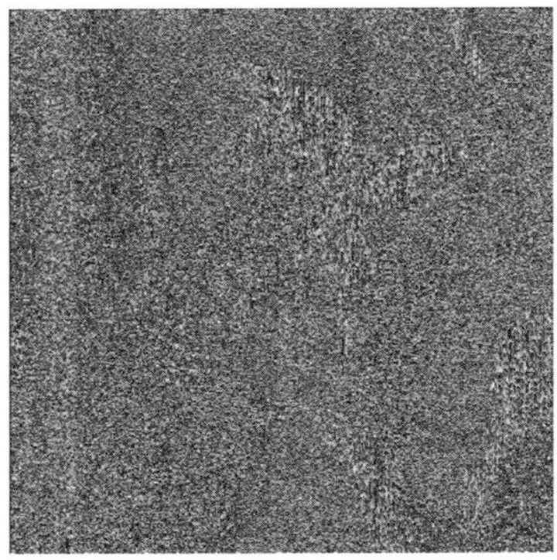

Fig. 6.7 Decoded image ($p = 0.3$)

where x_{ij} and y_{ij} are brightness values of the original and decoded image pixels respectively and $\bar{x}$, $\bar{y}$ are average brightness values of the images. Calculations of the correlation values of the recovered images in Figs. 6.6 and 6.7, are $C = 0.81$ and $C = 0.03$, respectively.

Simulations have shown (e.g., Fig. 6.3–6.7) that the recovered image cannot be visually recognized, if the correlation, C, is below 5 %. Therefore, depending on the application one could determine the security needed in relation to information value as well as operational speed and computational cost. In practice only military applications have strict security terms. There are several examples of applications that do not have very strict security requirements such as the electronic signature and the image databases where only the ones paid have access to high resolution images [17].

In the presented scheme such value is attained when the parameter difference $\Delta p = |p - p^{'}| > 0.7$. Therefore, the range (0, 1) of the parameter p may be divided into two clusters at most which implies that the system has two keys.

With just one parameter p, an attacker can recover an image very close to the original with $|p - p^{'}| < \epsilon$ where ϵ can be quite large, and of course this is inadmissible from the viewpoint of conventional cryptography. This is equivalent to saying that the system possesses very few possible keys, and thus an enemy can break the system by brute force after very few trials. However, as it is explained later in the chapter, increasing the number of parameters p and/or increasing the number of map iterations, the sensitivity of the system to parameter mismatch increases severely. This is equivalent to saying that the number of keys increases and thus an enemy would have to perform a very large number of trials to break the system by brute force. The number of trials needed as a function of number of system parameters and map iterations rises according to (6.6).

6.3 Encoding Systems Cascading the Basic Building Block and Increased Number of Parameters

It is known that many cryptographic schemes are designed by cascading their basic building blocks and thus repeating the encoding procedures. For instance, *DES* uses 16 repetitions, *IDEA* 8, *LOKI* 16, *Blowfish* 16, *GOST* 32, *Khufu* and *Khafre* 24, etc. [21]. Analysis of the properties of chaotic systems allows us to make the assumption that repetitions can also be efficient in the encoding scheme with nonlinear mixing.

One such property is *the sensitivity to initial conditions*. A map f is sensitive to initial conditions, if there exists $\delta > 0$, such that for all $x \in X$ and for all neighborhoods, H_x of point x, there exists a point in this neighborhood $y \in H_x$ and an integer k, such that $|f^k(x) - f^k(y)| > \delta$ [5].

Specifically, in chaotic systems the trajectories diverge with time, no matter how close they were at first. The degree of divergence is described and measured by the *Lyapunov* exponent λ. For chaotic systems $\lambda > 0$, therefore, the introduction of repetitions in the encoding scheme based on the chaotic synchronous response (which is equivalent to an increase of system iterations) naturally leads to a corresponding increase in the number of potential keys.

The encoding system, cascading the basic encrypting module is depicted in Fig. 6.8. The encoder/decoder of the system is composed of a chain of elements, each being the above-described circuit for encoding/decoding based on chaotic synchronous response.

Consequently, we consider the scheme with repetitions, using the 1D piecewise linear tent map of (6.1). The parameter value p is set to be the same along all the encoder chain. Similarly, all the decoder elements have the same parameter p.

An encoding test image with the scheme with repetitions (Fig. 6.8) showed that the size of clusters of the parameter p decreases with each repetition approximately by a factor of 2. This factor depends on the *Lyapunov* exponent λ, which has maximum value $\lambda_{max} = \ln 2$ at $p = 0.5$ and tends to zero when p moves towards the edges of the unit interval. So, the clusters in the center of the unit interval of the parameter p shrink more rapidly, than those near the edges and yet the average factor is only slightly less than $\exp(\lambda_{max})$. As a result, the number of keys increases, e.g., after 30 repetitions it becomes to 2^{30}. Therefore, the repetition of the encoding procedure is an efficient way of increasing the number of keys.

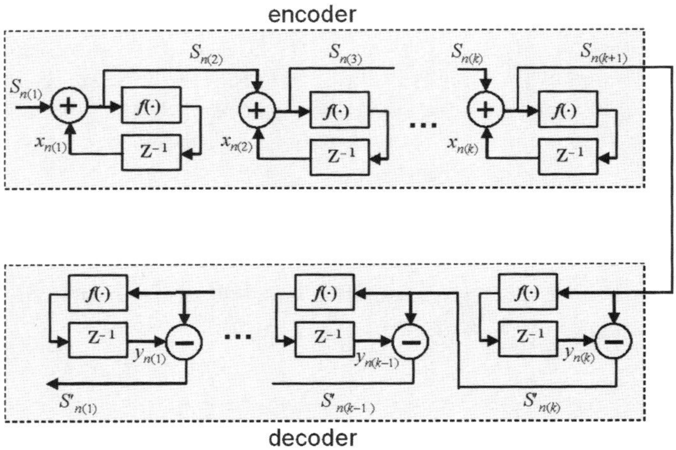

Fig. 6.8 Encoding scheme designed by cascading the basic building block

Fig. 6.9 Two parameter piecewise linear map

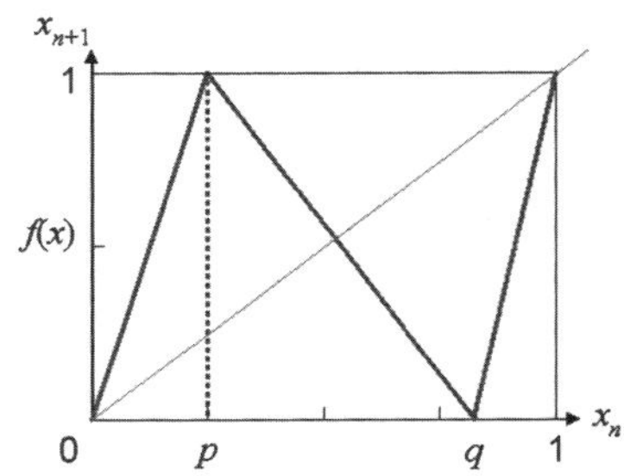

Another approach in order to increase the number of keys is to increase the number of the system parameters. This can be done by means of increasing the number of segments of the piecewise-linear encoding function $f(x)$, as is shown in Fig. 6.9. Note that introducing new segments in the map function makes the map strongly chaotic (i.e., increases the *Lyapunov* exponent) because the maximum *Lyapunov* exponent is given by

$$\lambda = \ln n \tag{6.4}$$

where n is the number of map segments (Fig. 6.9).

A chaotic attractor of such map occupies the entire phase space, and its invariant measure (distribution of the system variable x) is approximately even.

Using the map with parameters p and q (Fig. 6.9) we encoded the same test image as before and estimated the size of the corresponding parameter clusters; the number of repetitions is $k = 1, 2, \ldots, 9$. For fixed parameter values p and q, the number of keys rapidly grows with k (the size of clusters decreases), much faster than in the case of the scheme with the one-parameter piecewise-linear map of (6.1). After 5 and 9 repetitions the sizes of clusters for each of the parameters p and q are approximately 6×10^{-3} and 2×10^{-4}, respectively, i.e., the sizes of 2D clusters are 3.6×10^{-5} and 4×10^{-8}. Consequently, the number of keys is $1/(3.6 \times 10^{-5}) \approx 2.8 \times 10^4$ and $1/(4 \times 10^{-8}) \approx 2.5 \times 10^7$, respectively. Comparing the number of keys in the scheme with one-parameter map of (6.1) after 5 and 9 repetitions the number of keys is 25 and 500, respectively.

Thus, if for one-parameter piecewise-linear map the rate of increase of the number of keys is $\nu \approx 2$ for each repetition, then for two-parameter piecewise-linear map the rate is $\nu \approx 8$ for each repetition.

6.4 Estimation of the Number of Keys

Since the average divergence rate of trajectories is $\exp(\lambda)$, the upper bound of the increase rate of the number of clusters (keys) is

$$\nu = \exp(\lambda m) = n^m = (m+1)^m \tag{6.5}$$

where λ is given by (6.4), n is the number of map segments, m is the number of map parameters, $m = n - 1$. In the presented scheme with repetitions, the number of keys is

$$N = (m+1)^{mk} \tag{6.6}$$

where k is the number of repetitions.

6.5 Estimation of the Length of the Key

In traditional cryptographic schemes instead of the number of keys, the notion of the key length (in bits) is often used. In order to compare the discussed encoding method with known methods, let us introduce an equivalent of the key length: if the key length is L, then the number of keys is 2^L. Then, as follows from (6.7), the key length in the scheme based on chaotic synchronization is

$$L = \log_2 N = \log_2 (m+1)^{mk} \tag{6.7}$$

Consequently, from (6.7) in the case of 16 repetitions (as in *DES*) and 20 parameters the key length reaches 1,400 bit.

An analysis of (6.5) allows us to estimate the efficiency of the proposed scheme in view of computational expenses. Evidently, the system efficiency is inversely proportional to the number of repetitions, and the time of encoding in the scheme with n repetitions is n times larger than the time of encoding in the scheme with one repetition. At the same time, the dependence of the system efficiency on the number of parameters (key length) is negligible. This means that for equal key length the algorithm efficiency increases with decreased number of iterations and correspondingly increased number of parameters. Moreover, larger number of repetitions requires respectively increased cost of electronic hardware used to implement the systems whereas the corresponding increase of the cost if one increases the number of parameters is smaller. In this sense, encoding systems with greater number of parameters and smaller number of repetitions (one, in the limit) are preferable.

6.6 Concluding Remarks

A scheme of synchronous chaotic response (system with nonlinear mixing) is considered from the viewpoint of cryptographic data encoding. Unlike traditional cryptographic schemes, it operates with continuous-value variables and with floating-point arithmetic and thus appropriate for implementation with analog circuits.

An upper bound of the number of keys available in this scheme is obtained. Though (6.5) and (6.6) were obtained for model encoding functions, the authors believe that the obtained results are universal and pertain to the majority of piecewise-linear 1D chaotic maps.

The increase of the repetitions of the basic building block has some disadvantages. The main drawback is the manufacturing cost which is increased with every repetition. However, increasing the number of system parameters does not suffer from this disadvantage and moreover, it has similar if not better effects on system security.

In terms of computational performance and simplicity in the design of the system the scheme presented in this chapter is exceptionally efficient.

References

1. Andrecut, M.: Logistic map as a random number generator. Int. J. Mod. Phys. B **12**, 921–930 (1998)
2. Baptista, M.S.: Cryptography with chaos. Phys. Lett. A **240**, 50–54 (1998)
3. Cheng, H., Li, X.B.: Partial encryption of compressed image and videos. IEEE Trans. Signal Process. **48**, 2439–2451 (2000)
4. Dachselt, F., Schwartz, W.: Chaos and cryptography. IEEE Trans. Circ. Syst.-I **48**, 1498–1509 (2001)
5. Devaney, R.L.: An Introduction to Chaotic Dynamical Systems, 2nd edn. Addison-Wesley, Reading (1989)
6. Dmitriev, A.S., Panas, A.I., Starkov, S.O.: Experiments on speech and music signals transmission using chaos. Int. J. Bifurcat. Chaos **5**, 1249–1254 (1995)
7. Dmitriev, A.S., Panas, A.I., Starkov, S.O., Kuzmin, L.V.: Experiments on RF band communications using chaos. Int. J. Bifurcat. Chaos **7**, 2511–2527 (1997)
8. Dmitriev, A.S., Kuzmin, L.V., Panas, A.I.: Communication circuit with modulo summation of the chaotic and information signals. Radiotekhnika I Electronika **44**, 988–996 (in Russian) (1999)

9. Dmitriev, A.A., Andreyev, Yu.V., Bulushev, A.G.: Data encoding based on dynamic chaos. Uspekhi sovremenn. Radioelektron. **11**, 27–33 (in Russian) (2000)
10. Fischer, I., Liu, Y., Davis, P.: Synchronization of chaotic semiconductor laser dynamics on subnanosecond time scales and its potential for chaos communication. Phys. Rev. A **62**, 011801(R) (2000)
11. Fridrich, J.: Symmetric ciphers based on two-dimensional chaotic maps. Int. J. Bifurcat. Chaos **8**, 1259–1284 (1998)
12. Gonzailez, J.A., Pino, R.: Random number generator based on unpredictable chaotic functions. Comp. Phys. Comm. **120**, 109–114 (1999)
13. Habutsu, T., Nishio, Y., Sasase, I., Mori, S.: A secret key cryptosystem by iterating a chaotic map. In: D.W. Davies (ed.) Advances in Cryptology—EUROCRYPT '91. LNCS, vol. 547, pp. 127–140. Springer, Berlin (1991)
14. Ismail, LA., Amin, M., Diab, H.: An Efficient Image Encryption Scheme Based Chaotic Logistic Maps. Int. J. Soft Comput. **2**, 285–291 (2007)
15. Kwok, H.S., Tang, W.K.S.: A fast image encryption system based on chaotic maps with finite precision representation. Chaos Solitons Fractals **32**, 1518–1529 (2007)
16. Lu, H., Wang, S., Li, X.: A new spatiotemporally chaotic cryptosystem and its security and performance analyses. Chaos **14**, 617–629 (2004)
17. Mao, Y., Chen, G.: Chaos-based image encryption. In: E.B. Corrochano (ed.) Handbook of Geometric Computing: Applications in Pattern Recognition, Computer Vision, Neuralcomputing, and Robotics, 1st edn, pp 231–265. Springer. http://www.open-image.org/725publication/journal/CBIE.pdf (August 23, 2005). ISBN-10: 3540205950; ISBN-13: 978-3540205951
18. Pareek, N.K., Patidar, V., Sud, K.K.: Image encryption using chaotic logistic map. Image Vis. Comput. **24**, 926–934 (2006)
19. Pecora, L.M., Carroll, T.L.: Synchronization in chaotic systems. Phys. Rev. Lett. **64**, 821–824 (1990)
20. Pisarchik, A.N., Flores-Carmona, N.J., Carpio-Valadez, M.: Encryption and decryption of images with chaotic map lattices. Chaos **16**, 033118 (2006)
21. Schneier, B.: Applied Cryptography. Wiley, New York (1996)
22. Volkovskii, A.R., Rulkov, N.F.: Synchronous chaotic response of a nonlinear oscillator system as a principle for the detection of the information component of chaos. Tech. Phys. Lett. **19**, 97–99 (1993)
23. Wang, S., Kuang, J.: Chaos-based secure communications in a large community. Phys. Rev. **E66**, 065202 (2002)
24. Wong, W.K., Lee, L.P., Wong, K.W.: A modified chaotic cryptographic scheme. Comp. Phys. Comm. **138**, 234–236 (2001)
25. Yen, J.C., Guo, J.I.: A new chaotic key based design for image encryption and decryption. Proc. IEEE Int. Symp. Circ. Syst. **4**, 49–52 (2000)
26. Yu, X.Y., Zhang, J.: Chaotic Image Scrambling Algorithm Based on S-DES. J. Phys.: Conf. Ser. **48**, 349–353 (2006)

Part V
Chaos Synchronization: Systems and Circuits

Chapter 7
Chaos Synchronization of the Modified Autonomous Van der Pol-Duffing Circuits via Active Control

Ahmed Sadek Hegazi and Ahmed Ezzat Matouk

Abstract In this work, we study the dynamics and synchronization of a chaotic system describes the Modified Autonomous Van der Pol-Duffing (MAVPD) circuit. The detailed bifurcation diagrams are given to show the rich dynamics of the proposed system. Lyapunov exponents are calculated to verify the existence of chaos in this system. Chaos synchronization of MAVPD system is obtained using active control method. According to the qualitative theory of fractional differential equations, the existence and uniqueness of solutions for a class of commensurate fractional-order MAVPD systems are investigated. Furthermore, based on the stability theory of fractional-order systems, the conditions of local stability of linear fractional-order system are discussed. Moreover, the existence of chaotic behaviors in the fractional-order MAVPD system is shown. A necessary condition for this system to remain chaotic is obtained. It is found that chaos exists in this system with order less than three. Phase synchronization of the fractional-order MAVPD system is also achieved using an active control technique. Numerical simulations show the effectiveness of the proposed synchronization schemes.

7.1 Introduction

Chaos is an important dynamical phenomenon which has been extensively studied and developed by scientists since the work of Lorenz [1]. Lorenz chaotic system consists of three-dimensional autonomous integer-order differential equations.

A.S. Hegazi
Faculty of Science, Mathematics Department, Mansoura University, Mansoura 35516, Egypt
e-mail: hegazi@mans.edu.eg

A.E. Matouk (✉)
Faculty of Science, Mathematics Department, Mansoura University, Mansoura 35516, Egypt

College of Prep Year, Mathematics Department, Hail University, Hail 2440, Saudi Arabia
e-mail: aematouk@hotmail.com

S. Banerjee and L. Rondoni (eds.), *Applications of Chaos and Nonlinear Dynamics in Science and Engineering 3*, Understanding Complex Systems,
DOI 10.1007/978-3-642-34017-8_7,

A chaotic system has complex dynamical behaviors such as the unpredictability of the long-term future behavior and irregularity. Chaos has great potential applications in many disciplines such as, fluid mixing, chaotic heating of plasma for a nuclear fusion reactor and secure communications.

Fractional calculus is a 300 years old topic [2]. Recently, it has been found that differential equations has many applications in many fields of science like engineering [2], physics [3], finance [4], social sciences [5], mathematical biology [6,7] and game theory [8]. Hence, fractional differential equations have been utilized to study dynamical systems in general and applications of chaos in particular. There are many definitions of the fractional derivative; one of the most common definitions is the Caputo definition of fractional derivatives [9]:

$$\frac{d^\alpha}{dt^\alpha} f(t) = D^\alpha f(t) = I^{m-\alpha} f^{(m)}(t), \quad \alpha > 0, \tag{7.1}$$

where $f^{(m)}$ represents the m-order derivative of $f(t)$, $m = \lceil \alpha \rceil$ is the value α rounded up to the nearest integer, and the left sided Riemann-Liouville integral of order θ is given by

$$I^\theta u(t) = \frac{1}{\Gamma(\theta)} \int_0^t (t-s)^{\theta-1} u(s) ds, \quad \theta > 0, \tag{7.2}$$

where $\Gamma(\theta)$ is the gamma function. The operator D^α is called "Caputo differential operator of order α". According to the Poincaré-Bendixon theorem, chaos in fractional-order autonomous systems can occur for orders less than three and this can not happen in their integer-order counterparts.

Chaos synchronization has also attracted increasing attention over the past 2 decades since the pioneering work of Pecora and Carroll [10]. Afterwards, many effective methods have been presented for synchronizing identical chaotic systems like one-way coupling method [11], active control [12], adaptive control [13] and synchronization via feedback control techniques [14–17]. Indeed, chaos synchronization has potential applications in secure communications [18], physical, chemical and biological systems [19–21]. Moreover, synchronization can occur in many real systems even though the oscillators have different order. For example, in the synchronization between heart and lung, one can observe that both, circulatory and respiratory systems behave in synchronous way.

Recently, nonlinear electronic circuits have become a vehicle to study non-linear phenomena and chaos. Many chaotic circuits have been appeared like the chaotic circuits of Chua [22], Lü [23] and Autonomous Van der Pol-Duffing [24]. Meanwhile, the applications of chaos synchronization to some nonlinear electronic circuits have been studied by authors [25–27]. Consequently, our prime interest in this work is to study and investigate chaos synchronization in nonlinear electronic circuits via active control techniques.

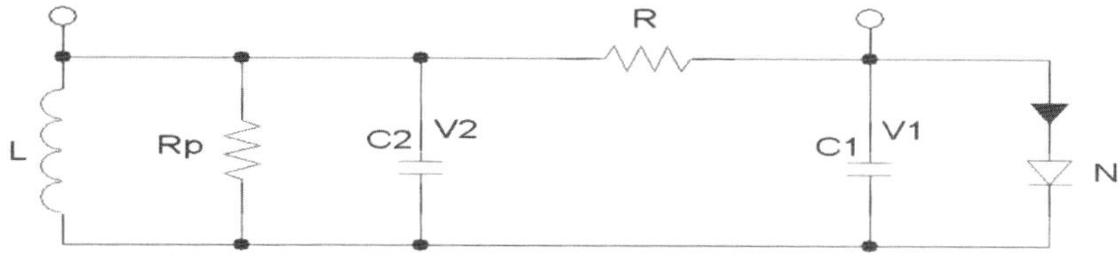

Fig. 7.1 The MAVPD chaotic circuit

7.2 The Integer-Order MAVPD System

The Modified Autonomous Van der Pol-Duffing (MAVPD) system describes the dynamics of nonlinear circuit whose nonlinear element has the following cubic form [28]:

$$i_N = F(v_1) = av_1 + bv_1^3, \quad (a < 0,\, b > 0). \tag{7.3}$$

By applying Kirchhoff's laws to the circuit in Fig. 7.1, the governing equations for the circuit elements (the voltages v_1 and v_2 across the capacitors c_1 and c_2, the current i_N through the nonlinear diode N and the current i_L through the inductor L) are represented by the set of three first-order autonomous differential equations:

$$\begin{aligned} c_1\dot{v}_1 &= -[bv_1^3 + av_1 + (1/R)(v_1 - v_2)], \\ c_2\dot{v}_2 &= (1/R)(v_1 - \gamma v_2) - i_L, \\ \dot{i}_L &= v_2/L, \end{aligned} \tag{7.4}$$

where c_1, c_2 are the capacitances of the two capacitors, L is the inductance, R is linear resistor and $\gamma = (R + R_P)/R_P$, R_P is the parallel resistor.

To study the qualitative behavior of the system (7.4), the variables are rescaled as follow:

$$\begin{aligned} &x = \sqrt{bR}v_1,\ y = \sqrt{bR}v_2,\ z = \sqrt{bR^3}i_L, \quad \tau = t/Rc_2, \\ &\mu = -(1 + aR),\ \nu = c_2/c_1 \quad and \quad \beta = c_2R^2/L. \end{aligned} \tag{7.5}$$

Then the system (7.4) is given as follows

$$\begin{aligned} \dot{x} &= -\nu(x^3 - \mu x - y), \\ \dot{y} &= x - \gamma y - z, \\ \dot{z} &= \beta y, \end{aligned} \tag{7.6}$$

where $. \equiv \frac{d}{d\tau}$ and γ, ν, β are all positive real numbers. System (7.6) shows double scroll chaotic attractor with the parameter values $\beta = 200$, $\mu = 0.1$, $\nu = 100$ and $\gamma = 1.6$ (see Fig. 7.2a). For these values of the parameters and $\gamma = 2.85$, one

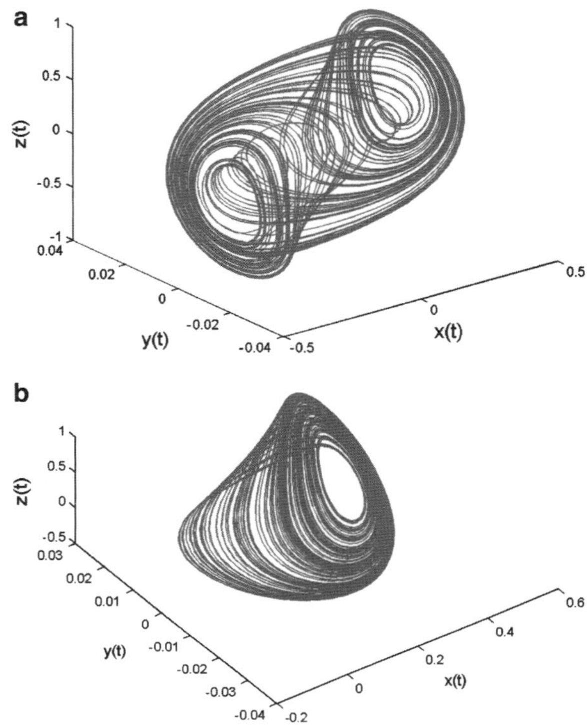

Fig. 7.2 Phase portraits of MAVPD system with the parameter values $\beta = 200$, $\mu = 0.1$, $\nu = 100$; (**a**) 3-D view of the double scroll attractor at $\gamma = 1.6$, (**b**) 3-D view of the one scroll chaotic attractor at $\gamma = 2.85$

scroll chaotic attractor exists (see Fig. 7.2b). Furthermore, using the parameter values $\beta = 200$, $\mu = 0.1$, $\nu = 100$ and $\gamma = 1.6$, Lyapunov exponents have been calculated using the efficient algorithm given in [29] and they approximately are; 2.22,0,−21.77.

The equilibrium points of system (7.6) are:

$$E_0 = (0,0,0),\ E_+ = (\sqrt{\mu}, 0, \sqrt{\mu}) \quad \text{and} \quad E_- = (-\sqrt{\mu}, 0, -\sqrt{\mu}).$$

System (7.6) has rich variety of dynamical behaviors including the chaotic and periodic behaviors which can be easily depicted from the following bifurcation diagrams (Fig. 7.3):

7.2.1 *Chaos Synchronization in the Integer-Order MAVPD System via Active Control*

In the following, we apply active control method to achieve chaos synchronization between two identical MAVPD drive and response chaotic systems. They are given as follow:

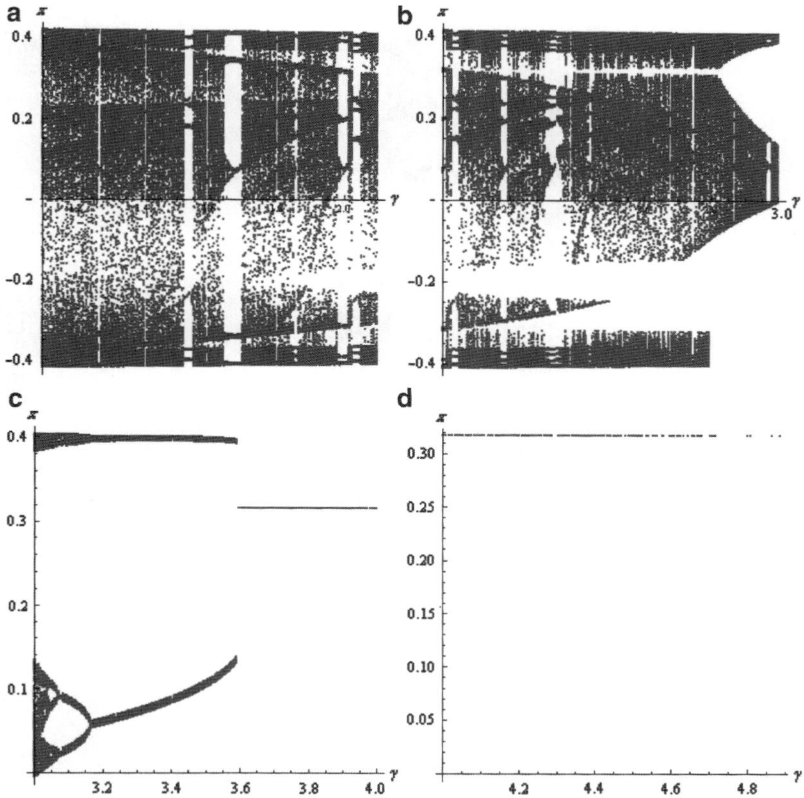

Fig. 7.3 (**a–d**): The bifurcation diagrams of MAVPD system

$$\begin{aligned}\frac{dx_1}{dt} &= -\nu(x_1^3 - \mu x_1 - y_1),\\ \frac{dy_1}{dt} &= x_1 - \gamma y_1 - z_1,\\ \frac{dz_1}{dt} &= \beta y_1,\end{aligned} \tag{7.7}$$

and

$$\begin{aligned}\frac{dx_2}{dt} &= -\nu(x_2^3 - \mu x_2 - y_2) + \mu_a(t),\\ \frac{dy_2}{dt} &= x_2 - \gamma y_2 - z_2 + \mu_b(t),\\ \frac{dz_2}{dt} &= \beta y_2 + \mu_c(t),\end{aligned} \tag{7.8}$$

where $\mu_a(t),\ \mu_b(t),\ \mu_c(t)$ are the control functions to be determined later. Let us define the synchronization errors between the drive and response systems as follow:

$$x_3 = x_2 - x_1,\ y_3 = y_2 - y_1,\ z_3 = z_2 - z_1. \tag{7.9}$$

Subtracting (7.7) from (7.8) and using the notation (7.9) yields

$$\begin{aligned}\frac{dx_3}{dt} &= -\nu((x_2^3 - x_1^3) - \mu x_3 - y_3) + \mu_a(t),\\ \frac{dy_3}{dt} &= x_3 - \gamma y_3 - z_3 + \mu_b(t),\\ \frac{dz_3}{dt} &= \beta y_3 + \mu_c(t).\end{aligned} \tag{7.10}$$

Now, we define the active control functions $\mu_a(t),\ \mu_b(t),\ \mu_c(t)$ as

$$\mu_a(t) = \nu(x_2^3 - x_1^3) + V_a(t),\ \mu_b(t) = V_b(t),\ \mu_c(t) = V_c(t). \tag{7.11}$$

Thus, we get

$$\begin{aligned}\frac{dx_3}{dt} &= \mu\nu x_3 + \nu y_3 + V_a(t),\\ \frac{dy_3}{dt} &= x_3 - \gamma y_3 - z_3 + V_b(t),\\ \frac{dz_3}{dt} &= \beta y_3 + V_c(t).\end{aligned} \tag{7.12}$$

Equation (7.12) describe the error dynamics and can be considered in terms of a control problem where the system to be controlled is a linear system with a control input $V_a(t),\ V_b(t)$ and $V_c(t)$ as functions of $x_3,\ y_3$ and z_3 which are to be determined for the purpose of synchronizing the two identical MAVPD systems. As long as these feedbacks stabilize the system (7.12), $x_3,\ y_3$ and z_3 converge to zero as time t goes to infinity. This implies that the two MAVPD systems are synchronized with feedback control. Thus we choose

$$\begin{pmatrix} V_a(t) \\ V_b(t) \\ V_c(t) \end{pmatrix} = A \begin{pmatrix} x_3 \\ y_3 \\ z_3 \end{pmatrix}, \tag{7.13}$$

where A is a 3×3 constant matrix. For proper choice of the elements of the matrix A, the feedback system must have all of the eigenvalues with negative real parts. In this case, the closed loop system will be stable. If we choose the matrix A in the form

$$A = \begin{pmatrix} -(\mu\nu + 1) & -\nu & 0 \\ -1 & \gamma - 1 & 1 \\ 0 & -\beta & -1 \end{pmatrix},$$

then the closed loop system has eigenvalues that are found to be −1, −1 and −1. Therefore, the zero equilibrium point of system 7.12) is locally asymptotically stable. Numerical integrations of the drive and response systems with the above-mentioned parameter values and using the initial conditions; $(x_1(0), y_1(0), z_1(0)) = (0.1, 0.01, -0.1)$, $(x_2(0), y_2(0), z_2(0)) = (0.2, 0.02, 0.1)$ show that the synchronization of two MAVPD systems is obtained using the previous active controller (see Fig. 7.4a–c).

7.3 The Fractional-Order MAVPD Circuit

Recently, generating chaos by fractional-order circuits has become a focal topic for research, many fractional-order chaotic circuits have been developed such as; the fractional-order Chua's circuit [30], the fractional-order Chaotic Liu circuit [31], the fractional-order unified chaotic system's circuit [32], and the fractional-order hyper-chaotic circuit [33]. Furthermore, it has been shown that chaos synchronization in some fractional-order systems has better results than the corresponding integer-order counterparts [14,34,35]. Thus, we will study chaos and phase synchronization in the fractional-order MAVPD circuit.

7.3.1 Modeling the Fractional-Order MAVPD Circuit

In the following, we are going to show the basic laws that can be applied to obtain the fractional-order MAVPD circuit.

7.3.1.1 Fractional Capacitor Theory

In 1994, S. Westerlund and L. Ekstam, proposed a new linear capacitor model. It is based on Curie's empirical law of 1889 [2]:

$$i(t) = \frac{U_0}{h_1 t^\alpha}, \quad (0 < \alpha < 1, t > 0), \tag{7.14}$$

where $i(t)$ is the current through the capacitor, h_1 is constant related to the capacitance of the capacitor and type of dielectric, α is a constant related to the losses of the capacitor, and U_0 is the dc voltage applied at $t = 0$.

For a general input voltage $u(t)$ the current is

$$i(t) = C\frac{d^\alpha u(t)}{dt^\alpha}, \tag{7.15}$$

where C is the capacitance of the capacitor. It is related to the type of dielectrics.

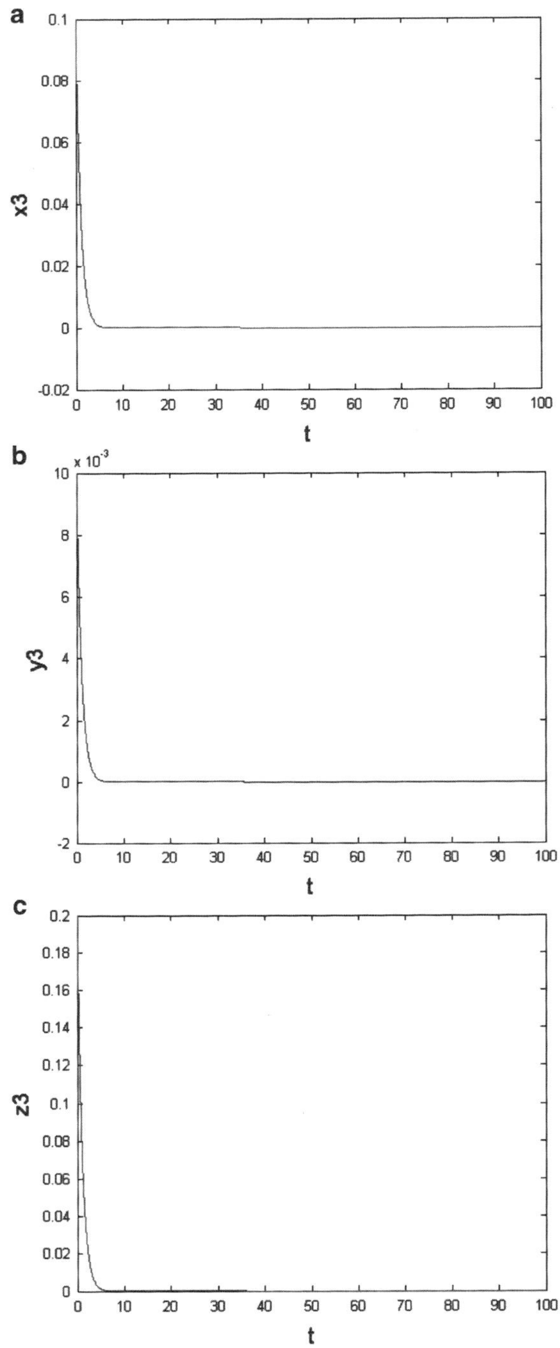

Fig. 7.4 The synchronization errors x_3, y_3 and z_3 of system (7.10) converge to zero after the active control (7.11) is activated; (**a**) x_3 tends to zero, (**b**) y_3 tends to zero, (**c**) z_3 tends to zero

In 2002, S. Westerlund investigated the behavior of a real inductor. He showed that if $i(t)$ is the current in the inductor, then the voltage is

$$u(t) = L\frac{d^{\alpha}i(t)}{dt^{\alpha}}, \tag{7.16}$$

where L is the inductance of the inductor and the constant α is related to the "proximity effect" [30]. Applying Kirchhoff's laws and the relations (7.15), (7.16) into the circuit shown in Fig. 7.1, we obtain:

$$\begin{aligned} c_1\frac{d^{\alpha_1}v_1}{dt^{\alpha_1}} &= -[bv_1^3 + av_1 + (1/R)(v_1 - v_2)], \\ c_2\frac{d^{\alpha_2}v_2}{dt^{\alpha_2}} &= (1/R)(v_1 - \gamma v_2) - i_L, \\ \frac{d^{\alpha_3}i}{dt^{\alpha_3}} &= v_2/L, \end{aligned} \tag{7.17}$$

where c_1, c_2 are the capacitances of the two capacitors, L is the inductance, α_1 is real order of the capacitor C_1, α_2 is real order of the capacitor C_2, and α_3 is real order of the inductor L. Assuming that $\alpha_1 = \alpha_2 = \alpha_3 = \alpha$, and using the rescaling relations (7.5), the commensurate fractional-order MAVPD system is given as follows [35]:

$$\begin{aligned} \frac{d^{\alpha}x}{dt^{\alpha}} &= -\nu(x^3 - \mu x - y), \\ \frac{d^{\alpha}y}{dt^{\alpha}} &= x - \gamma y - z, \\ \frac{d^{\alpha}z}{dt^{\alpha}} &= \beta y, \end{aligned} \tag{7.18}$$

where α is the fractional-order satisfying $\alpha \in (0, 1]$. When $\alpha = 1$, system (7.18) is the original integer-order MAVPD system.

The equilibrium points of fractional-order system (7.18) are the same as its integer-order form and are given as:

$$E_0 = (0,0,0),\ E_+ = (\sqrt{\mu}, 0, \sqrt{\mu}) \quad \text{and} \quad E_- = (-\sqrt{\mu}, 0, -\sqrt{\mu}). \tag{7.19}$$

7.3.2 Fractional Calculus

Consider the initial value problem:

$$D^{\alpha}X(t) = f(t, X(t)),\ 0 \le t \le T,\ X^{(k)}(0) = X_0^{(k)},\ k = 0, 1, \ldots, l-1. \tag{7.20}$$

Theorem 3.1. *(**Existence** [36]). Assume that* $\mathrm{E} := [0,\ \chi^*] \times [\mathrm{X}_0^{(0)} - \varepsilon,\ \mathrm{X}_0^{(0)} + \varepsilon]$ *with some* $\chi^* > 0$ *and some* $\varepsilon > 0$ *and let the function* $f : \mathrm{E} \to R$ *be continuous. In addition, let* $\chi := \min\{\chi^*,\ (\varepsilon\Gamma(\alpha+1)/\|f\|_\infty)^{1/\alpha}\}$ *Then, there exists a function* $\mathrm{X} : [0,\ \chi] \to R$ *solving the initial value problem (7.20).*

Theorem 3.2. *(**Uniqueness** [36]). If* $\mathrm{E} := [0,\ \chi^*] \times [\mathrm{X}_0^{(0)} - \varepsilon,\ \mathrm{X}_0^{(0)} + \varepsilon]$ *with some* $\chi^* > 0$ *and some* $\varepsilon > 0$. *Moreover, let the function* $f : \mathrm{E} \to R$ *be bounded on* E *and satisfy a Lipschitz condition with respect to the second variable, i.e.* $|f(t,\ \mathrm{X}) - f(t,\ \mathrm{Y})| \le \rho|\mathrm{X} - \mathrm{Y}|$ *with some constant* $\rho > 0$ *independent of* t, X *and* Y. *Then, denoting* χ *as in theorem 3.1, there exists at most one function* $\mathrm{X} : [0,\ \chi] \to R$ *solving the initial value problem (7.20).*

In [36] Diethelm et al. proved that if the function f is continuous, then the initial value problem (7.20) is equivalent to Volterra integral equation of the second kind, given as

$$\mathrm{X}(t) = \sum_{k=0}^{l-1} \frac{t^k}{k!} \mathrm{X}_0^{(k)} + \frac{1}{\Gamma(\alpha)} \int_0^t (t-\zeta)^{\alpha-1} f(\zeta, \mathrm{X}(\zeta)) d\zeta. \tag{7.21}$$

Theorem 3.3. *The initial value problem of the commensurate fractional-order MAVPD system (7.18) can be given as follows:*

$$D^\alpha \mathrm{X}(t) = A_1 \mathrm{X}(t) + x(t) A_2 \mathrm{X}(t), \quad \mathrm{X}(0) = \mathrm{X}_0, \tag{7.22}$$

where $0 < t \le \chi$, $\mathrm{X}(t) = (x(t),\ y(t),\ z(t))^T \in R^3$, $\mathrm{X}_0 = (x_0,\ y_0,\ z_0)^T$, $A_1 = \begin{pmatrix} \mu\nu & \nu & 0 \\ 1 & -\gamma & -1 \\ 0 & \beta & 0 \end{pmatrix}$, $A_2 = \begin{pmatrix} -\nu x & 0 & 0 \\ 0 & 0 & 0 \\ 0 & 0 & 0 \end{pmatrix}$, *for some constant* $\chi > 0$, *then it has a unique solution.*

Proof. Consider the function $f(\mathrm{X}(t)) = A_1\mathrm{X}(t) + x(t)A_2\mathrm{X}(t)$ which is continuous and bounded on the interval $[\mathrm{X}_0 - \varepsilon,\ \mathrm{X}_0 + \varepsilon]$ for any $\varepsilon > 0$. Moreover, we have

$$\left|f(\mathrm{X}(t)) - f(\mathrm{X}'(t))\right| = \left|A_1[\mathrm{X}(t) - \mathrm{X}'(t)] + x(t)A_2\mathrm{X}(t) - x'(t)A_2\mathrm{X}'(t)\right|.$$

Now, since

$$\left|x(t) - x'(t)\right| \le \left|\mathrm{X}(t) - \mathrm{X}'(t)\right|, \text{ we get}$$

$$\left|A_2[x(t) - x'(t)]\mathrm{X}(t) + A_2[x'(t)\{\mathrm{X}(t) - \mathrm{X}'(t)\}]\right| \le \|A_2\| \cdot [|\mathrm{X}(t)| + \left|x'(t)\right|](\left|\mathrm{X}(t) - \mathrm{X}'(t)\right|).$$

Thus,

$$\left|f(\mathrm{X}(t)) - f(\mathrm{X}'(t))\right| \le \|A_1\| \cdot \left|\mathrm{X}(t) - \mathrm{X}'(t)\right| + \|A_2\| \cdot [|\mathrm{X}(t)| + \left|x'(t)\right|](\left|\mathrm{X}(t) - \mathrm{X}'(t)\right|),$$

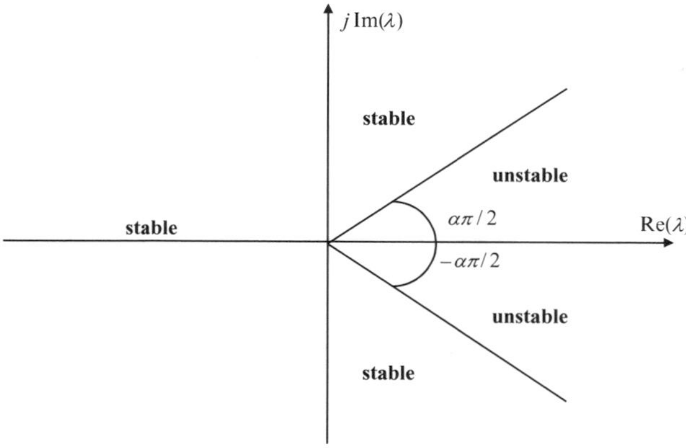

Fig. 7.5 Stability region of a linearized fractional-order system

which implies that

$$\begin{aligned}\left|f(\mathrm{X}(t))-f(\mathrm{X}'(t))\right| &\leq \|A_1\|\cdot\left|\mathrm{X}(t)-\mathrm{X}'(t)\right|+\|A_2\|\cdot[2\,|\mathrm{X}_0|+2\varepsilon](\left|\mathrm{X}(t)-\mathrm{X}'(t)\right|)\\ &\leq (\|A_1\|+\|A_2\|\cdot[2\,|\mathrm{X}_0|+2\varepsilon])\left|\mathrm{X}(t)-\mathrm{X}'(t)\right|\\ &\leq \rho\left|\mathrm{X}(t)-\mathrm{X}'(t)\right|,\end{aligned} \tag{7.23}$$

where one sets $\rho = \|A_1\|+\|A_2\|\cdot[2\,|\mathrm{X}_0|+2\varepsilon] > 0, \mathrm{X}'(t) \in R^3$, $\|.\|$ and $|.|$ represent matrix norm and vector norm respectively. Now, owing to the attractiveness of the attractor, there exists $\eta > 0$ such that $|x(t)| \leq \eta < \infty$ which implies that $\|A_2\|$ is bounded. Thus, Lipschitz condition is satisfied. Therefore, the results of Theorems 3.1 and 3.2 imply that the initial value problem of the commensurate fractional order MAVPD system has a unique solution. □

On the other hand, the local stability of the equilibrium points of a linear fractional-order system is governed by the following Matignon's results [37]:

$$|\arg(\lambda_i)| > \alpha\pi/2, (i = 1, 2, 3), \tag{7.24}$$

where $\lambda_1, \lambda_2, \lambda_3$ are the eigenvalues of the equilibrium points, and the stability region of these equilibrium points is obtained from Fig. 7.5 (in which $j = \sqrt{-1}$).

Now, consider the autonomous system (7.18) in the following form:

$$\frac{d^\alpha \mathrm{X}(t)}{dt^\alpha} = f(\mathrm{X}(t)),\ \mathrm{X}(0) = \mathrm{X}_0, \tag{7.25}$$

where $\mathrm{X}(t) = (x, y, z)^T = (x_1, x_2, x_3)^T \in R^3$, $f : R^3 \to R^3$ is a nonlinear vector function in terms of X. Let $J(\mathrm{X}^*) = \left(\frac{\partial f_i}{\partial x_j}\right)_{ij}\Big|_{\mathrm{X}=\mathrm{X}^*}$ be the Jacobian matrix at the equilibrium point X^*.

7.3.3 Chaos in the Fractional-Order MAVPD System

7.3.3.1 A Necessary Condition in Fractional-Order Systems to Remain Chaotic

Consider the commensurate fractional-order system $\frac{d^\alpha X}{dt^\alpha} = f(X)$, where $\alpha \in (0,1)$ and $X \in R^n$. According to (7.24), the equilibrium points of the previous system are locally asymptotically stable if all their eigenvalues satisfy $|\arg(\lambda(J))| > \alpha\pi/2$, where $J = \partial f/\partial X$, and λ is an eigenvalue of J. When $n = 3$, a saddle point is an equilibrium point on which the corresponding linearized system has at least one eigenvalue in the region of stability and one eigenvalue in the region of unstability. Moreover, if one of the eigenvalues is unstable and the other eigenvalues are stable, the equilibrium point is said to be "saddle point of an index 1". However, if two of the eigenvalues are unstable and one eigenvalue is stable, then the equilibrium point is said to be "saddle point of an index 2" [38]. From [38], we conclude that scrolls are generated only around the saddle points of an index 2, however saddle points of an index 1 are responsible only for connecting scrolls. Now, consider three dimensional chaotic system $\dot{X} = f(X)$ has only three equilibrium points. So, if this system has double scroll attractor, then one of its equilibrium points is saddle point of an index one and others are saddle points of an index 2. Assume that λ is an unstable eigenvalue of one of the saddle points of an index 2. Thus, a necessary condition for the commensurate fractional-order system $\frac{d^\alpha X}{dt^\alpha} = f(X)$, to remain chaotic is keeping the eigenvalue λ in the unstable region (see Fig. 7.5). This implies that

$$\tan(\alpha\pi/2) > \frac{|\mathrm{Im}(\lambda)|}{\mathrm{Re}(\lambda)} \Rightarrow \alpha > \frac{2}{\pi}\tan^{-1}(\frac{|\mathrm{Im}(\lambda)|}{\mathrm{Re}(\lambda)}). \tag{7.26}$$

At the parameter values $\beta = 200$, $\mu = 0.1$, $\nu = 100$ and $\gamma = 1.6$, the equilibrium points and their eigenvalues are given as:

$$E_0 = (0,0,0): \quad \lambda_1 = 13.3409, \quad \lambda_{2,3} = -2.4704 \pm 11.9922j,$$
$$E_+ = (0.3162, 0, 0.3162): \quad \lambda_1 = -23.3020, \quad \lambda_{2,3} = 0.8510 \pm 13.0742j,$$
$$E_- = (-0.3162, 0, -0.3162): \quad \lambda_1 = -23.3020, \quad \lambda_{2,3} = 0.8510 \pm 13.0742j, \quad j = \sqrt{-1}.$$

The equilibrium point $E_0 = (0,0,0)$ is a saddle point of an index 1. However, the equilibrium points E_+ and E_- are saddle points of an index 2. Thus, using (7.26), the necessary condition for the commensurate fractional-order MAVPD system (7.18) to remain chaotic is $\alpha > (2/\pi)\tan^{-1}(|\mathrm{Im}(\lambda_{2,3})|/\mathrm{Re}(\lambda_{2,3}))$. Consequently, the lowest fractional-order α for which the fractional-order MAVPD system (7.18) demonstrates chaos using the above-mentioned parameters is given by the inequality $\alpha > 0.96$. Since the order of the fractional-order chaotic system is the sum of the orders of all involved derivatives, then the lowest order for the fractional-order

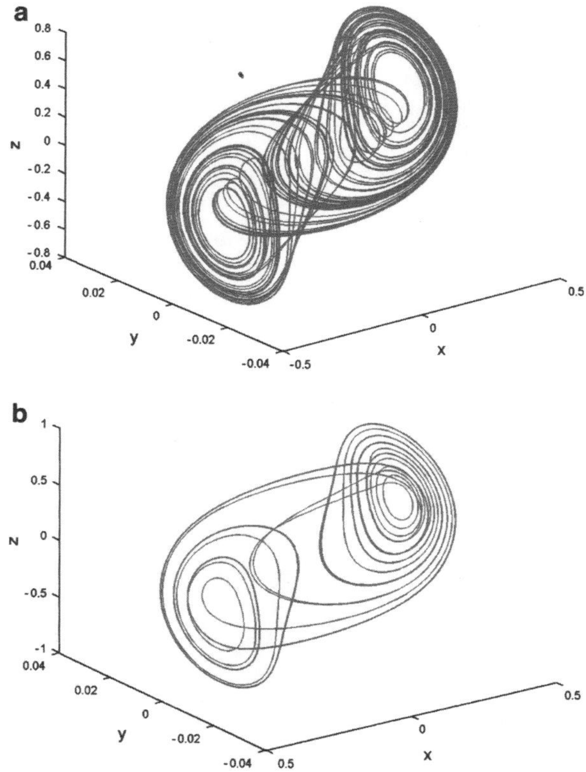

Fig. 7.6 The double scroll attractor of system (7.18) using the parameter values $\beta = 200$, $\mu = 0.1$, $\nu = 100$, $\gamma = 1.6$: (**a**) at $\alpha = 0.98$, (**b**) at $\alpha = 0.99$

MAVPD system (7.18) to yield chaos is 2.91. Applying the numerical technique given in [39, 40] to a discretization form of the initial value problem for (7.18), simulation results show that the double scroll attractor is observed when 0.99 and 0.98 (see Fig. 7.6).

7.3.4 Phase Synchronization of the Fractional-Order MAVPD System

In the following, we study phase synchronization between two identical fractional-order MAVPD systems using an active control technique. Based on the stability analysis of dynamical systems, if all eigenvalues are negative, then the system will tend to zero, yielding complete synchronization. However, phase synchronization occurs if there is any zero eigenvalue. In the case of phase synchronization, the difference between various states of synchronized systems may not necessarily approach zero, but will stay less than or equal to a constant.

The two identical drive and response systems are given as follow:

$$\begin{aligned}\frac{d^\alpha}{dt^\alpha}x_1 &= -\nu(x_1^3 - \mu x_1 - y_1),\\ \frac{d^\alpha}{dt^\alpha}y_1 &= x_1 - \gamma y_1 - z_1,\\ \frac{d^\alpha}{dt^\alpha}z_1 &= \beta y_1,\end{aligned} \tag{7.27}$$

and

$$\begin{aligned}\frac{d^\alpha}{dt^\alpha}x_2 &= -\nu(x_2^3 - \mu x_2 - y_2) + u_1(t),\\ \frac{d^\alpha}{dt^\alpha}y_2 &= x_2 - \gamma y_2 - z_2 + u_2(t),\\ \frac{d^\alpha}{dt^\alpha}z_2 &= \beta y_2 + u_3(t),\end{aligned} \tag{7.28}$$

where $u_1(t)$, $u_2(t)$, $u_3(t)$ are the control functions. Define the synchronization errors as follow:

$$e_1 = x_2 - x_1,\ e_2 = y_2 - y_1,\ e_3 = z_2 - z_1. \tag{7.29}$$

By subtracting (7.27) from (7.28), we get

$$\begin{aligned}\frac{d^\alpha}{dt^\alpha}e_1 &= -\nu((x_2^3 - x_1^3) - \mu e_1 - e_2) + u_1(t),\\ \frac{d^\alpha}{dt^\alpha}e_2 &= e_1 - \gamma e_2 - e_3 + u_2(t),\\ \frac{d^\alpha}{dt^\alpha}e_3 &= \beta e_2 + u_3(t).\end{aligned} \tag{7.30}$$

Now, we define the control functions $u_1(t)$, $u_2(t)$ and $u_3(t)$ as

$$u_1(t) = \nu(x_2^3 - x_1^3) + V_1(t), \quad u_2(t) = V_2(t), \quad u_3(t) = V_3(t). \tag{7.31}$$

This leads to

$$\begin{aligned}\frac{d^\alpha}{dt^\alpha}e_1 &= \mu\nu e_1 + \nu e_2 + V_1(t),\\ \frac{d^\alpha}{dt^\alpha}e_2 &= e_1 - \gamma e_2 - e_3 + V_2(t),\\ \frac{d^\alpha}{dt^\alpha}e_3 &= \beta e_2 + V_3(t).\end{aligned} \tag{7.32}$$

Fig. 7.7 (**a–c**): Phase synchronization between the drive and response systems (7.27), (7.28)

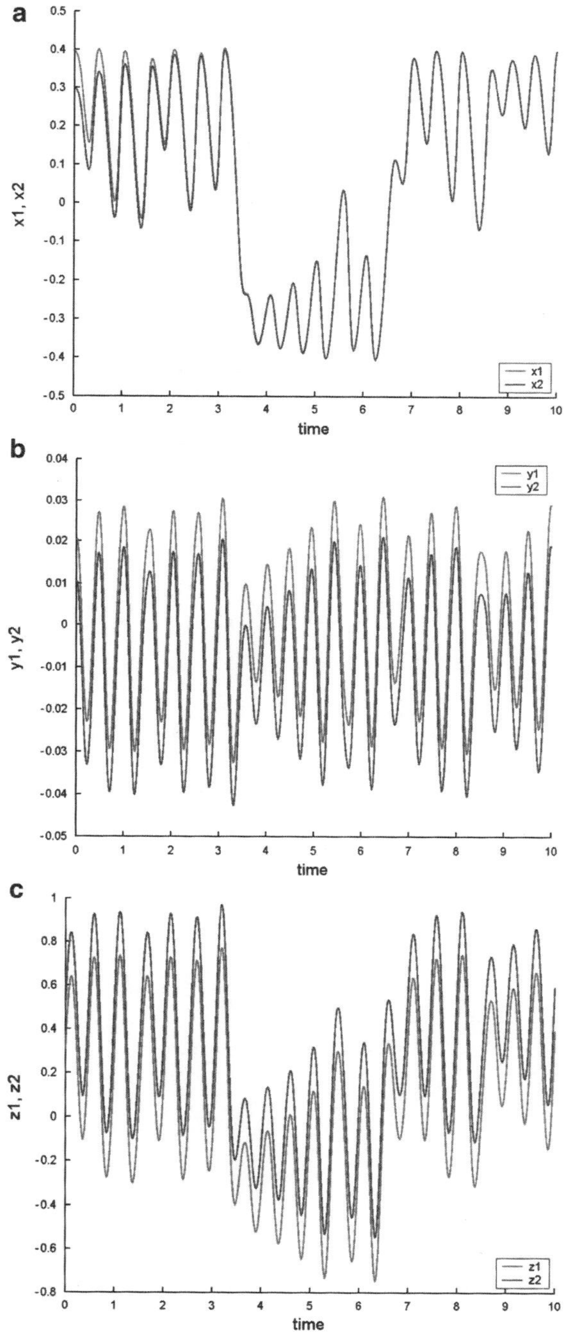

Thus, the linear active control functions $V_1(t)$, $V_2(t)$ and $V_3(t)$ can be chosen as

$$\begin{pmatrix} V_1 \\ V_2 \\ V_3 \end{pmatrix} = A \begin{pmatrix} e_1 \\ e_2 \\ e_3 \end{pmatrix}, \tag{7.33}$$

where the matrix A is given by

$$A = \begin{pmatrix} -(\mu\nu + 1) & -\nu & 0 \\ -1 & \gamma - 1 & 1 \\ 0 & -\beta & -1 \end{pmatrix}.$$

The eigenvalues of the linear system (7.32) are -1, -1 and -1. Using the Matignon's conditions (7.24), it follows that all the eigenvalues λ_i of the matrix A satisfy $|\arg(\lambda_i)| > \alpha\pi/2$. Consequently, the zero equilibrium point of the linear system (7.32) is locally asymptotically stable and the synchronization between the drive and response systems (7.27), (7.28) is achieved.

Denoting λ_1, λ_2 and λ_3 as the eigenvalues of system (7.32), then (7.33) is rewritten as follows:

$$V_1(t) = -\mu\nu e_1 - \nu e_2 + \lambda_1 e_1,\ V_2(t) = -e_1 + \gamma e_2 + e_3 + \lambda_2 e_2,\ V_3(t) = -\beta e_2 + \lambda_3 e_3. \tag{7.34}$$

Now, the systems (7.27) and (7.28) are numerically integrated using the above-mentioned parameter values and commensurate fractional-order $\alpha = 0.98$. The initial conditions are chosen as follow; $(x_1(0), y_1(0), z_1(0)) = (0.4, 0.02, 0.3)$, and $(x_2(0), y_2(0), z_2(0)) = (0.3, 0.01, 0.5)$. Now, by choosing $\lambda_1 = -1$, $\lambda_2 = 0$, and $\lambda_3 = 0$, the control functions (7.34) can be determined and phase synchronization between the drive and response systems can also be achieved. The corresponding numerical results are shown in Fig. 7.7a–c.

7.4 Conclusion

We have studied chaotic dynamics of a Modified Autonomous Van der Pol-Duffing (MAVPD) circuit and its fractional-order counterpart. Moreover, we have studied chaos synchronization in these systems via active control techniques.

On the other hand, studying the periodic solutions and numerical verifications of Hopf bifurcations in the fractional-order systems are still open problems and need further investigations in a future work.

Acknowledgements The authors wish to thank Prof. E. Ahmed and Prof. H.N. Agiza for discussion and help. The corresponding author 'Matouk' wishes to thank the editor, Dr. S. Banerjee, and Dr. Eid AL-Haisouni, the Dean of Prep-Year college for their kind help and support.

References

1. Lorenz, E.N.: Deterministic non-periodic flows. J. Atmos. Sci. **20**, 130–141 (1963)
2. Podlubny, I.: Fractional Differential Equations. Academic, New York (1999)
3. Hilfer, R. (ed.): Applications of Fractional Calculus in Physics. World Scientific, New Jersey (2000)
4. Laskin, N.: Fractional market dynamics. Phys. A **287**, 482–492 (2000)
5. Ahmad, W.M., El-Khazali, R.: Fractional-order dynamical models of love. Chaos Solitons and Fractals **33**, 1367–1375 (2007)
6. Ahmed, E., Elgazzar, A.S.: On fractional order differential equations model for nonlocal epidemics. Phys. A **379**, 607–614 (2007)
7. El-Sayed, A.M.A., El-Mesiry, A.E.M., El-Saka, H.A.A.: On the fractional-order logistic equation. Appl. Math. Lett. **20** 817–823 (2007)
8. El-Sayed, A.M.A., Ahmed, E., Herzallah, M.A.E.: On the fractional-order games with non-uniform interaction rate and asymmetric games. J. Fractional Calculus Appl. **1**(1), 1–9 (2011)
9. Caputo, M.: Linear models of dissipation whose Q is almost frequency independent-II. Geophys. J. R. Astron. Soc. **13**, 529–539 (1967)
10. Pecora, L.M., Carroll, T.L.: Synchronization in chaotic systems. Phys. Rev. Lett. **64**, 821–824 (1990)
11. Lakshmanan, M., Murali, K.: Chaos in Nonlinear Oscillators: Controlling and Synchronization. World Scientific, Singapore (1996)
12. Bai, E.W., Lonngren, K.E.: Synchronization of two Lorenz systems using active control. Chaos Solitons and Fractals **8**, 51–58 (1997)
13. Hegazi, A.S., Agiza, H.N., El-Dessoky, M.M.: Synchronization and adaptive synchronization of nuclear spin generator system. Chaos Solitons and Fractals **12**, 1091–1099 (2001)
14. Hegazi, A.S., Matouk, A.E.: Dynamical behaviors and synchronization in the fractional order hyperchaotic Chen system. Appl. Math. Lett. **24**, 1938–1944 (2011)
15. Matouk, A.E.: Dynamical analysis feedback control and synchronization of Liu dynamical system. Nonlinear Anal.: TMA **69**, 3213–3224 (2008)
16. Matouk, A.E.: Chaos synchronization between two different fractional systems of Lorenz family. Math. Probl. Eng. **2009**, 11 (2009). doi:10.1155/2009/572724.
17. Matouk, A.E.: Dynamical behaviors, linear feedback control and synchronization of the fractional order Liu system. J. Nonlinear Syst. Appl. **1**(3), 135–140 (2010)
18. Carroll, T.L., Pecora, L.M.: Synchronizing chaotic circuits. IEEE Trans. Circ. Syst. I **38**, 453–456 (1991)
19. Blasius, B., Huppert, A., Stone, L.: Complex dynamics and phase synchronization in spatially extended ecological systems. Nature **399**, 354–359 (1999)
20. Li, Y., Chen, L., Cai, Z., Zhao, X.: Study on chaos synchronization in the Belousov-Zhabotinsky chemical system. Chaos Solitons and Fractals **17**, 699–707 (2003)
21. Uchida, A., Kinugawa, S., Yoshimori, S.: Synchronization of chaos in two microchip lasers by using incoherent feedback method. Chaos Solitons and Fractals **17**, 363–368 (2003)
22. Chua, L.O.: The genesis of Chua's circuit. AEU. Int. J. Elecron. Comm. **46**, 187–257 (1992)
23. Han, F., Wang, Y., Yu, X., Feng, Y.: Experimental confirmation of a new chaotic attractor. Chaos Solitons and Fractals **21**, 69–74 (2004)
24. King, G.P., Gaito, S.T.: Bistable chaos. I. Unfolding the cusp. Phys. Rev. A **46**, 3092–3099 (1992)
25. Agiza, H.N., Matouk, A.E.: Adaptive synchronization of Chua's circuits with fully unknown parameters. Chaos Solitons and Fractals **28**, 219–227 (2006)
26. Chua, L.O., Kocarev, L.J., Eckert, K., Itoh, M.: Experimental chaos synchronization in Chua's circuit. Int. J. Bifurcat. Chaos **2**, 705–708 (1992)
27. Cuomo, K.M., Oppenheim, V.: Circuit implementation of synchronized chaos with application to communication. Phys. Rev. Lett. **71**, 65–68 (1993)

28. Matouk, A.E., Agiza, H.N.: Bifurcations, chaos and synchronization in ADVP circuit with parallel resistor. J. Math. Anal. Appl. **341**, 259–269 (2008)
29. Wolf, A., Swift, J.B., Swinney, H.L., Vastano, J.A.: Determining Lyapunov exponents from a time series. Phys. D **16**, 285–317 (1985)
30. Petras, I.: A note on the fractional-order Chua's system. Chaos Solitons and Fractals **38**, 140–147 (2008)
31. Chen, X.R., Liu, C.X., Wang, F.Q., et al.: Study on the fractional-order Liu Chaotic system with circuit experiment and its control. Acta Phys. Sin. **57**, 1416–1422 (2008)
32. Chen, X.R., Liu, C.X., Wang, F.Q.: Circuit realization of the fractional-order unified chaotic system. Chin. Phys. B **17**, 1664–1669 (2008)
33. Liu, C.X., Ling, L.: Circuit implementation of a new hyperchaos in fractional-order system. Chin. Phys. B **17**, 2829–2836 (2008)
34. Hegazi, A.S., Ahmed, E., Matouk, A.E.: The effect of fractional order on synchronization of two fractional order chaotic and hyperchaotic systems. J. Fractional Calculus Appl. **1**(3), 1–15 (2011)
35. Matouk, A.E.: Chaos, feedback control and synchronization of a fractional-order modified Autonomous Van der Pol-Duffing circuit. Comm Nonlinear Sci Numer Simulat **16**, 975–986 (2011)
36. Diethelm, K., Ford, N.J.: Analysis of fractional differential equations. J. Math. Anal. Appl. **265**, 229–248 (2002)
37. Matignon, D.: Stability results for fractional differential equations with applications to control processing. In: IEEE-SMC Proceedings of the Computational Engineering in Systems and Application Multiconference, IMACS, Lille, France, vol. 2, pp. 963–968 (1996)
38. Tavazoei, M.S., Haeri, M.: A necessary condition for double scroll attractor existence in fractional-order systems. Phys. Lett. A **367**, 102–113 (2007)
39. Diethelm, K.: An algorithm for the numerical solution of differential equations of fractional order. Electron. Trans. Numer. Anal. **5**, 1–6 (1997)
40. Diethelm, K., Ford, N.J., Freed, A.D.: A predictor-corrector approach for the numerical solution of fractional differential equations. Nonlinear Dynam. **29**, 3–22 (2002)

Chapter 8
Outer and Inner Synchronization in Networks on Rössler Oscillators: An Experimental Verification

Rajarshi Middya, Shankar Kumar Basak, Anirban Ray, and Asesh Roychowdhury

Abstract This chapter deals with the synchronization within the network and among different networks. Synchronization within a network is called 'inner synchronization' and synchronization among different networks is named as 'outer synchronization'. First, the 'inner synchronization' for different network topologies are studied numerically and experimentally. Next outer 'synchronization' amongst two different networks are studied when they are connected through similar nodes(homogenous situation). Again the study of 'outer synchronization' among two different network are repeated when they are connected through different nodes(heterogenous case). Though second situation occurs often in nature, it has been studied very little.

8.1 Introduction

The process of synchronization holds an important place in the present day study of nonlinear system and complex network. Knowledge about complex network and its synchronization is growing through rapid growth [1–9]. According to definition, a complex network is a large set of inter connected nodes, in which a node is a fundamental unit with specific contents [10].

Random graph, small world effect and scale free characteristics are mostly noticeable among different complex networks. After its construction in early 1960 by Erdös and Renyi [11], random graph dominated mathematical research of

R. Middya
Department of Electronics and Telecommunication Engineering, Jadavpur University, Kolkata 700032, West Bengal, India
e-mail: rajarshi.middya@gmail.com

S.K. Basak · A. Ray · A. Roychowdhury (✉)
Department of Physics, Jadavpur University, Kolkata 700032, West Bengal, India
e-mail: basak.sankar8@gmail.com; anirban.chaos@gmail.com; asesh_r@yahoo.com

S. Banerjee and L. Rondoni (eds.), *Applications of Chaos and Nonlinear Dynamics in Science and Engineering 3*, Understanding Complex Systems,
DOI 10.1007/978-3-642-34017-8_8, © Springer-Verlag Berlin Heidelberg 2013

complex network for half a century. This was partly for the absence of supercomputaional power and partly for the absence of detailed topological information about different large scale real world networks. Small world effect was introduced by Strogatz and Watts [1] to investigate the transition from regular networks to random ones. Such networks behave with a high degree of clustering as in regular networks and small average distance among them. After that Barabasi and Albert [2] brought forward a power law distribution and more nodes having few connections, but only few nodes have many connections and hub.

Networks mostly focus on modeling, dynamical analysis and control. Network synchronization can be considered in two ways (a)Synchronization arising inside a network composed of coupled dynamical systems i.e. 'inner synchronization' and (b) synchronization occurring between two coupled complex networks in mutual coupling configuration i.e. 'outer synchronization'. Early works on 'inner synchronization' has attracted much attention till now [10]. In 'inner synchronization', a network of the following ideal structure is usually considered.

$$\dot{x}_i(t) = f(x_i(t)) + c' \sum_{j=1}^{N} a_{ij} \Gamma x_j(t), i = 1, 2, \ldots\ldots, N, \tag{8.1}$$

where $x_i = (x_{i1}, x_{i2}, \cdots x_{iN})^T \in R^n$ are the state variables of the node i and N is the number of network nodes. $f : R^n$ is a continually differentiable function which determines the dynamical behavior of the nodes. $c' > 0$ is a coupling strength and $\Gamma \in R^{n \times n}$ is the constant matrix linking coupling variables. $A = a_{ij\,N \times N}$ represents coupling among different nodes of an entire network. Entries of A are defined as follows, $a_{ij} = 1$ if connection exist between different nodes i and j; otherwise $a_{ij} = 0$. Diagonal elements of A are defined as $a_{ii} = -\sum_{j=1}^{N}$ and clearly, if degree of node i is k_i, then $a_{ii} = -k_i$, $i = 1, 2, \cdots, N$. In the paper [10], in all nodes belonging to the network studied, the synchronous state is determined by $\dot{x}(t) = f(x(t))$. Since they considered the synchronization within the network, we referred it as 'inner synchronization'. Later, this approach was expanded and improved by introducing weighted connection, time dependent coupling matrix, non-linear coupling function and time delays [12–19]. A different approach where synchronous state is different from $\dot{x}(t) = f(x(t))$ is also studied in [20].

This gives rise to a natural question, whether two similar network synchronizes with each other. There are real life examples where outer synchronization exists between different networks. We will describe them later. This kind of synchronization aptly named as 'outer synchronization', is answered in [21]. In all of the above cases, nodes of a network are assumed to show same dynamics. But there are cases where different nodes exhibit different dynamics. This kind of network do exist in real life. Does a synchronization between two such networks exist? We cite two such examples here. From the angle of sociology our world can be divided into two networks: developed network constructed by developed countries and under developed network constructed of under developed countries. Now from a finer point, all developed countries are not developed to same extant.

Same goes for under developed network. With the gradual increase of international exchange, the two networks will be synchronized. In animal world, synchronization phenomena universally exists if no external intervention appears. In predator and prey community, the number of predator and prey remains invariant under no outside intervention. But not all predators hunt same prey. So the study of the 'outer synchronization' between coupled networks with different nodes becomes imperative.

In the present chapter we study this interesting topic. For more details, synchronization of a network under different network topologies are given in Sect. 8.2. First numerical results are shown. Then experimental results are presented. In the next section, we find the situation when two such networks synchronizes with each other. Two different conditions are considered (a) when networks are connected through similar nodes(homogeneous coupling) and (b) when networks are connected through different nodes(heterogeneous coupling).

8.2 Inner Synchronization

Each node of the network consists of a Rössler oscillator. In this situation, we have considered the variety of possible topological arrangements and also considered the different ways to connect them. However, it sould be noted that instead of the standard Rössler system, we have considered a slightly modified Rössler form [22],

$$\begin{aligned} \dot{x} &= -\alpha x - \beta y - z \\ \dot{y} &= x + \gamma y \\ \dot{z} &= g(x) - z \end{aligned} \tag{8.2}$$

Here $g(x)$ stands for a step function which is written as

$$\begin{aligned} g(x) &= 0 \qquad\qquad x \leq 3 \\ &= \mu(x-3) \quad x > 3 \end{aligned} \tag{8.3}$$

where α, β, γ and μ are parameters. The system given in (8.9) has two families of fixed points

$$P_1 \implies (0,0,0) \tag{8.4}$$

$$P_2 \implies (\frac{3\mu\gamma}{\sigma}, -\frac{3\mu}{\sigma}, \frac{3\mu(\beta-\alpha\gamma)}{\sigma}) \tag{8.5}$$

with $\sigma = (\alpha+\mu)\gamma - \beta$ This system was used as it is easier to implement in the electronic circuit. In the experimental setup, the parameter values used were $\alpha = 0.05$, $\beta = 0.5$, $\gamma = 0.15$ and $\mu = 15$, whence the stability of fixed points become evident from the eigenvalues of the Jacobian. The origin has eigenvalues $(-1, 0.05 \pm 0.7i)$

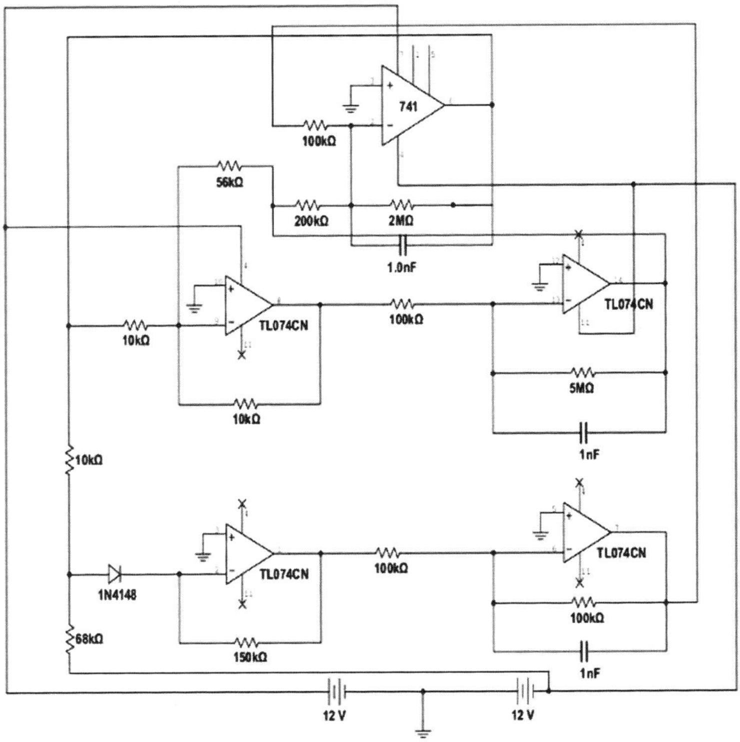

Fig. 8.1 Electronic Circuit for Rossler system

and the other has the set (0.1847, 3.2788, −4.363514). The corresponding circuit is given in Fig. 8.1 and the attractor obtained from this equation is given in Fig. 8.2.

Now, we connected such Rössler systems in a network with different topologies. In the next section we discuss the different network topologies that we have used to make a single network.

8.2.1 Different Network Topologies

We have used four Rössler oscillators to make a single network and used four network topologies.

- Uni-directional Line Topology
- Bi-directional Line Topology
- Star Topology
- Global Topology

The schematics for the above mentioned networks are shown in Figs. 8.3, 8.4, 8.5 and 8.6 respectively.

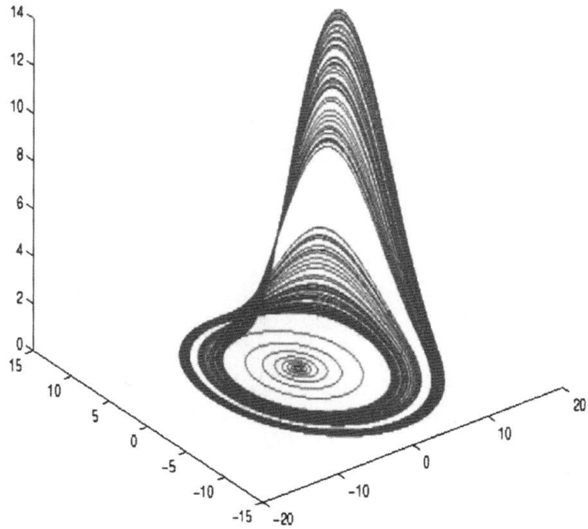

Fig. 8.2 Phase space for Rossler attractor

Fig. 8.3 Uni-directional line network

Fig. 8.4 Bi-directional line network

Fig. 8.5 Star network

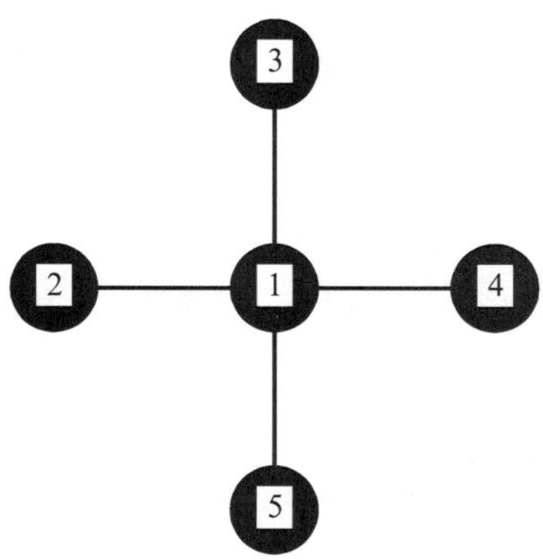

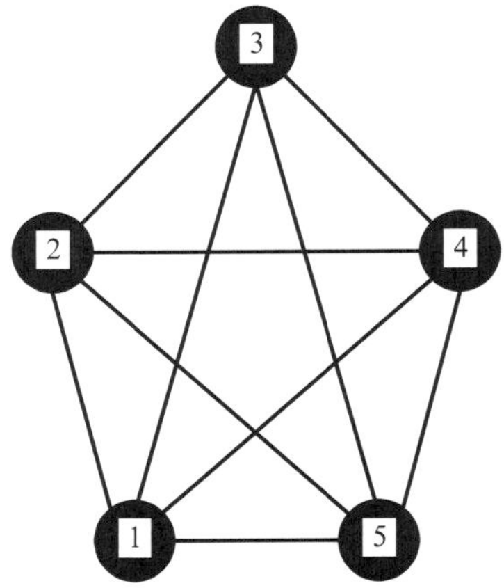

Fig. 8.6 Global network

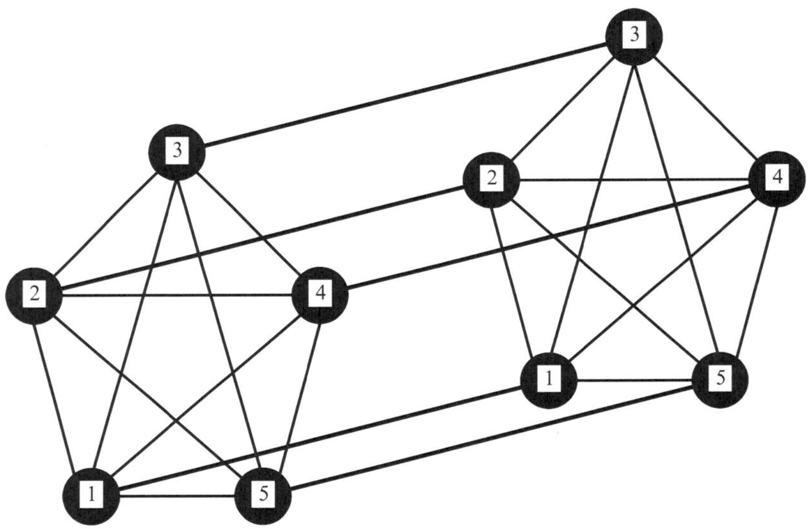

Fig. 8.7 Two global networks coupled with one another homogeneously

8.2.2 Outer Synchronnization

From the different network topologies that are shown in previous section, we take the Global Topology and make connections among two such networks. Now, two networks can be connected in two ways—they can be connected through similar nodes (this type of connection is called 'homogeneous connection') as shown in Fig. 8.7 or, through different nodes (the connection type is heterogeneous) as shown in Fig. 8.8. In both cases, two networks are connected bi-directionally.

Now, the work has been done in two steps (a) numerically and (b) experimentally. First we will describe the numerical approach and then discuss the experimental one.

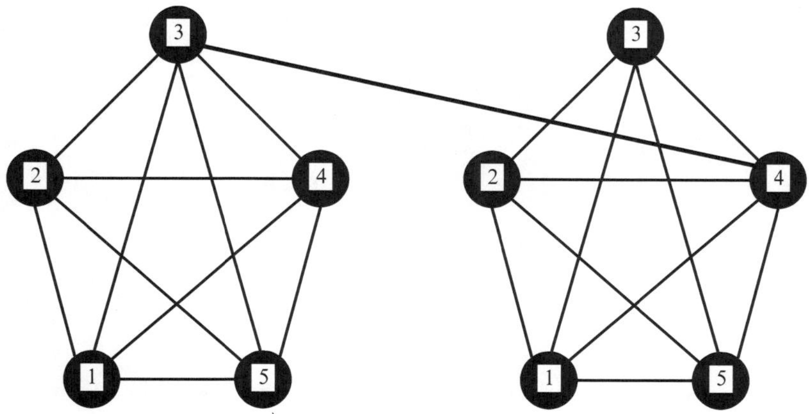

Fig. 8.8 Two global networks coupled with one another heterogeneously

8.3 Numerical Analysis

8.3.1 Global Network

The numerical analysis for each network is done independently. Dynamical equation of each nodes is given by

$$\dot{x}_i(t) = f(x_i(t)) + c' \sum_{j=1}^{N} a_{ij} \Gamma x_j(t) \tag{8.6}$$

Here x_i is the vector representing node variables, a_{ij} is the element of adjacency matrix A and Γ is the coupling matrix of each variable. Coupling strength is given by c'. At first, we had done the analysis of the Global Network. The adjacency matrix of a five node global network is given as

$$A = \begin{pmatrix} -4 & 1 & 1 & 1 & 1 \\ 1 & -4 & 1 & 1 & 1 \\ 1 & 1 & -4 & 1 & 1 \\ 1 & 1 & 1 & -4 & 1 \\ 1 & 1 & 1 & 1 & -4 \end{pmatrix} \tag{8.7}$$

The linking matrix Γ is given as

$$\begin{pmatrix} 0 & 0 & 0 \\ 0 & 1 & 0 \\ 0 & 0 & 0 \end{pmatrix} \tag{8.8}$$

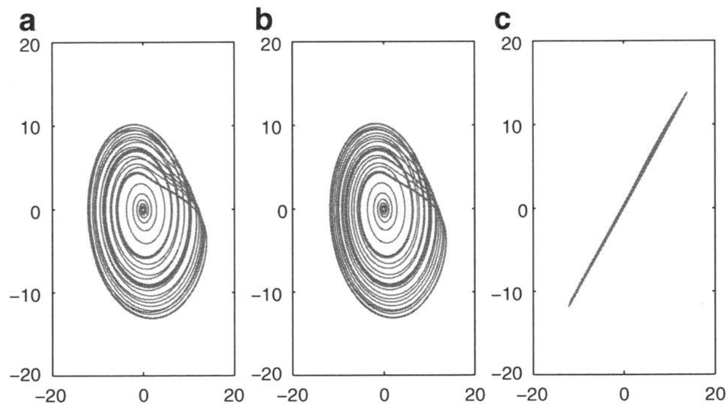

Fig. 8.9 (**a**) Attractor of Rossler 3, (**b**) Attractor of Rossler 5, (**c**) Phase space of Y for Rossler 3 and 5

i.e. we have chosen 'y' variable for connection. The system of equations for the 'i'th node of global network is given by

$$\begin{aligned}
\dot{x}_i &= -\alpha x_i - \beta y_i - z_i \\
\dot{y}_i &= x_i + \gamma y_i + c'(\sum_{j \neq i} y_j - 4 * y_i) \\
\dot{z}_i &= g(x_i) - z_i
\end{aligned} \tag{8.9}$$

Here $g(x_i)$ is a step function described (8.3). All parameter values are as described in previous section.

On solving Eg. (8.9) the phase space looks like Fig. 8.9.

To quantify the chaos, we also measured the maximum Lyapunov Exponents and plotted them against the different values of the coupling constant given by c in the above shown equations. This is shown in Fig. 8.10

8.3.2 Outer Synchronization Between Global Network

8.3.2.1 Heterogenous Coupling

This type of network is shown in Fig. 8.8. Such a network can be described by the following set of equations,

$$\dot{x}_i(t) = f(x_i(t)) + c' \sum_{j=1}^{N} a_{ij} \Gamma x_j(t) + C \delta_{mi} \Gamma (y_l - x_m) \tag{8.10}$$

$$\dot{y}_i(t) = f(y_i(t)) + c'' \sum_{j=1}^{N} a_{ij} \Gamma y_j(t) + C \delta_{li} \Gamma (x_m - y_l) \tag{8.11}$$

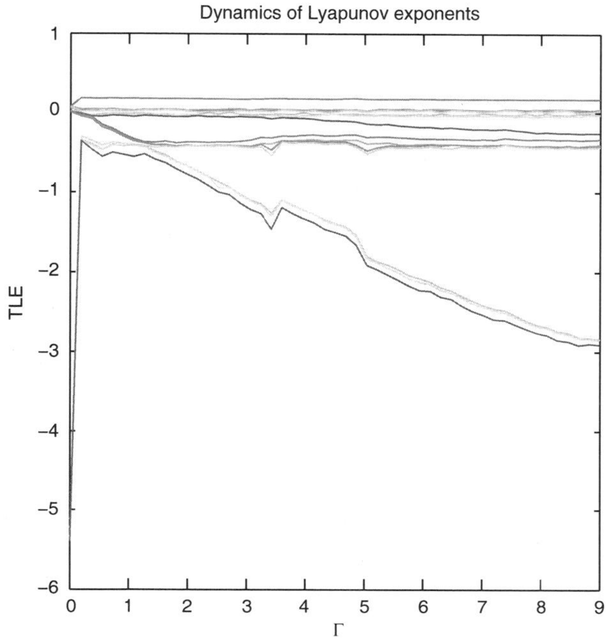

Fig. 8.10 Maximum Lyapunov exponents as a function of the coupling constant

where c' and c'' are coupling of the two networks respectively and $i = 1, \cdots, 5$. C is the coupling coefficient in between the two global networks. Here Γ has the expression shown earlier, and δ_{ij} is the Kronecker delta function. In the present case $m = 3$ and $l = 1$ i.e. third node of the first network is connected to first node of the second network. Explicitly, we write the expression for the connecting nodes as

$$\dot{x}_{12} = ax_{11} + x_{12} + c'(x_{22} + x_{32} + x_{42} + x_{52} - 4x_{12})$$
$$\dot{x}_{22} = ax_{21} + x_{22} + c'(x_{12} + x_{32} + x_{42} + x_{52} - 4x_{22})$$
$$\dot{x}_{32} = ax_{31} + x_{32} + c'(x_{12} + x_{22} + x_{42} + x_{52} - 4x_{32}) + C(y_{12} - x_{32})$$
$$\dot{x}_{42} = ax_{41} + x_{42} + c'(x_{12} + x_{22} + x_{32} + x_{52} - 4x_{42})$$
$$\dot{x}_{52} = ax_{51} + x_{52} + c'(x_{12} + x_{22} + x_{32} + x_{42} - 4x_{52})$$
$$\dot{y}_{12} = ay_{11} + y_{12} + c''(y_{22} + y_{32} + y_{42} + y_{52} - 4y_{12}) + C(x_{32} - y_{12})$$
$$\dot{y}_{22} = ay_{21} + y_{22} + c''(y_{12} + y_{32} + y_{42} + y_{52} - 4y_{22})$$
$$\dot{y}_{32} = ay_{31} + y_{32} + c''(y_{12} + y_{22} + y_{42} + y_{52} - 4y_{32})$$
$$\dot{y}_{42} = ay_{41} + y_{42} + c''(y_{12} + y_{22} + y_{32} + y_{52} - 4y_{42})$$
$$\dot{y}_{52} = ay_{51} + y_{52} + c''(y_{12} + y_{22} + y_{32} + y_{42} - 4y_{52})$$

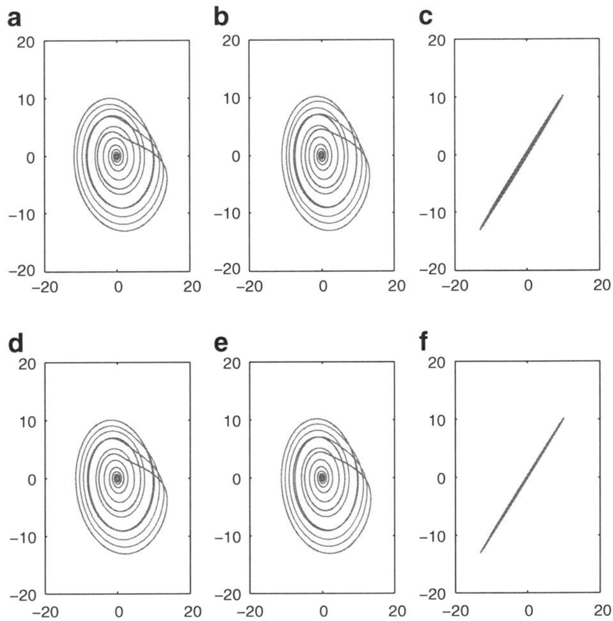

Fig. 8.11 Phase space output for 2-global network with five nodes each under heterogenous coupling (**a**) Attractor of Rossler 3, (**b**) Attractor of Rossler 4, (**c**) Phase space of Y for Rossler 2 and 5, (**d**) Attractor of Rossler 2, (**e**) Attractor of Rossler 5, (**f**) Phase space of Y for Rossler 3 and 4

Here For such a network the phase space is given by Figs. 8.11 and 8.12 shows the variation of the maximum Lyapunov Exponents with respect to the Coupling constant C in between the networks.

8.3.2.2 Homogenous Coupling

This type of network is shown in Fig. 8.7. Such a network can be described by the following set of equations.

$$\dot{x}_i(t) = f(x_i(t)) + c' \sum_{j=1}^{N} a_{ij} \Gamma x_j(t) + C\Gamma(y_i - x_i) \tag{8.12}$$

$$\dot{y}_i(t) = f(y_i(t)) + c'' \sum_{j=1}^{N} a_{ij} \Gamma y_j(t) + C\Gamma(x_i - y_i) \tag{8.13}$$

Here meaning of all variables remain same as above section. Explicitly, equations representing the dynamics for the connecting nodes are given as

Fig. 8.12 Maximum Lyapunov exponents as a function of the coupling constant for 2-global network with five nodes each under heterogenous coupling

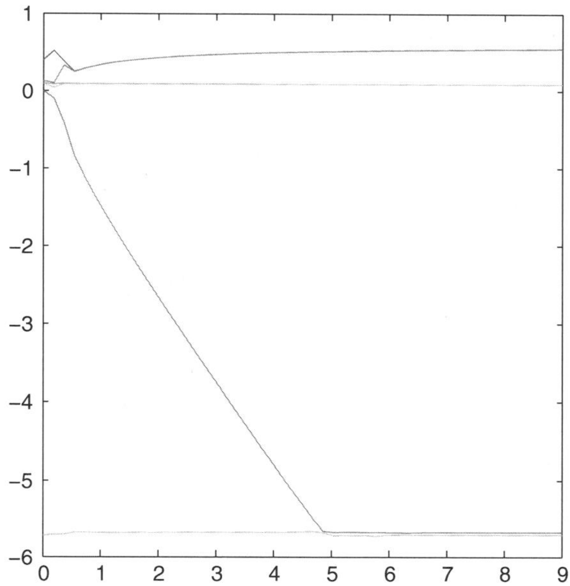

$$\dot{x}_{12} = ax_{11} + x_{12} + c'(x_{22} + x_{32} + x_{42} + x_{52} - 4x_{12}) + C(y_{12} - x_{12})$$
$$\dot{x}_{22} = ax_{21} + x_{22} + c'(x_{12} + x_{32} + x_{42} + x_{52} - 4x_{22}) + C(y_{22} - x_{22})$$
$$\dot{x}_{32} = ax_{31} + x_{32} + c'(x_{12} + x_{22} + x_{42} + x_{52} - 4x_{32}) + C(y_{32} - x_{32})$$
$$\dot{x}_{42} = ax_{41} + x_{42} + c'(x_{12} + x_{22} + x_{32} + x_{52} - 4x_{42}) + C(y_{42} - x_{42})$$
$$\dot{x}_{52} = ax_{51} + x_{52} + c'(x_{12} + x_{22} + x_{32} + x_{42} - 4x_{52}) + C(y_{52} - x_{52})$$
$$\dot{y}_{12} = ay_{11} + y_{12} + c''(y_{22} + y_{32} + y_{42} + y_{52} - 4y_{12}) + C(x_{12} - y_{12})$$
$$\dot{y}_{22} = ay_{21} + y_{22} + c''(y_{12} + y_{32} + y_{42} + y_{52} - 4y_{22}) + C(x_{22} - y_{22})$$
$$\dot{y}_{32} = ay_{31} + y_{32} + c''(y_{12} + y_{22} + y_{42} + y_{52} - 4y_{32}) + C(x_{32} - y_{32})$$
$$\dot{y}_{42} = ay_{41} + y_{42} + c''(y_{12} + y_{22} + y_{32} + y_{52} - 4y_{42}) + C(x_{42} - y_{42})$$
$$\dot{y}_{52} = ay_{51} + y_{52} + c''(y_{12} + y_{22} + y_{32} + y_{42} - 4y_{52}) + C(x_{52} - y_{52})$$

where c' and c'' are the coupling coefficients of the networks and C is the coupling coefficient in between the two global networks. For such a network the phase space is given by Figs. 8.13 and 8.14 shows the variation of the maximum Lyapunov Exponents with respect to the Coupling constant C in between the networks.

8.3.3 Further Numerical Study

Now we discuss our numerical results obtained from changing Γ matrix to take other variables. As for example, we take the x component instead of the y (as we had

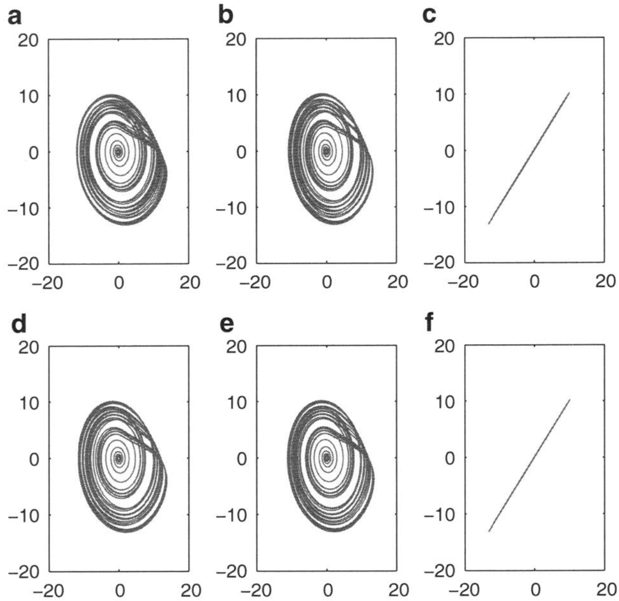

Fig. 8.13 Phase space output for 2-global network with five nodes each under homogenous coupling (**a**) Attractor of Rossler 3, (**b**) Attractor of Rossler 4, (**c**) Phase space of Y for Rossler 2 and 5, (**d**) Attractor of Rossler 2, (**e**) Attractor of Rossler 5, (**f**) Phase space of Y for Rossler 3 and 4

done in our work), and couple it with the *y* component of the second node of Fig. 8.6, and *z* component of the third node of Fig. 8.6, and going on, such type of coupling can be termed as a heterogeneous one. This heterogeneous coupling has been studied numerically for all the networks for which numerical and experimental analysis had been done in the previous sections.

8.3.3.1 Global Topology

The Global Topology is shown in Fig. 8.6. The coupling is done using the *x, y, z, x, y*—component of node *1, 2, 3, 4, 5* respectively. For such a coupling the chaotic state is present only when the value for the coupling constant is small enough. This is seen in Fig. 8.15.

As can be observed from the Fig. 8.15, as we increase the coupling between the different nodes (which is different from one another), the chaotic state no longer exists.

8.3.3.2 Outer Synchronization of Global Networks Under Heterogenous coupling

This is the coupling between two global networks as shown in Fig. 8.8. Here the coupling is done using the *x, y, z, x, y* and *z*—component of node *1, 2, 3, 4, 5* and *6*

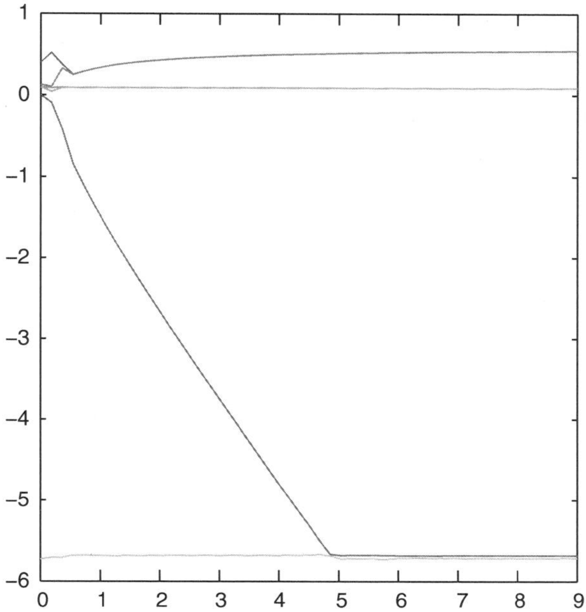

Fig. 8.14 Maximum Lyapunov exponents as a function of coupling constant for 2-global network with five nodes each under homogenous coupling

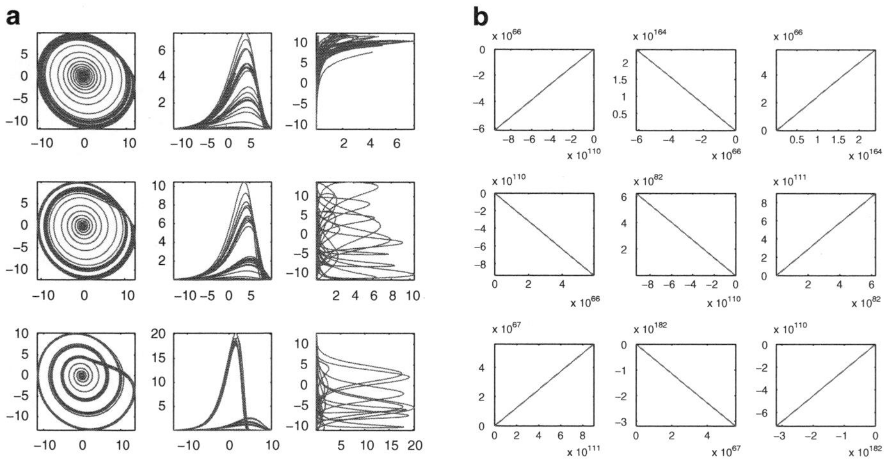

Fig. 8.15 (**a**) Heterogeneous coupling with very low value of C and (**b**) Heterogeneous coupling with a high value of C

respectively. For such a coupling the chaotic state is present only when the value for the coupling constant is small enough. This is seen in Fig. 8.16.

As can be observed from the Fig. 8.16, as we increase the coupling between the two networks (which is different from one another), irrespective of the changes made to the coupling inside each, the chaotic state no longer exists. In order

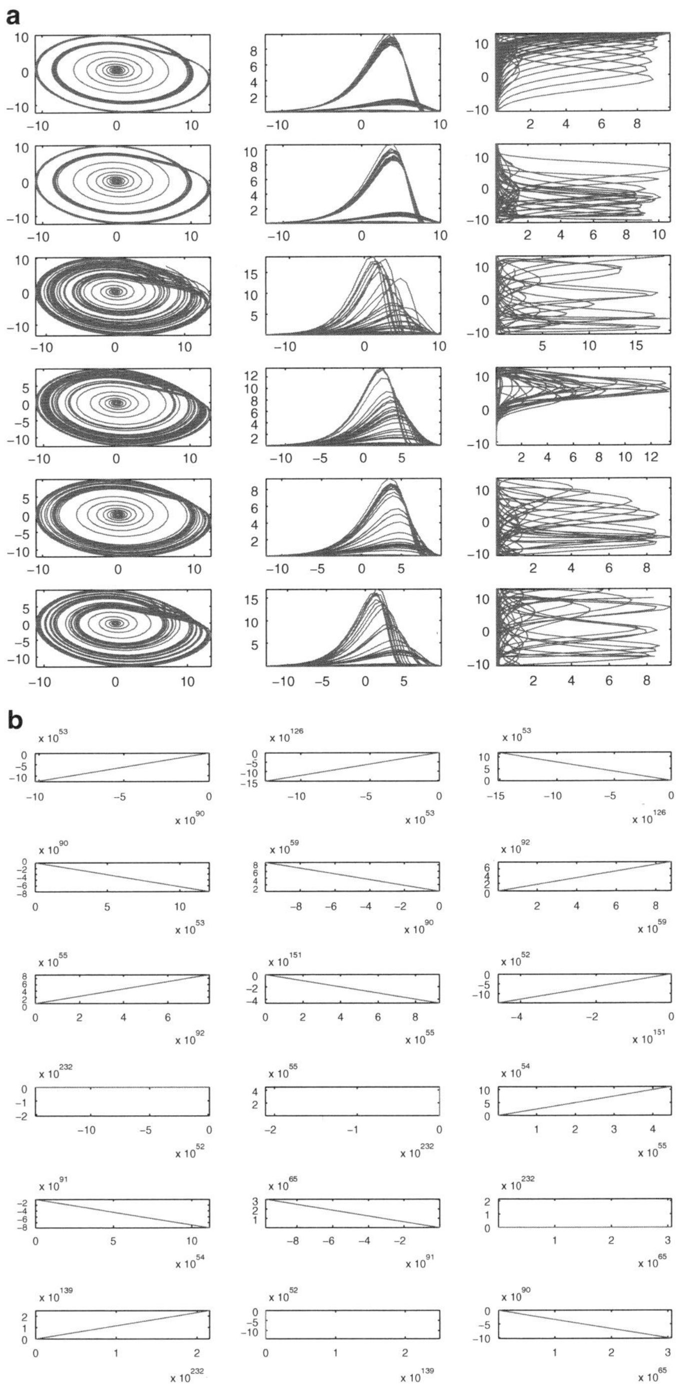

Fig. 8.16 (**a**) Heterogeneous coupling with very low value of C and (**b**) Heterogeneous coupling with a high value of C

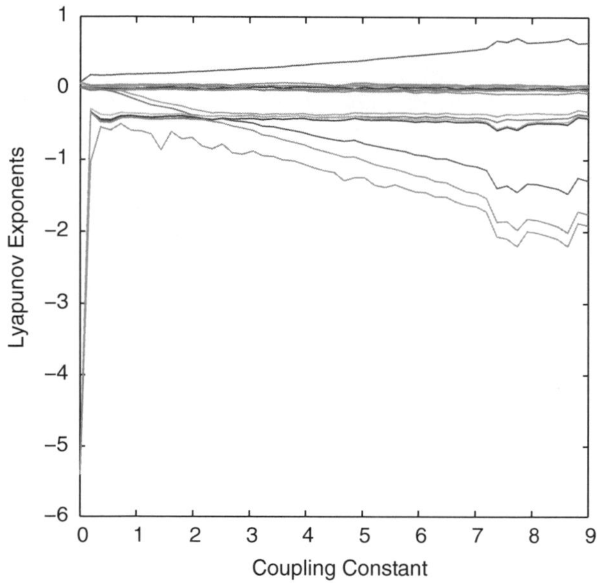

Fig. 8.17 Maximum Lyapunov exponents as a function of the coupling constant

to measure or rather quantify the chaos in the system, the maximum Lyapunov exponents as a function of the coupling constant has been computed for this system as shown in Fig. 8.17.

8.3.3.3 Outer Synchronization of Global Networks Under Homogeneous coupling

This is the coupling between two global networks as shown in Fig. 8.7. Here the coupling is done using the *x, y, z, x, y* and *z*—component of node *1, 2, 3, 4, 5* and *6* respectively. For such a coupling the chaotic state is present only when the value for the coupling constant is small enough. This is seen in Fig. 8.18.

As can be observed from the Fig. 8.18, as we increase the coupling between the two networks (which is different from one another), irrespective of the changes made to the coupling inside each, the chaotic state no longer exists. Fig. 8.19 shows the variation of the maximum Lyapunov exponents with the coupling constant. As can be observed, the majority of them are negative apart from one which is greater than zero.

In the above subsections, the difference between the three types of network topologies does influence their coupling. In the first one with only five nodes, the coupling between each node was the constraint, whereas, for the next two networks, it was the coupling between the networks that played a major role in deciding whether the oscillators in each node are in chaotic state or not.

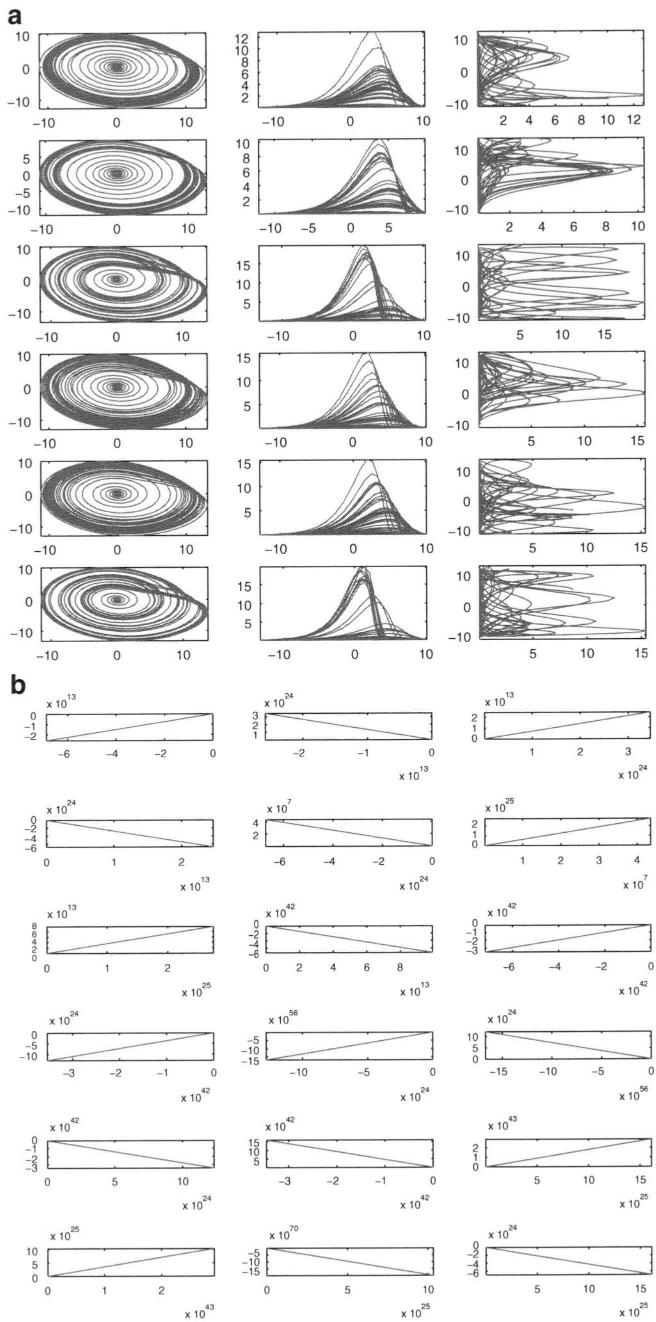

Fig. 8.18 (**a**) Heterogeneous coupling with very low value of C and (**b**) Heterogeneous coupling with a high value of C

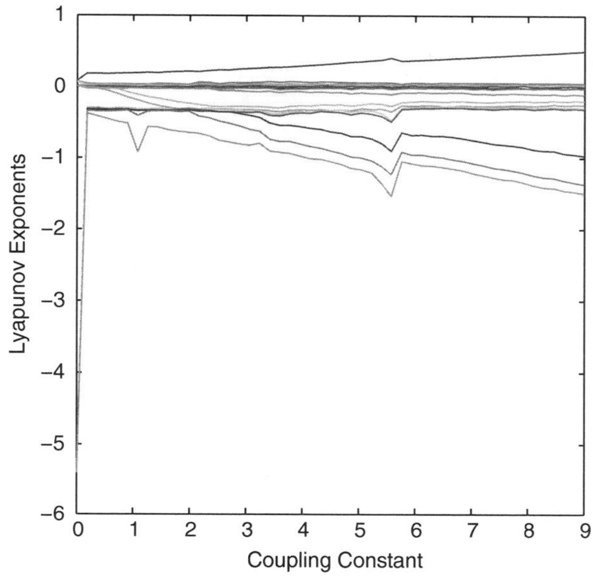

Fig. 8.19 Maximum Lyapunov exponents as a function of the coupling constant

As can be seen from Figs. 8.10, 8.12 and 8.14, for each case, one of the Lyapunov Exponents is positive, one is near zero and the others are negative.

8.4 Experimental Verification

The experiments were conducted in three stages. At first we looked into the different network topologies as shown in Figs. 8.3–8.6 and analyzed them to find out the value of the coupling constant for which they synchronize. Then we constructed the network shown in Fig. 8.8 and used the same method to determine the coupling constant for synchronization. And the last stage was to implement another network as shown in Fig. 8.7 and determine the coupling constant for that.

8.4.1 Inner Synchronization

8.4.1.1 Uni-directional Line Topology

As mentioned earlier four different topologies of networks—line, bidirectional line, star and global—are used here. The uni directional line network has five Rössler oscillators connected in series such that data flows in only one direction i.e the communication between them is one way as shown in Fig. 8.3. The coupling circuit required for such a network has been developed and is shown in Fig. 8.20. The pins

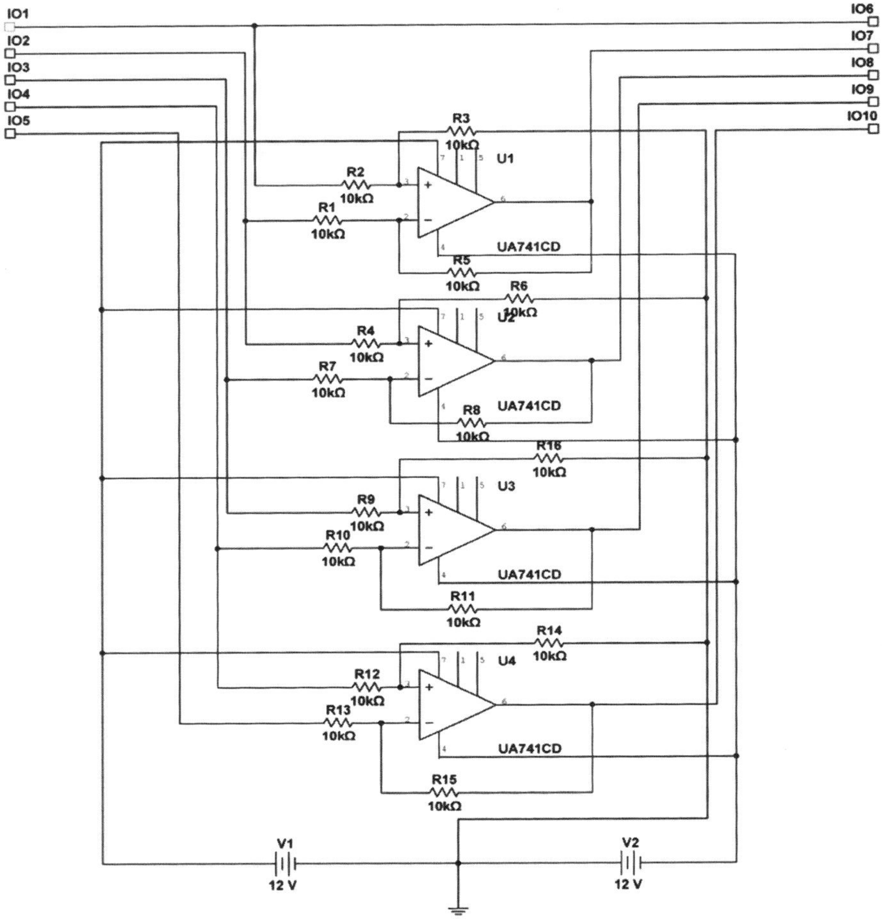

Fig. 8.20 Uni-directional network coupling

on the L.H.S of the figure denotes the output of each Rössler oscillator which is fed into this coupling circuit, which consists of four differential amplifiers. The output of these are fed into the input of the respective Rössler circuits via a resistance known as coupling resistance. This coupling resistance is decided upon by the value of the coupling constant and it is this resistance whose variation takes us to the state of synchronization in the network.

8.4.1.2 Bi-directional Line Topology

The bidirectional line network as shown in Fig. 8.4 has the coupling circuit Fig. 8.21. In this network, the nodes are connected serially i.e connection is between node 1

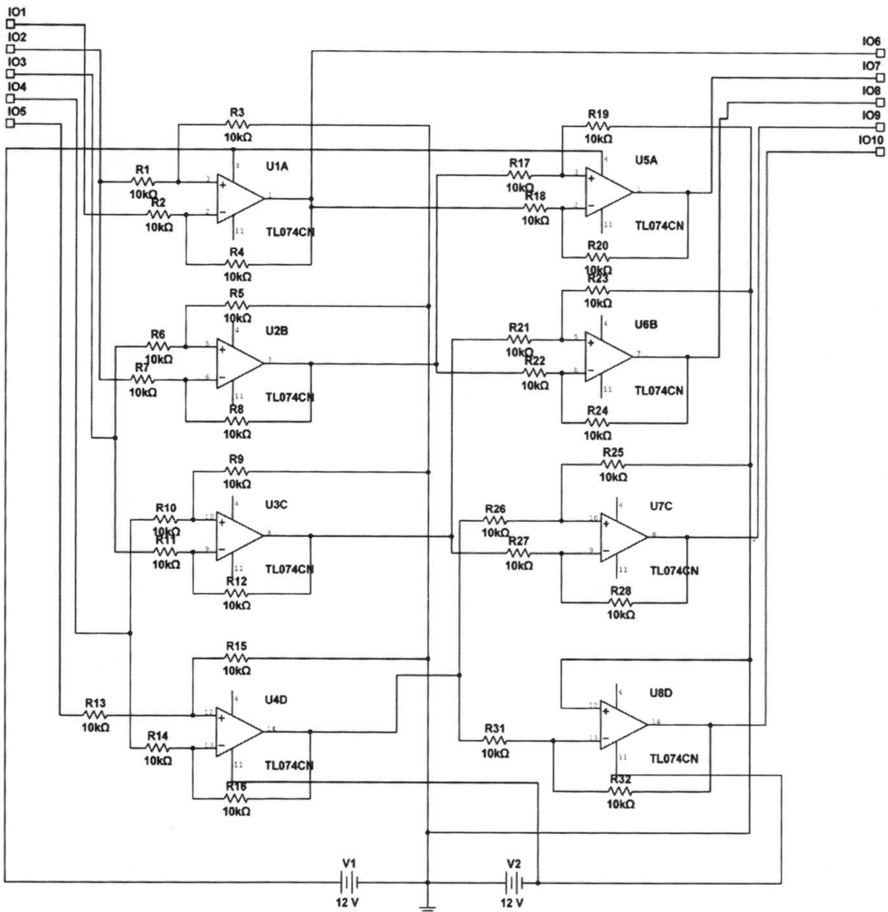

Fig. 8.21 Bi-directional line network coupling

and 2 then 2 and 3 then 3 and 4 and 4 and 5. The data exchange between the nodes is bidirectional, i.e both of them receive and transmit. For this type of network, the output of the coupling circuit given by the pins (IO6 → IO10) are fed into the input of each node (Róssler oscillator) via a coupling resistance whose variation changes the synchronization of the network.

8.4.1.3 Star Topology

Next in line is the Star network as shown in Fig. 8.5. In this network, all the nodes apart from node 1 are connected to node1. So data exchange takes place between 1 → 2, 1 → 2, 1 → 3, 1 → 4 and 1 → 5. The coupling circuit for this type of network is shown in Fig. 8.22. As can be seen, the circuits consists of five differential

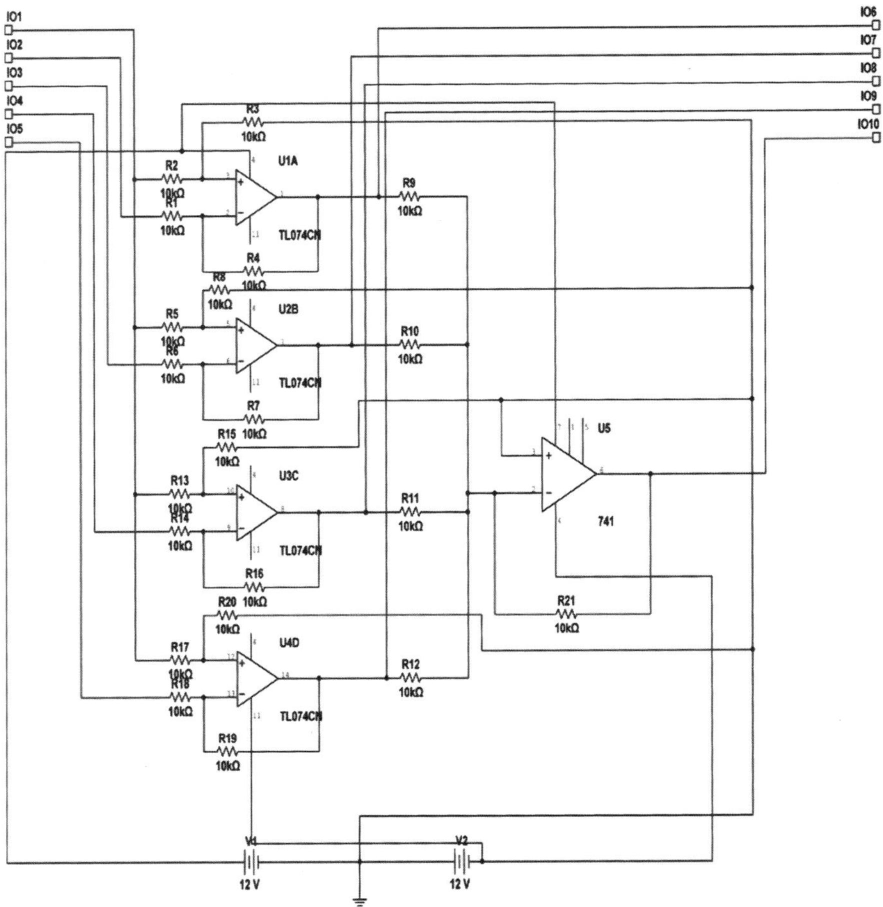

Fig. 8.22 Star network coupling

amplifiers and their output, denoted by pins IO6 → IO10 is fed into the y—input of the corresponding node (Róssler oscillator) via a coupling resistance whose variation changes the state of synchronization of the circuit as shown in Fig. 8.23.

8.4.1.4 Global Topology

Finally it is the Global Network where every node is connected to four others in the network as shown in Fig. 8.6. The coupling circuit is a bit complicated and is shown in Fig. 8.24. As is the case for the other previously mentioned networks, the outputs denoted by pins IO6 → IO10 is fed into the y—input of the corresponding node (Róssler oscillator) via a coupling resistance whose variation changes the state of synchronization of the circuit as shown in Fig. 8.25.

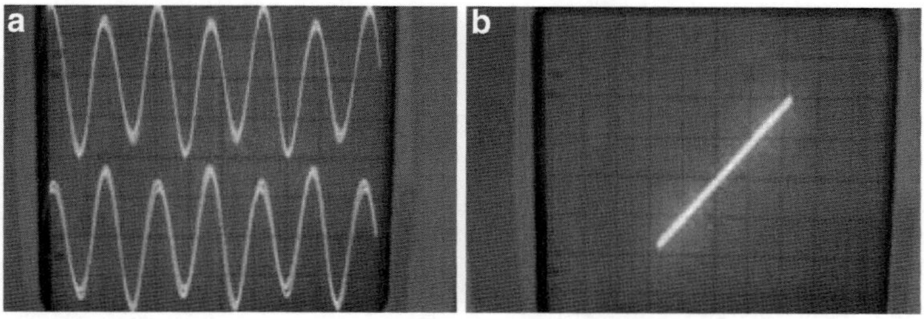

Fig. 8.23 (**a**) Time series of signals from node 4 and 1 and (**b**) Phase space of signals from node 4 and 1

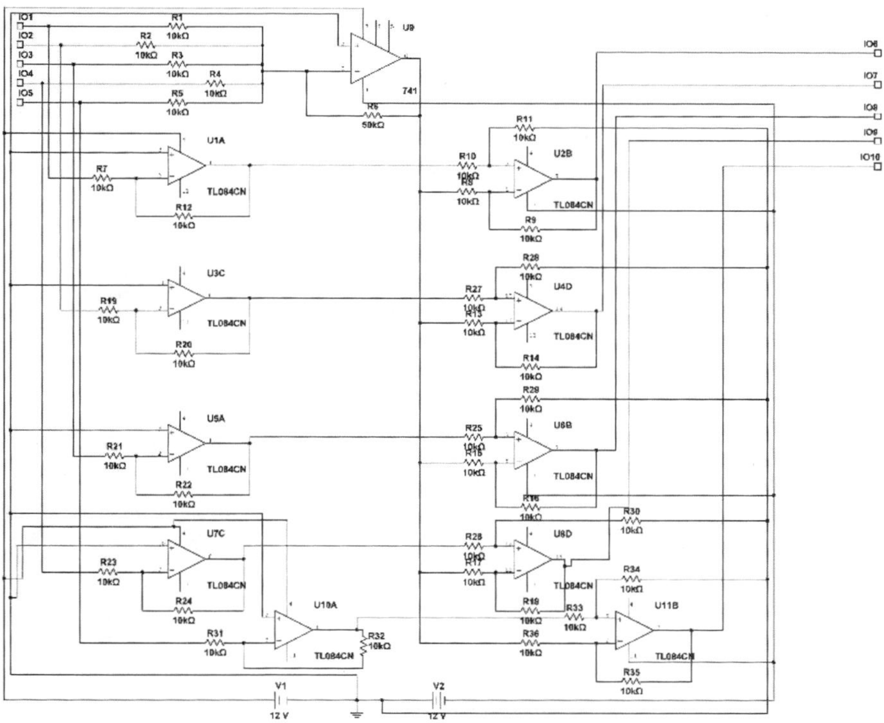

Fig. 8.24 Global network coupling

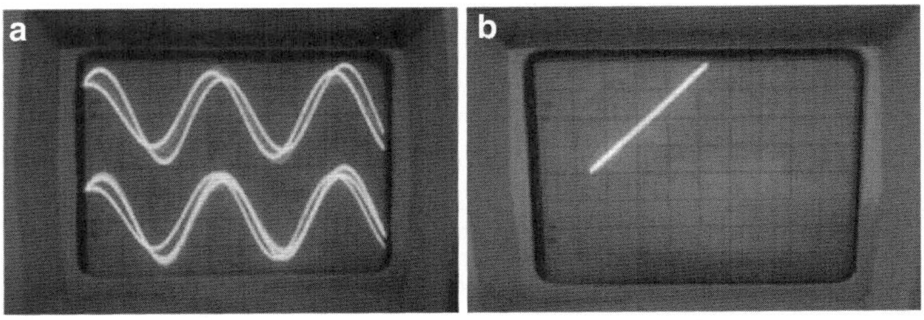

Fig. 8.25 (**a**) Time series of signals from node 3 and 5 and (**b**) Phase space of signals from node 3 and 5

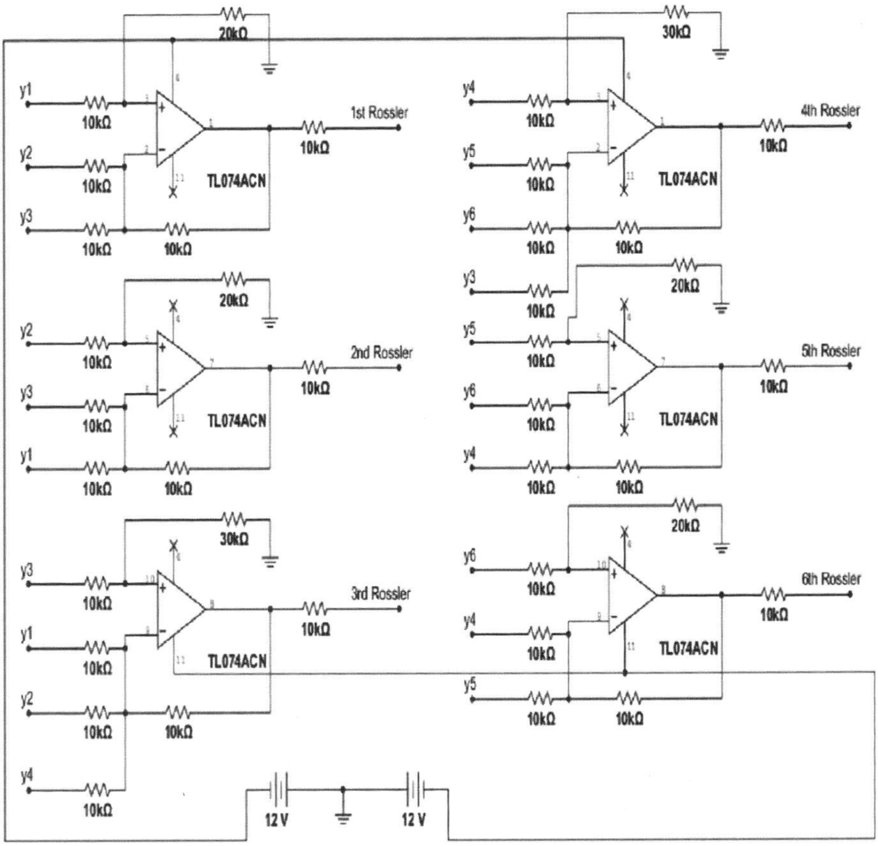

Fig. 8.26 Coupling between two global networks

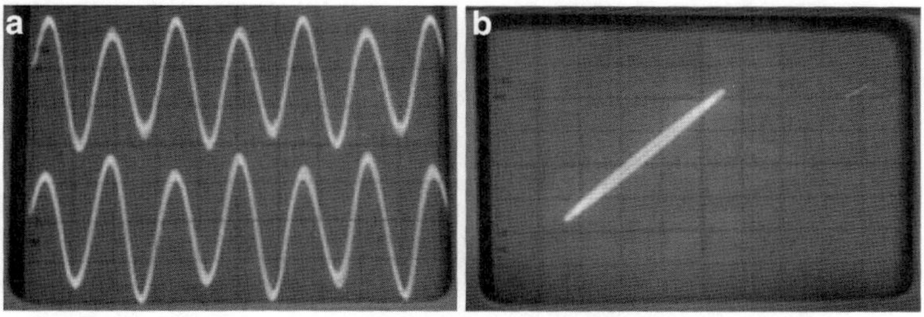

Fig. 8.27 (**a**) Time series of signals from node 1 and 4 and (**b**) Phase space of signals from node 1 and 4

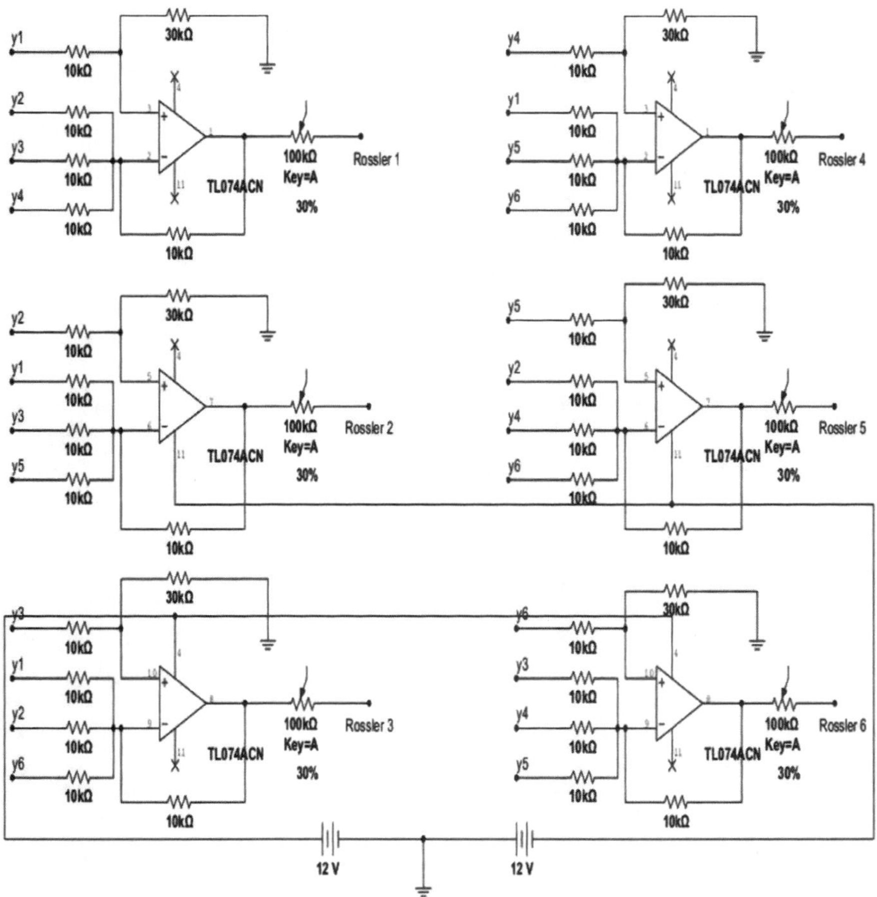

Fig. 8.28 Coupling between two global networks

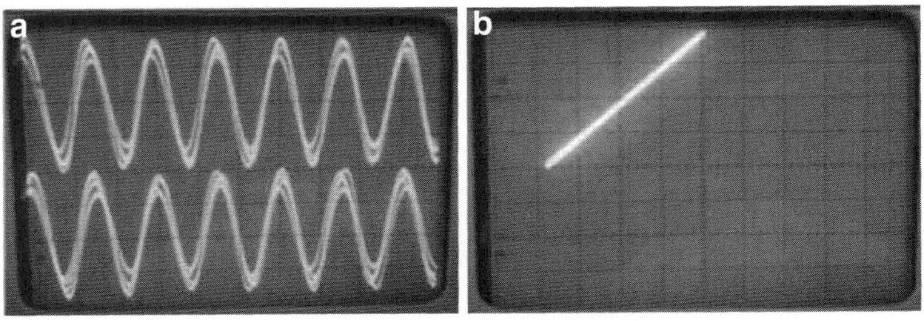

Fig. 8.29 (**a**) Time series and (**b**) Phase space of signals from node 3 of network 1 and node 2 of network 2

8.4.2 Outer Synchronization

8.4.2.1 Heterogenous coupling

Here we took two global networks consisting of three nodes each and then we see the effect of their coupling on their synchronization. The schematic for this network is shown in Fig. 8.8. The two global networks can be coupled using the circuit in Fig. 8.26. The coupling is changed by varying the potentiometers and finally when the value of the coupling resistances (potentiometers) becomes around 33 K, the entire network achieves synchronization as shown in Fig. 8.27.

8.4.2.2 Homogenous Coupling

Finally, we arrive at one to one coupling between two networks, where each node from one network is connected to its counterpart in the second network i.e node 1 of network 1 is connected to node 1 of network 2, node 2 of network 1 is connected to node 2 of network 2 etc. as shown in Fig. 8.7. For this type of coupling, the coupling circuit is given by Fig. 8.28. The variation of the coupling resistance gives us the synchronized state at the value of 10 K. For this, the corresponding output is shown in Fig. 8.29.

8.5 Conclusion

We have studied the synchronization of chaotic networks in three ways (a) a typical global network with 5-nodes, (b) the case of two global networks coupled with two nodes, one form each, and (c) the case of two global networks coupled in a prism like way, as shown in the above figures. Now, these three ways showed us that each coupling attains the state of synchronization while being chaotic. This result is confirmed both numerically and experimentally. During the numerical analysis, we found that among all the Lyapunov exponents for each network, one is positive, one lies near zero and the rest are negative. This result reconfirmed the phenomena—the systems synchronize while each being chaotic. Future investigations will be concerned with the analysis of the systems in presence of noise and effect of heterogenous coupling in derails.

Acknowledgements Author AR is thankful to Council of Scientific and Industrial Research for providing with Senior Research Fellowship and ARC is thankful to University Grant Commission-BSR grant for upgrading the computational facility. He is also grateful for D.S. Kothari Fellowship.

References

1. Watts, D.J., Strogatz, S.H.: Collective dynamics of 'small-world' networks. Nature **393**, 440 (1998)
2. Barabasi, A.L., Albert, R.: Emergence of scaling in random networks. Science **286**, 509 (1999)
3. Albert, R., Barabasi, A.L.: Statistical mechanics of complex networks. Rev. Mod. Phys. **74**, 47 (2002)
4. Newman, M.E.J.: The structure and function of complex networks. SIAM Rev. **45**, 167 (2003)
5. Boccaletti, S., Latora, V., Morenoy, Y., Chavez, M., Hwang, D.U.: Complex networks: structure and dynamics. Phys. Rep. **424**, 175 (2006)
6. Barahona, M., Pecora, L.M.: Synchronization in small-world systems. Phy. Rev. Lett. **89**, 054101 (2002)
7. Jalan, S., Amritkar, R.E.: Self-organized and driven phase synchronization in coupled maps. Phy. Rev. Lett. **90**, 014101 (2003)
8. Nishikawa, T., Motter, A.E., Lai, Y.C., Hoppensteadt, F.C.: Heterogeneity in oscillator networks: are smaller worlds easier to synchronize? Phy. Rev. Lett. **91**, 014101 (2003)
9. Zhou, C.S., Motter, A.E., Kurths, J.: Universality in the synchronization of weighted random networks. Phy. Rev. Lett. **96**, 034101 (2006)
10. Wang, X.F., Chen, G.: Complex networks: topology, dynamics and synchronization. Int. J. Bifurcat. Chaos Appl. Sci. Eng. **12**, 885 (2002)
11. Erdös, P., Renyi, A.: The evolution of random graphs. Publ. Math. Inst. Hung. Acad. Sci **5**, 17 (1960)
12. Zhou, C.S., Kurths, J.: Dynamical weights and enhanced synchronization in adaptive complex networks. Phy. Rev. Lett. **96**, 164102 (2006)
13. Lu, J.H., Yu, X., Chen, G.: Chaos synchronization of general complex dynamical networks. Phys. A **334**, 281 (2004)
14. Li, C.P., Sun, W.G., Xu, D.: Synchronization of complex dynamical network with nonlinear inner-coupling functions and time delays. Prog. Theor. Phys. **114**, 749 (2005)
15. Li, C.P., Sun, W.G., Kurths, J.: Synchronization of complex dynamical networks with time delays. Phys. A **361**, 24 (2006)
16. Li, C.G., Chen, G.: Synchronization in general complex dynamical networks with coupling delays. Phys. A **343**, 263 (20064)
17. Masoller, C., Marti, A.C., Zanette, D.H.: Synchronization in an array of globally coupled maps with delayed interactions. Phys. A **325**, 186 (2003)
18. Marti, A.C., Masoller, C.: Delay-induced synchronization phenomena in an array of globally coupled logistic maps. Phys. Rev. E **67**, 056219 (2003)
19. Marti, A.C., Masoller, C.: Synchronization of globally coupled non-identical maps with inhomogenous delayed interactions. Phys. A **342**, 344 (2004)

20. Sun, W.G., Chen, Y., Li, C.P., Fang, J.Q.: Synchronization and bifurcation of general complex dynamical networks. Comm. Theor. Phys. **47**, 1073 (2007)
21. Li, C.P., Sun, W.G., Kurths, J.: Synchronization between two coupled complex networks. Phys. Rev. E **76**, 046204 (2007)
22. Heagy, J.F., Carroll, T.L., Pecora, L.M.: Synchronous chaos in coupled oscillator systems. Phys. Rev. E **50**, 1874 (1994). doi:10.1103/PhysRevE.50.1874. http://link.aps.org/doi/10.1103/PhysRevE.50.1874

Chapter 9
The Route from Synchronization to Desynchronization of Chaotic Operating Circuits and Systems

Stavros G. Stavrinides and Antonios N. Anagnostopoulos

Abstract In a world of an ever increasing information-trafficking, ultra wideband information transmission, as well as security are main issues. Chaotic oscillators inherently possess these two properties. Consequently, chaotic transmitter-receiver synchronization emerges as a key-topic in networked communication. The issues of synchronization robustness and synchronization stability are of great importance, if one means to turn in advantage properties of chaotic operating circuits and systems. As a result, not only synchronization accomplishment but also the ways a system desynchronizes is of great importance, if not crucial; however, these are not, usually, thoroughly studied. In this contribution a short overview of experimental work dedicated to the ways chaotic-synchronized communication systems desynchronize, is provided.

9.1 Introduction

Both words of *chaos* and *synchronization* are of Greek origin. Chaos "$\chi\acute{\alpha}o\varsigma$" originally meant the state of the universe before the creation of cosmos, later it came to mean a state of confusion and any lack of order. On the other hand synchronization "$\sigma\upsilon\gamma\chi\rho o\nu\iota\sigma\mu\acute{o}\varsigma$" means having a common timing. Both meanings have remained unchanged till nowadays and were destined to characterize a whole section in sciences and engineering; that of nonlinear system behavior and the resulting complexity [1].

S.G. Stavrinides (✉)
Department of Electrical and Computer Engineering, University of Cyprus, P.O. Box 20537, 1687 Nicosia, Cyprus
e-mail: stavros@physics.auth.gr

A.N. Anagnostopoulos
Physics Department, Aristotle University of Thessaloniki, 54124 Salonica, Greece
e-mail: anagnost@physics.auth.gr

S. Banerjee and L. Rondoni (eds.), *Applications of Chaos and Nonlinear Dynamics in Science and Engineering 3*, Understanding Complex Systems,
DOI 10.1007/978-3-642-34017-8_9, © Springer-Verlag Berlin Heidelberg 2013

The significance of broadband and/or secure communication is obvious in a world increasingly dependent on information transmission [2–4]. The introduction of chaos theory in telecommunication applications provided this whole branch with a new vision and has already offered several new applications and performance enhancements to existing communication systems [5–7]. Inherent properties of chaotic oscillating circuits are very beneficial to broadband data transmission, wired or wireless, lending at the same time security advantages [8].

A chaotic electronic oscillator is capable of producing non-repeating sequences that possess frequencies with a wide spectral distribution; therefore broadband transmission is an inherent property of such systems [9]. Since, it is very hard to predict chaotic behavior and signal sequences even when the chaotic function is known to the interceptors, secure data transmission is another inherent property of such systems. This is because slightly different estimation of the initial condition and control parameter sensitivity, would lead to a very different chaotic sequence. Two key features of chaos are a fairly "noise-like" time-series and a crucial dependence on initial conditions and control parameters, which allow for low probability of detection and interception, in the case that transmission utilizes chaotic information-bearing signals, respectively [10–14].

On the other hand, chaotic communication systems are simpler, concerning circuit realization, compared to traditional spread spectrum systems [6, 15, 16]. The key fact in chaos-based communication systems is that the procedure used by the transmitter to generate the chaotic waveform is deterministic; the knowledge of this procedure by an authorized receiver allows him to replicate, or *synchronize*, the chaotic waveform, and then to recover the message by subtracting the chaotic carrier [17, 18]. The confidentiality of the encryption technique is based on the difficulty to reproduce the chaotic carrier signal if an intruder does not know the particular dynamical system used [19].

Such a telecommunication system, consisting of two nonlinear electronic circuits in synchrony, must be endowed with two additional attributes:

The first is that the synchronization property has to be stable. This means that it has to be immune against small deviations of the matching between transmitter and receiver and additionally, it has to be robust against the influence of certain noise levels, unavoidable in any communication system. It should be noted that circuit or system matching, although desired, must be kept within some limits to avoid easy synchronization of even non-identical circuits, damaging thus the security property of the system.

The latter is robustness and it is of great importance not only because noise is unavoidable, but also because the system has to parry intentional attacks by malevolent signals trying not to listen in but to destroy the transmission. As a result, the transitions (routes) from complete synchronization to complete loss of it, termed complete desynchronization, in chaotic communication circuits and systems have to be studied.

In this contribution, a presentation of real chaotic synchronized circuits demonstrating, next to their synchronized state, an interesting behavior of synchronization to desynchronization routes, appears. The chapter structure is as follows: At the first part, a rather abstractive presentation of nonlinear circuits and synchronization is provided. The rest of the chapter is devoted to the description of circuits exhibiting synchronization to desynchronization routes. Finally, the last part deals with the presentation of experimental results, the study and the evaluation of phenomena regarding these routes, such as "*On-Off*" and "*In-Out*" intermittency.

9.2 Nonlinear Circuits

Most electrical and electronic circuits could establish, under certain conditions, either a periodic—semi periodic or a chaotic behavior. This is due to the fact, that almost all passive and active circuit components demonstrate some kind of nonlinear feature, giving rise to what is called strange behavior (complex behavior) [7,20–22].

Such strange behaviors have been faced from the very beginning of electronic engineering and have been tackled as undesirable ones; in fact, this meant that circuit design had to avoid such unexplainable operational behaviors. The first ever reported abnormal circuit behavior was in 1927 in the Van der Pol—Van der Mark oscillator. This externally driven neon vacuum tube based oscillator exhibited an *irregular noise*, according to the circuit's designers [23].

The study of complex phenomena and chaos in real electronic circuits (not modeling of nonlinear differential equations by analog computers [24]) started during the 1980s. In the beginning, the study regarded simple non-autonomous circuits, such as R-L-Diode circuits and the first phenomenon reported was the period doubling route to chaos [25–27]. Autonomous circuits appeared later on, like the famous Chua circuit; a well-known circuit that provides the scientific community with a swarm of chaotic behaviors [28, 29].

An important issue in nonlinear circuits is the reason why an electronic circuit (a dynamical system in general) exhibits a complex behavior. The answer is that this is due to nonlinear behavior of active (diodes, transistors, saturation operating op amps etc.) and/or passive components (capacitors, inductors) [30].

Electronic chaotic oscillator design and their synchronization is a very motivating part of chaotic electronics, which nowadays, attracts the interest of the engineering community. Chaotic oscillators are either 3rd order autonomous oscillators or 2nd order non-autonomous ones. In most cases, of course, greater order oscillators could also exhibit chaos and sometimes hyper-chaos [31]. The first are operating by themselves, while the latter need an external stimulus. In any case a simple harmonic oscillator could come to exhibit a chaotic behavior by introducing a "local annoyance" in terms of importing in the circuit design a nonlinear element or a kind of memory or some form of hysteresis [32].

Nowadays, nonlinear circuits and systems and the study of all related (chaotic) phenomena, unveils a very rigorous potential, with a perspective of many real-world applications in the area of electronics, especially telecommunications, targeting etc. [18, 33]. As a result there is great interest in designing electronic circuits capable of producing chaotic signals, in an easily controlled way, suitable for data modulation and demodulation over a deterministic chaotic carrier [34].

Electronic nonlinear circuits exhibiting chaotic behavior are not only studied for their real and potential application in wideband and secure data transmission, but they also represent very good experimental implementations of nonlinear equations or dynamical systems with chaotic behavior. They are cheap, easy to construct and if damaged easy to repair. Moreover, just a power supply and an oscilloscope are sufficient for monitoring their performance, in most of the cases. Among their other advantages are the convenience to fine-tune their control parameters and the reproducibility of the circuits. All these merits are not to be underestimated, since even the simplest mechanical oscillators cannot be realized so easily, not to mention that they cannot be precisely controlled as the electronic circuits [31].

Experimental synchronization for communication purposes and its reverberation on chaotic electronic circuit design and design optimization accordingly, are a very attractive issue in the scientific and engineering world [35].

9.3 Synchronization

Systems exhibiting periodic oscillations were the first that the phenomenon of synchronization was observed. Motion synchronization between weakly coupled clocks (hanged from the same wooden rod) was first reported by Huygens in 1665 and it was considered as "a phenomenon of sympathy between two clocks" [36]. Later on, it was revealed that synchronization also existed in the case of systems with irregular (i.e. chaotic) oscillations [37].

In general, tuning up the oscillations of two or more interacting dynamical systems is called synchronization. More specifically, chaotic synchronization is a process in which two or more chaotic dynamical systems adjust their oscillations or a given property of their oscillations, to a common behavior, due to a coupling or to a forcing. The synchronized systems could be either identical or equivalent or different [35, 37]. It should be mentioned that synchronization differs from resonance, since in synchronized systems (periodically or irregularly-chaotically oscillating) each system retains its own dynamics [37].

Nonlinear oscillator synchronization is a process that is frequently encountered in nature. The ability that nonlinear dynamical systems possess, to synchronize, is a significant property. As a consequence, chaotic system synchronization is encountered in a variety of scientific fields from sciences and engineering (astronomy, meteorology, electronic engineering etc.) to biological, social and economic sciences (econophysics, epidemiology etc.) [38–40].

Chaotic synchronization between nonlinear dynamical systems is somehow unanticipated, taking into account their main characteristic, that of sensitive dependence on the initial conditions. This feature implies exponential growth of initial state deviations, even in the case of identical dynamical systems, getting them uncorrelated in the course of time [41]. However, it has been experimentally shown that synchronization is possible for chaotic dynamical systems [42, 43].

From the most general aspect, there are two coupling configurations, namely unidirectional and bidirectional one. Each one could lead both the coupled systems to synchronized states, by using quite different mechanisms. These mechanisms are different so far that, till now, no method has ever been proposed to link the two cases or to reduce one process to another [37, 44].

The first case, i.e. unidirectional coupling, is a typical master-slave configuration, where two subsystems form a system. One subsystem acts as the drive while the other one acts as the response; meaning that the driving system evolves freely, nevertheless driving the evolution of the response system.

In the second case, that of bidirectional coupling, both subsystems are coupled with each other in such a way that they interact and mutually influence their behavior. The main idea behind this kind of coupling is that the coupling factor induces an adjustment of the orbits onto a common manifold, where the synchronization manifestoes itself; thus, inducing a mutual synchronization behavior in both subsystems. Occasional escapes of the trajectories from this manifold correspond to synchronization loss, representing the cases of incomplete synchronization.

Both coupling schemes are used in electronic chaotic circuits with typical applications in ultra-wideband communications with chaos or cryptography. However, the bidirectional coupling is encountered very often in nonlinear optics e.g. coupled laser system with feedback and biology [45], e.g. between interacting neurons with feedback [46].

Synchronization ranges from complete agreement of trajectories, where coordinates of different subsystems coincide, to locking of phases. During the last 20 years, many different synchronization states have been reported and studied such as Complete and Generalized Synchronization or Phase and Lag Synchronization [37, 41].

Complete Synchronization was the first discovered and is the simplest form of synchronization in chaotic systems. This type of behavior can usually be expected when identical subsystems are involved. In this case, although, each subsystem operates in a chaotic way, there is complete agreement of trajectories and corresponding coordinates, of different subsystems coincide. This is achieved by means of a coupling signal, in such a way that they remain in step with each other in the course of the time. This mechanism was first shown to occur when two identical chaotic systems are coupled unidirectionally, provided that the conditional Lyapunov exponents of the subsystem to be synchronized are all negative [8].

If the produced oscillation trajectories are totally different, up to even in the slightest detail, then the so called *complete desynchronization* state arises and the oscillators are considered as *completely desynchronized*. However, between these two extreme states a lot of other cases exist. Some of the most common and extensively studied are cited below.

Generalized Synchronization, as a notion, goes further in using different physical systems. The idea of Generalized Synchronization has been introduced in 1995, aiming in treating synchronization between non-identical subsystems. In this case the output of one system is associated to a given function of the output of the other system and this functional relation holds for all time [47].

Phase Synchronization appears when coupled non-identical oscillatory systems can reach an intermediate mode of operation, where a locking of the phases is established, while correlation in the amplitudes remains weak. Transition to Phase Synchronization was first reported in two coupled Rössler dynamical subsystems [37].

Lag Synchronization is an intermediate stage between Phase and Complete Synchronization. The two outputs of each subsystem lock their phases and amplitudes, but with the presence of a time lag. Practically, output signals are identical in their form, but shifted in respect to each other by a constant time interval [41, 48].

Inverse Synchronization appears when signals are identical in shape but with their amplitudes inverted [48].

Projective synchronization arises when signals from each sub-system are identical in shape but their amplitudes differ by a constant factor. In this case there is a similitude under scale [49].

In the case that two nonlinear oscillating systems are forced to an interruptible coupling i.e. in the case of electronic circuits via a switch opening and closing according to some given law, then *impulsive synchronization* is established [50, 51].

Another aspect in the synchronization of chaotic systems is the way this is established or lost. This happens by varying a system control parameter, which could either be the synchronized-system coupling factor or a parameter mismatch between the coupled sub-systems. When the control parameter falls short of a critical value, the synchronized state becomes unstable and characteristic dynamics are observed. The difference between two corresponding variables (signals) from each subsystem blows occasionally from its almost zero value, exhibiting desynchronization. This state is a state of *incomplete synchronization*, an intermediate state declaring the system's synchronization stability.

The mechanism behind the incomplete synchronization of chaotic systems is the competition between the inherent instability of trajectories, due to nonlinear elements and the synchronization tendency, due to coupling. As a result, coupling strength plays a crucial role in the evolution of the phenomenon [51, 52].

Depending on the way that this incomplete synchronization behaves, two similar, but not identical kinds of transition—intermittent synchronization—namely the "*On-Off*" and the "*In-Out*" intermittent synchronization, have been reported [52].

For the experimentalist, in the case of "*On-Off*" intermittent synchronization the synchronization phase portrait appears to escape from the diagonal in a rather random-like way, without favoring any particular direction. When full desynchronization is achieved then most of the synchronization phase space is covered by trajectories in a deterministic chaotic way.

On the other hand, for the theorist this kind of synchronization presents a laminar length distribution (for each value of the control parameter) that scales with the following law [53, 54]:

$$P(\tau) \propto a \cdot \tau^{\beta} \cdot e^{(-\delta\tau)} \tag{9.1}$$

and holding for exponent β the value of $\beta = -1.5$. As a result, (9.1) forms a straight line with a slope $\beta = -1.5$, in a double logarithmic plot. It should be noted that, in the case of incomplete synchronization, laminar lengths are defined as the time intervals that the system is synchronized. At the same time, scaling of the mean laminar lengths $< \tau >$ with the difference of the synchronization control parameter (q) from its critical value (q_{crit}) holds the following power law [53, 54]:

$$\langle \tau \rangle \propto (q - q_{crit})^{\gamma} \tag{9.2}$$

The exponent γ possesses the value of $\gamma = 1$. The control parameter critical value q_{crit} is defined as the one that onset of the incomplete synchronization phenomenon, in this case the "*On-Off*" intermittency, takes place [52–54].

The case of "*In-Out" intermittency* is a more general version of "*On-Off*". In this case trajectories also escape occasionally from the relatively steady synchronized state to irregular "burst" states and then quickly return back to synchronization. But this time the system attractor blows out from a lower dimensional subspace in a random-like way, due to transverse instability. As a result, the experimental synchronization phase portrait escapes from the diagonal, but this time not directing to almost all possible directions, but to a certain and discrete structure.

Theoretically, in this kind of incomplete synchronization, the power law that holds for the laminar length distribution (for each value of the control parameter) has the following form [52, 55]:

$$P(\tau) \propto \overbrace{\underbrace{\kappa \cdot \tau^{\beta} \cdot e^{(-\delta\tau)}}_{In-Off} + \lambda \cdot e^{(-\zeta\tau)}}^{In-Out} \tag{9.3}$$

As depicted on (9.3) in "*In-Out*" intermittency, laminar length distribution in double logarithmic plot, demonstrates the well-known linear part with a slope $\beta = -1.5$ (like in the "*On-Off*" case)—due to first term of (9.3); while the second term of (9.3) causes an identifiable shoulder developed at large values of laminar lengths. Regarding the mean laminar length $< \tau >$ scaling according to the difference of synchronization control parameter (q) from its critical value (q_{crit}), it follows the power law in (9.2) [52, 55, 56].

In electronic circuits, simple resistive coupling is the most common way of synchronizing them, thus this kind of coupling being bidirectional. Unidirectional coupling is established by using buffers or amplifiers in the coupling branch. Direct coupling is another case that is used in unidirectional coupling and is applied in synchronization of chaotic electronic circuits, through cable or RF. As a result numerous circuits suitable for applications in communications with the use of synchronized coupled chaotic circuits have appeared over the last 20 years.

9.4 Nonlinear Circuits Demonstrating Desynchronization Routes

As already mentioned, synchronization between nonlinear systems and consequently between nonlinear circuits is a topic of great interest and with a potential of real world applications. However incomplete circuit synchronization also attracts the attention of designers, since it is an issue that critically characterizes synchronization robustness, and thus the efficiency of communication systems.

In the lines that follow, a presentation of three real coupled chaotic circuit systems, experimentally demonstrating phenomena of incomplete synchronization, appears; namely, a third-order double-scroll circuit-based system [57–59], in two coupling schemes (uni- and bi-directional), a fourth order chaotic circuit-based system [60], also exhibiting a double-scroll attractor and a simple coupled PLL system [61].

These nonlinear circuits have been studied in the ways that they desynchronize i.e. the routes that they follow when they abandon synchronized operation to full desynchronization.

9.4.1 Double-Scroll Coupled Circuits

The system of two identical nonlinear, autonomous, double-scroll circuits, generally coupled either unidirectionally (master-slave configuration) or bidirectionally, is a noteworthy case, demonstrating both the previously mentioned synchronization to desynchronization phenomena: the "*On-Off*" and the "*In-Out*" intermittency [57–59].

The synchronized, coupled system, in general, appears in Fig. 9.1. It consists of identical third-order, double-scroll, chaotic circuits. The coupling parameter ξ, between the two circuits, can vary depending on the circuitry utilized in each case.

The normalized system equations are presented in (9.4):

$$\frac{dx_1}{dt} = y_1$$

$$\frac{dy_1}{dt} = z_1 + \xi_1(y_2 - y_1)$$

$$\frac{dz_1}{dt} = -ax_1 - ay_1 - bz_1 + c \cdot f(x_1)$$

$$\frac{dx_2}{dt} = y_2$$

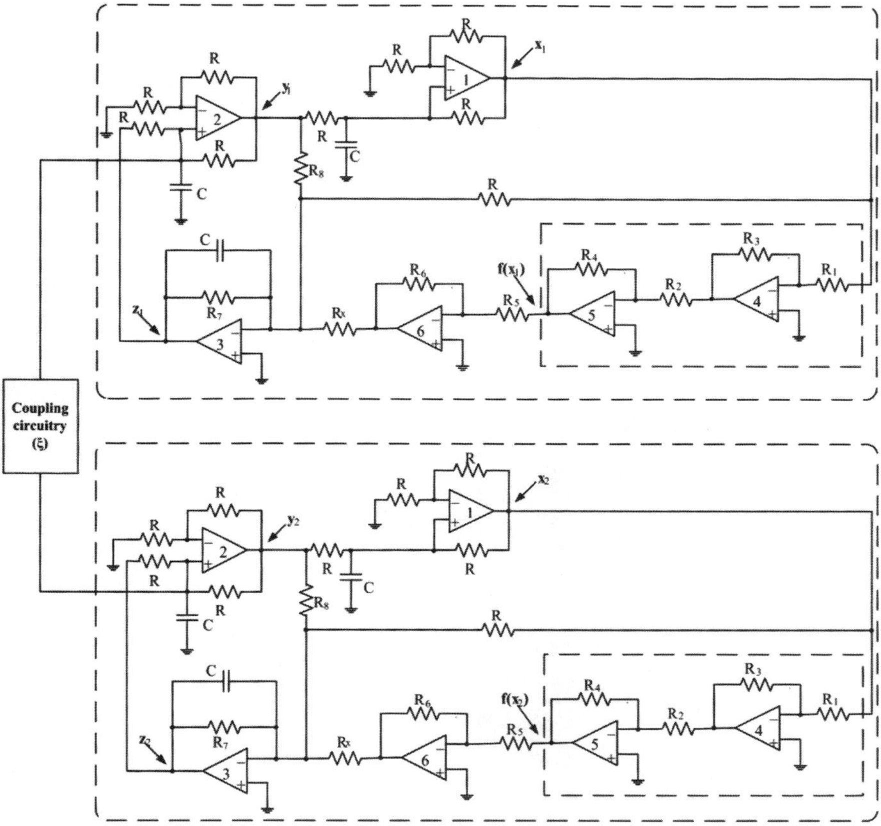

Fig. 9.1 The system of the coupled double-scroll circuits

$$
\begin{aligned}
\frac{dy_2}{dt} &= z_2 + \xi_2(y_1 - y_2) \\
\frac{dz_2}{dt} &= -ax_2 - ay_2 - bz_2 + c \cdot f(x_2)
\end{aligned}
\tag{9.4}
$$

where α, b and c, are the system parameters and are defined as follows:

$$
a = \frac{1}{RC}, \; b = \frac{1}{R_7C}, \; c = \frac{1}{R_xC}. \tag{9.5}
$$

It should be mentioned that signals $x_{1,2}$, $y_{1,2}$, and $z_{1,2}$, represent the voltages at the outputs of the operational amplifiers numbered 1, 2 and 3 respectively, as it is shown in Fig. 9.1. The first three equations of (9.4) describe the first of the two coupled identical, double-scroll circuits, while the last three describe the second one.

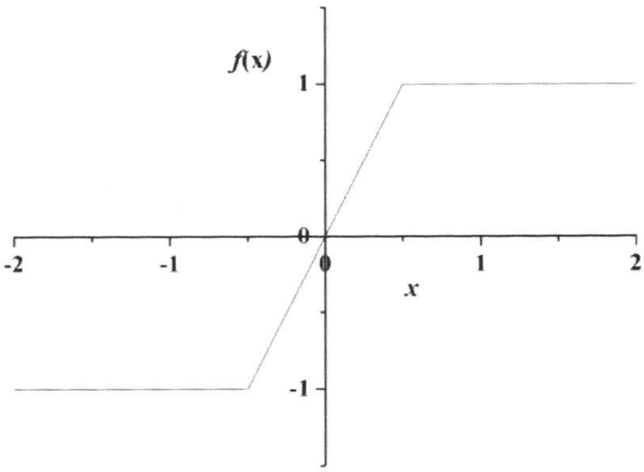

Fig. 9.2 The saturation circuit characteristic curve f(x) vs. x

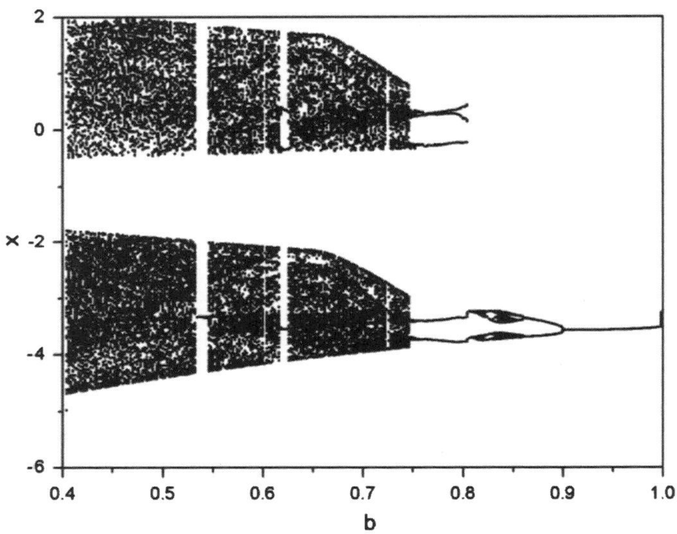

Fig. 9.3 Bifurcation diagram of x vs. b, for $\alpha = 0.5$ and $c = 1$, for each of the double-scroll circuits

The saturation functions f(x_1) and f(x_2) used—(9.4)—are defined by the following expression:

$$f(x) = \begin{cases} 1, & x \geq k \\ \frac{1}{k}x, & -k \leq x \leq k \\ -1, & x < k \end{cases} \tag{9.6}$$

In Fig. 9.2 the characteristic curve of f(x) vs. x is presented. This implementation differs from other similar ones, such as those proposed by Elwakil et al. [62, 63], and Lu et al. [64]. In the present implementation the characteristic demonstrates two saturation plateaus at ± 1 and an intermediate linear part with a slope that equals to 1/k. This implementation offers the advantage of a rather smooth transition between

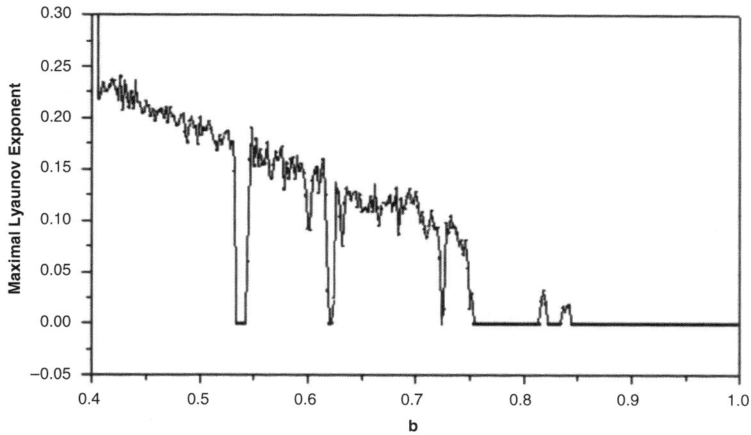

Fig. 9.4 Maximal Lyaunov exponent vs. b, for $\alpha = 0.5$ and c = 1, for each double-scroll circuit

the two plateaus avoiding premature and overdue jumps between them, as it is often the case in the precipitous transition between them. The circuitry implementing the saturation functions is inside the dashed frame in each sub-circuit (op amps 4 and 5). It should be mentioned that the value of parameter k (and therefore the slope value) is decided by the following equation:

$$k = \frac{R_8}{R_7}. \tag{9.7}$$

Finally, the coupling coefficients in each circuit are ξ_1 and ξ_2. They are present at the equations of both circuits (the second and the fifth equation) of equation system (9.4) depending on the kind of coupling i.e. unidirectional or bidirectional one. The value or the functional form of ξ is decided by the circuitry parts used i.e. resistor, memristor, etc.

As it is obvious, all parameter values are determined by the values of the passive components used, so each of the single double scroll circuits operates in a chaotic mode, demonstrating a double scroll attractor. These parameter values could be estimated by utilizing classic numerical analysis i.e. the circuit's bifurcation diagram or the maximal Lyapunov exponent spectrum. In Fig. 9.3 the bifurcation diagram of variable (signal) x vs. system parameter b is presented, while in Fig. 9.4 the corresponding maximal Lyapunov exponent spectrum, depending upon the same parameter b, appears. It is apparent from both figures that for $b = 0.5$ or $b = 0.7$ each double scroll circuit seems to exhibit a rich chaotic dynamic behavior [57–59].

9.4.1.1 Bidirectional Coupling

In the case of resistive bidirectional coupling between the two identical double-scroll circuits, the coupling circuitry (ξ) is just a linear resistor R_C. In this case,

Table 9.1 Parameter and component values in "*On-Off*" synchronization and bidirectional coupling

$\alpha = 0.5$	b = 0.5	c = 1	k = 0.5
$R_1 = 25\,k\Omega$	$R_3 = R_4 = R_5 = R_6 = 1\,k\Omega$	$R_8 = 20\,k\Omega$	$R = 20\,k\Omega$
$R_2 = 14.3\,k\Omega$	$R_7 = 20\,k\Omega$	$R_X = 10\,k\Omega$	$C = 1\,nF$

Table 9.2 Parameter and component values in "*In-Out*" synchronization and bidirectional coupling

$\alpha = 0.5$	b = 0.5	c = 0.8	k = 0.7
R1 = 25 kΩ	R3 = R4 = R5 = R6 = 1 kΩ	R8 = 14.4 kΩ	R = 20 kΩ
R2 = 14.3 kΩ	R7 = 20.4 kΩ	RX = 12.5 kΩ	C = 1 nF

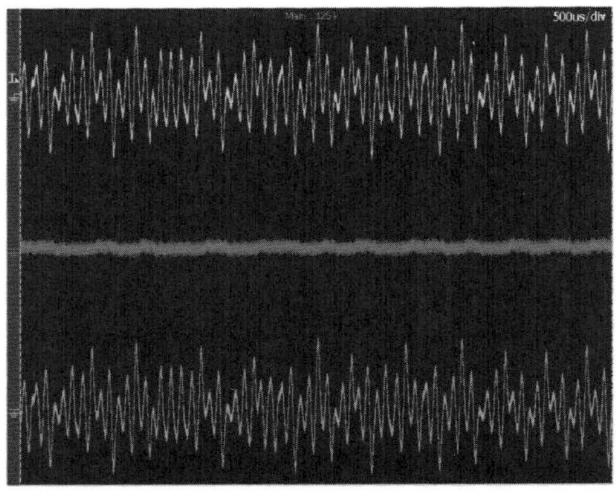

Fig. 9.5 A snapshot of the time-series of each circuit (*upper* and *lower*), in the case of bidirectional coupling. In the middle difference signal appears. Apparently, the system is synchronized

the coupling coefficient is present at the equations of both circuits $\xi_1 = \xi_2 = \xi$ and it is defined as follows:

$$\xi = \frac{R}{2R_C}. \tag{9.8}$$

By using resistance R_C in the denominator of coupling parameter ξ as a control parameter, the system appears to undergo a transition from full synchronization to full desynchronization, by traversing a region of incomplete synchronization [57, 59]. This region is a very interesting one, since various phenomena appear, for different combinations of parameter values.

For two different parameter value sets, the system of the two resistively coupled, double-scroll circuits exhibit either the "*On-Off*" or the "*In-Out*" intermittent synchronization, until finally it fully desynchronizes. In specific, for the values of all passive components and therefore parameters that appear in Table 9.1, the "*On-Off*" transition arises [57]. On the other hand, in Table 9.2 the corresponding component and parameter values, in the case of "*In-Out*" intermittent synchronization, are quoted [59]. It should be noted that in the realized circuits the op amps that were utilized were LF411, while a symmetrical power supply of ±15 V was used; thus the saturation voltages were measured to be ±14.3 V.

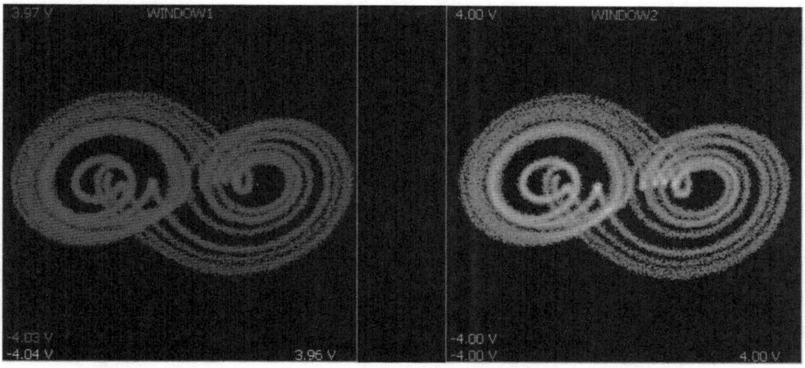

Fig. 9.6 The phase portrait of each double-scroll circuit (*left* and *right*), in the case of bidirectional coupling, for $R_C = 5\,\mathrm{k}\Omega$ ($\xi = 2$)—state of system synchronization

Fig. 9.7 The synchronization phase portrait for $R_C = 5\,\mathrm{k}\Omega$ ($\xi = 2$), with the attractor resting on the diagonal

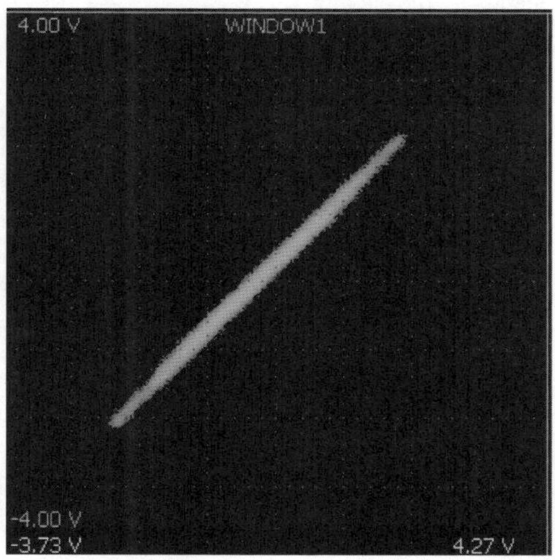

In both cases and as long as $R_C < 8.7\,\mathrm{k}\Omega$ (this means that the coupling parameter ξ possesses a value $\xi > 1.14$) the two double scroll circuits remained synchronized as it is illustrated, by their time-series (y_1 and y_2) and the corresponding difference signal (which is almost zero), in Fig. 9.5. In this case $R_C = 5\,\mathrm{k}\Omega$, thus coupling parameter $\xi = 2$. In Fig. 9.6 the corresponding phase portraits appear, while in Fig. 9.7 the synchronization phase portrait of y_1 vs. y_2 is presented. Synchronization attractor rests on the diagonal, as expected.

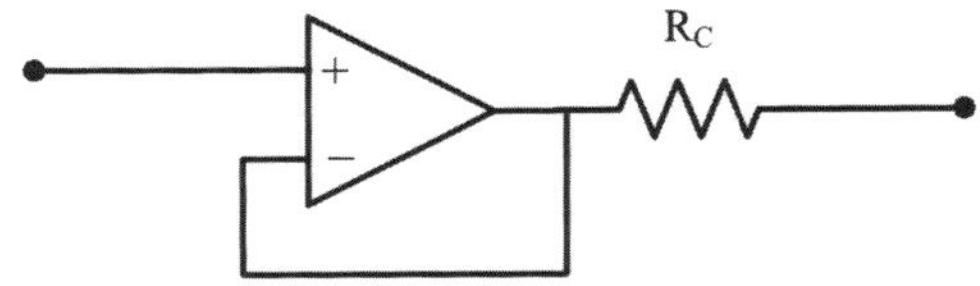

Fig. 9.8 The coupling circuitry of Fig. 9.1, in the case of unidirectional coupling

Table 9.3 Parameter and component values in "*On-Off*" synchronization and unidirectional coupling

$\alpha = 0.5$	b = 0.5	c = 1	k = 0.5
R1 = 25 kΩ	R3 = R4 = R5 = R6 = 1 kΩ	R8 = 20 kΩ	R = 20 kΩ
R2 = 14.3 kΩ	R7 = 20 kΩ	RX = 10 kΩ	C = 1 nF

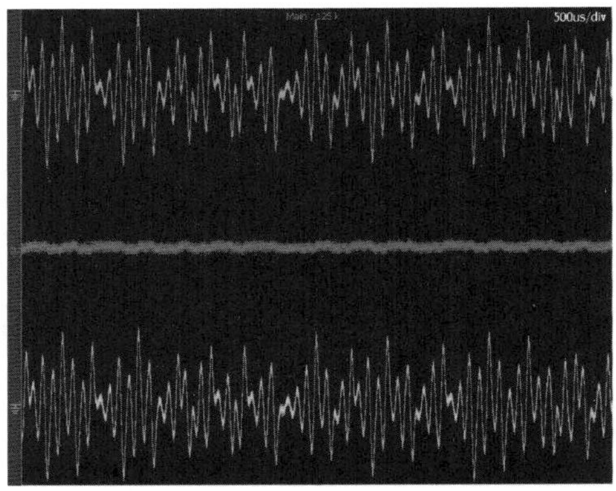

Fig. 9.9 A snapshot of each synchronized double-scroll circuit time-series (*upper* and *lower*), in the case of unidirectional coupling, for $R_C = 5\,\mathrm{k}\Omega (\xi = 2)$. In the middle, difference signal appears

9.4.1.2 Unidirectional Coupling

In the case of resistive unidirectional coupling between the two identical double-scroll circuits, the coupling circuitry (ξ) is not just a linear resistor R_C but there is a kind of insularity between the two double-scroll sub-circuits, by utilizing an op amp based buffer, as indicated in Fig. 9.8 [58]. Apparently the realized circuit system implements a master-slave configuration. In this case, the coupling coefficient is present in the equations of the second (slave) circuit. Thus, in the normalized state equations set—(9.1)—coupling factors ξ_1 and ξ_2 have the values $\xi_1 = 0$ and $\xi_2 = \xi$ and ξ is defined by an expression exactly the same to that of (9.8), in the case of bidirectional coupling.

The values of the elements of the two identical circuits, that ensure chaotic operation for both circuits, are listed in Table 9.3, together with the values of all normalized parameters (α, b, c and k). In the experimental circuit the op amps utilized were LF411, as well. Power supply voltages (positive and negative) were set to $\pm 15\,\mathrm{V}$.

In this kind of coupling, the system remains synchronized for larger values of coupling parameter ξ, compared to bidirectional coupling. Consequently, as long as

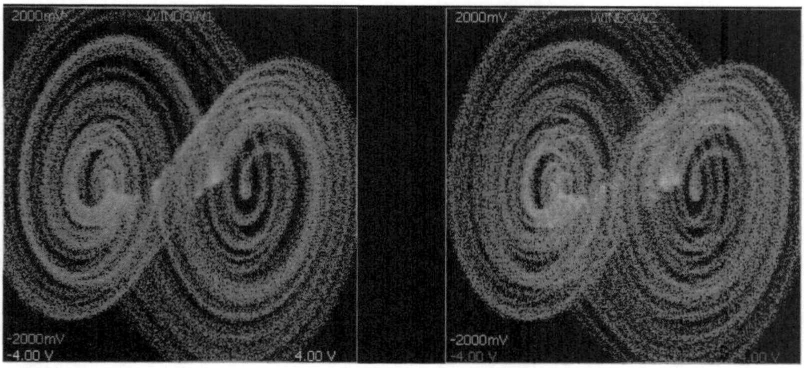

Fig. 9.10 The phase portrait of each double-scroll circuit (*left* and *right*), in the case of unidirectional coupling, for $R_C = 3\,\text{k}\Omega$ ($\xi = 2$)—state of system synchronization

Fig. 9.11 Synchronization phase portrait for $R_C = 3\,\text{k}\Omega$ ($\xi = 2$); the attractor rests on the diagonal

the coupling parameter ξ gets values $\xi > 1.74$—this happens for coupling resistance $R_C < 5.75\,\text{k}\Omega$—the system exhibits a synchronized behavior.

In Fig. 9.9, time-series y_1 and y_2 (upper and lower) and the corresponding difference signal are illustrated. Synchronization between the two double scroll circuits is evident by the almost zero difference signal. In Fig. 9.10 the corresponding (identical) phase portraits appear and in Fig. 9.11 synchronization phase portrait of y_1 vs. y_2 with its attractor resting again on the diagonal, depicting the system's synchronization state [58].

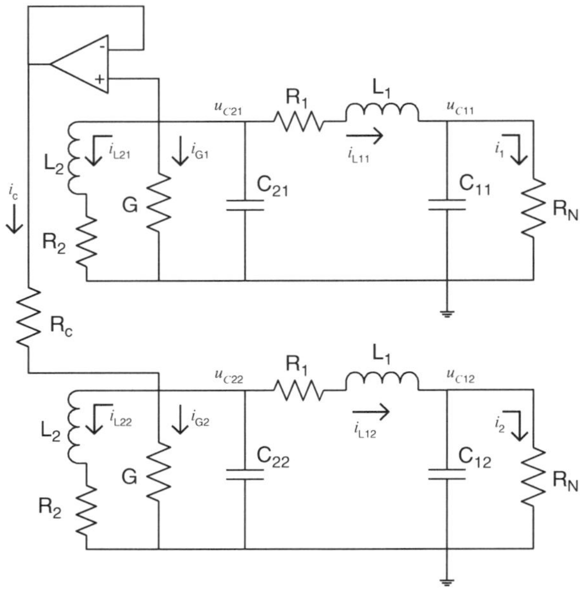

Fig. 9.12 The 4th order circuit based synchronized system

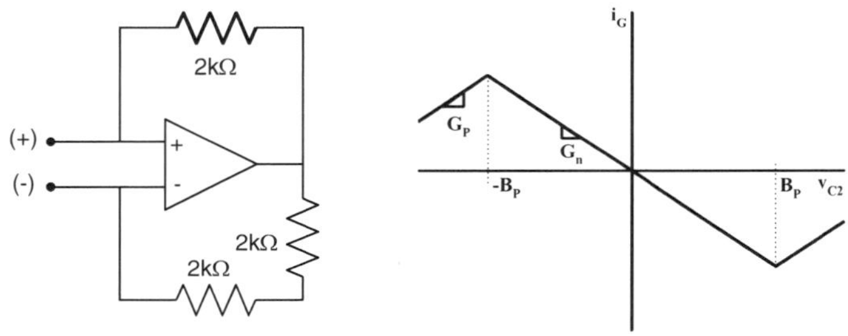

Fig. 9.13 The circuit implementation of negative conductance G and its real $i - v$ (*right*)

9.4.2 4th Order Autonomous Nonlinear Coupled Circuits

The "*In-Out*" intermittent synchronization has also been reported in the case of two initially identical, fourth-order, autonomous, nonlinear circuits, which are unidirectionally coupled, via a linear resistor, as shown in Fig. 9.12 [60]. This circuit also exhibits a double scroll attractor. The coupling is implemented by interconnecting the coupling circuitry between the C_2 nodes (υ_{C2} -coupling) of each sub-circuit. It should be noted that the interconnected circuits are based on a third order autonomous piecewise linear circuit, introduced by Chua and Lin [65, 66], that is capable of realizing every member of the Chua circuit family.

Each sub-circuit includes two active elements: a linear negative conductance G, which circuit implementation is presented in Fig. 9.13, together with the corresponding characteristic curve (on the right) and a nonlinear resistor of N-type R_N, with

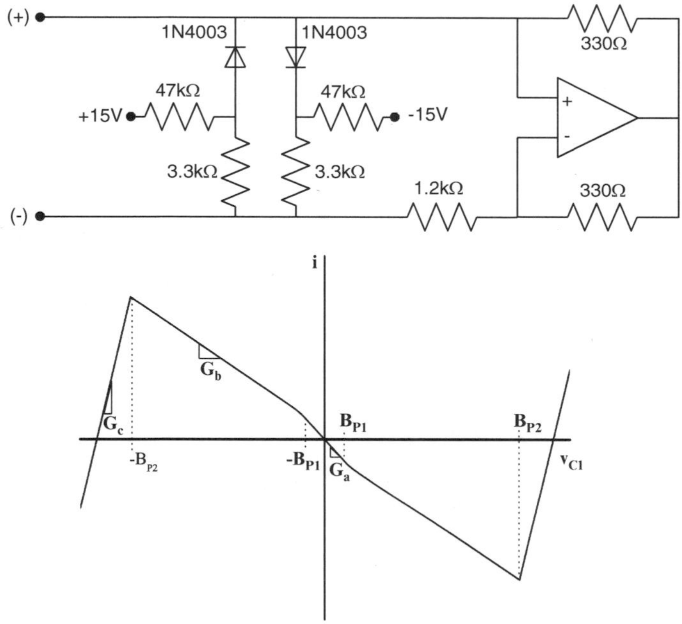

Fig. 9.14 The circuit implementation of nonlinear resistor R_N and its piecewise $i - v$ (*lower*)

Table 9.4 Component values of each synchronized 4th order circuit and the values of all the parameters regarding the active-circuit-elements

Component values	Parameter values
R11 = R12 = R1 = 2.0 KΩ	BP1 = 1.45 V
R21 = R22 = R2 = 109.0 KΩ	BP2 = 10.00 V
L11 = L12 = L1 = 10.2 mH	Ga = −0.83 mS
L21 = L22 = L2 = 21.5 mH	Gb = −0.5 mS
C11 = 6.6 nF	LR = 6.5 V
C21 = C22 = C2 = 5.2 nF	Gp = Gn = −0.5 mS

the corresponding circuit implementation appearing in Fig. 9.14, demonstrating a symmetrical piecewise linear $v - i$ characteristic curve (lower figure).

All circuit elements (regarding the synchronized system illustrated in Fig. 9.12) were considered to have the same values in both sub-circuits, except capacitors C_{11} and C_{12}, since capacitor C_{12} is destined to be used as a control parameter. The exact passive-element values appear in Table 9.4, together with the parameter values of both the active circuit elements (G and R_N). It should be noted that they were kept fixed and were obtained and experimentally verified for the resistance values that appear in Figs. 9.13 and 9.14. Moreover, it should be mentioned that all operational amplifiers used, were LF411 [60].

The coupled circuit system dynamics (Fig. 9.12) are described by set of state equations (9.9). The first four equations describe the driving circuit dynamics, while the other (last) four describe the dynamics of the slave circuit:

$$
\begin{aligned}
\frac{du_{C11}}{dt} &= \frac{1}{C_{11}}\left[i_{L11} - f(u_{C11})\right] \\
\frac{du_{C21}}{dt} &= \frac{1}{C_{21}}\left[-G \cdot u_{C21} - i_{L11} - i_{L21}\right] \\
\frac{du_{L11}}{dt} &= \frac{1}{L_1}\left[-u_{C11} - u_{C21} - R_1 \cdot i_{L11}\right] \\
\frac{du_{L21}}{dt} &= \frac{1}{L_2}\left[u_{C21} - R_2 \cdot i_{L21}\right] \\
\frac{du_{C21}}{dt} &= \frac{1}{C_{12}}\left[i_{L12} - f(u_{C12})\right] \\
\frac{du_{C22}}{dt} &= \frac{1}{C_{22}}\left[\frac{u_{C21} - u_{C22}}{R_C} - G \cdot u_{C22} - i_{L12} - i_{L22}\right] \\
\frac{du_{L12}}{dt} &= \frac{1}{L_1}\left[-u_{C12} - u_{C22} - R_1 \cdot i_{L12}\right] \\
\frac{du_{L22}}{dt} &= \frac{1}{L_2}\left[u_{C22} - R_2 \cdot i_{L22}\right]
\end{aligned}
\qquad (9.9)
$$

where $f(u_{\mathrm{C1j}})$ is provided by (9.10) and represents the current flowing through the nonlinear resistor R_{N}. (j refers to sub-circuit 1,2):

$$
\begin{aligned}
f(u_{\mathrm{C1j}}) = {} & G_C u_{\mathrm{C1j}} + 0.5(G_a - G_b) \cdot \left(\left|u_{\mathrm{C1j}} + E_1\right| - \left|u_{\mathrm{C1j}} - E_1\right|\right) + \\
& 0.5(G_b - G_c) \cdot \left(\left|u_{\mathrm{C1j}} + E_2\right| - \left|u_{\mathrm{C1j}} - E_2\right|\right)
\end{aligned}
\qquad (9.10)
$$

Finally, the unidirectional coupling appears in the 6th equation of (9.9). The coupling resistor possessed a value of $R_{\mathrm{C}} = 5\,\mathrm{k\Omega}$. In this case the synchronized system behavior was studied by bringing about a parameter mismatch between the two sub-circuits, while the coupling parameter remained unchanged. As long as capacitor C_{12} (slave circuit) had the same value to that of $C_{11} = 6.60\,\mathrm{nF}$ (master circuit), the two identical sub-circuits remain fully synchronized.

In Fig. 9.15 the time-series u_{C11} (upper) and u_{C12} (lower) of the driving and the driven sub-circuits, together with their difference (in the middle) are presented.

Note the zero difference between the corresponding u_{C1j} signals, indicating a state of synchronization. In Fig. 9.16 the corresponding identical phase portraits u_{C11} vs. u_{C21} (left), of the master circuit, and u_{C12} vs. u_{C22} (right), of the slave circuit, are shown, respectively. As expected in the case of complete synchronization the time-series and the attractors of both the sub-circuits are identical and the difference signal is continuously almost zero. This is further confirmed in Fig. 9.17, where the synchronization phase portrait (u_{C11} vs. u_{C21}) is illustrated [60].

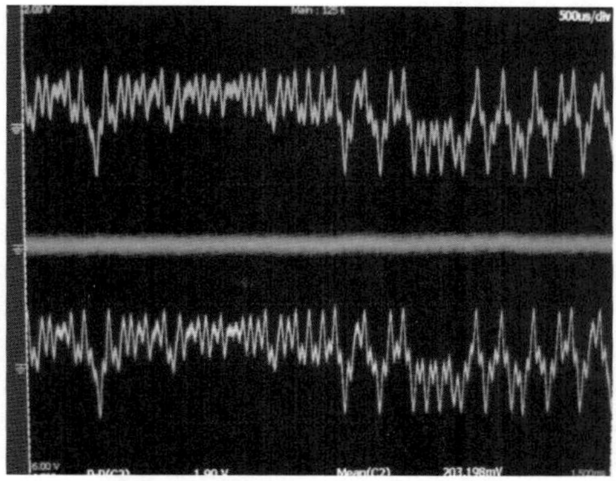

Fig. 9.15 A snapshot of each 4th order circuit time-series u_{C1j} (*upper* and *lower*), in the case of exact parameter matching. In the middle a zero difference signal appears, indicating a synchronized state of operation

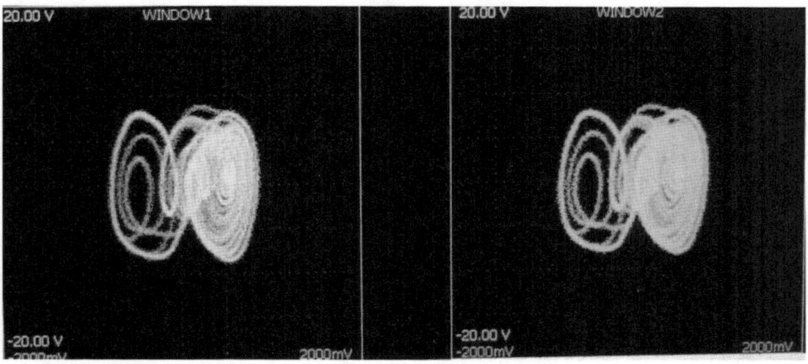

Fig. 9.16 Phase portraits of both the master (*left*) and the slave (*right*) circuits, in the case of exact parameter matching

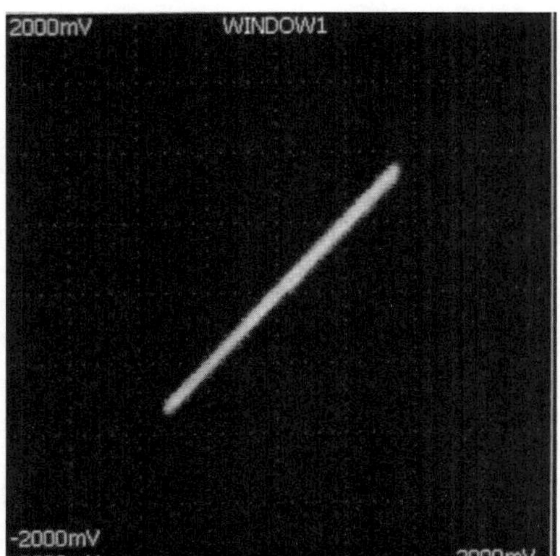

Fig. 9.17 The corresponding to Figs. 9.15 and 9.16 synchronization phase portrait; the attractor rests on the diagonal depicting synchronization

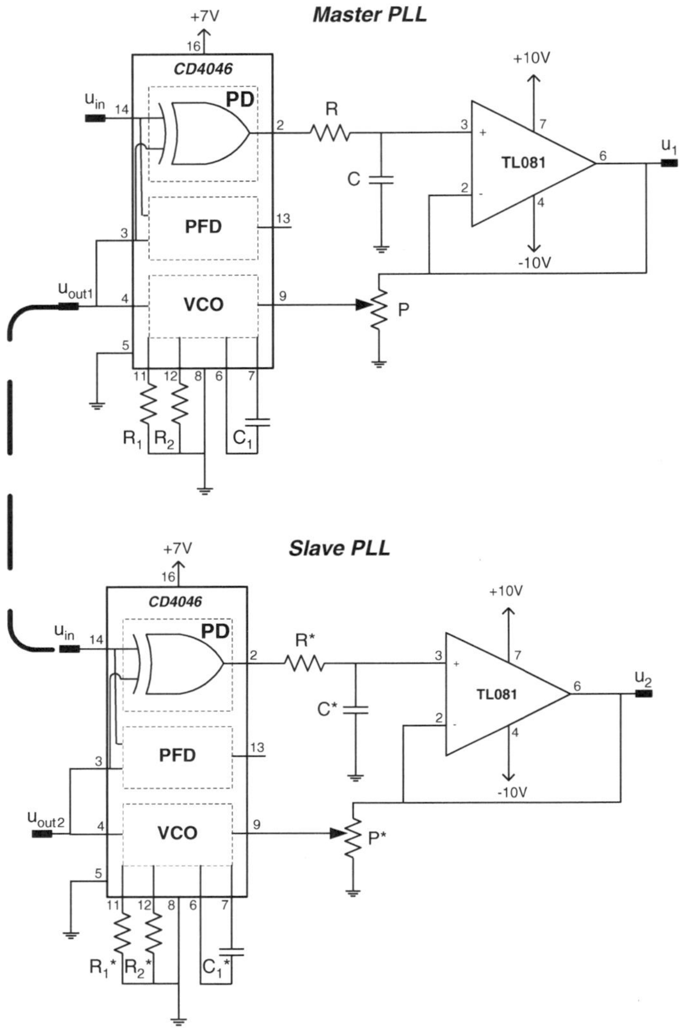

Fig. 9.18 Master-slave PLL circuit diagram

9.4.3 PLL Simple-Coupling Circuits

The system of two PLL's interconnected with the simplest possible way [67] i.e. the master-slave configuration (unidirectional coupling), is presented hereby [61]. Even in this simplest coupling, it is revealed that synchronization intermittency appears on the way from full synchronization between the PLLs to full desynchronization. The experimental setup was realized by two commercial PLL IC's i.e. CD4046; and it appears in Fig. 9.18. It is apparent that there is a single (and simple) coupling between the two PLLs. The first one is working in a normal, single PLL configuration, while the second one is driven by the output frequency of the first. This is the most simple and commonly used connection of two PLLs (the output frequency of a PLL is used as the reference frequency to the second, without the existence of any feedback path).

In order to show that the chaotic behavior is inherent in this configuration and it is not related to a particular kind of circuit, we used the simplest configuration for building the two PLLs. The EX-OR phase detectors (PD) and the voltage controlled oscillators (VCO) used, were those built-in the PLL ICs. It should be noted that most commercial PLLs, including CD4046 that we used in our experiments, have two types of built-in phase detectors (an EX-OR simple phase detector and a phase-frequency detector). Both low-pass filters (LPF), plugged at the phase detector outputs, have a cut-off frequency of 3.9 kHz and are of a simple (widely used) circuit configuration. The role of these analog filters is to suppress the undesired higher frequency harmonics and enhance the low frequency voltage signal which represents the mean phase error value. Simple voltage dividers (potentiometers noted as (P) in the circuit diagram) are used as loop-gain attenuators.

Finally, the center frequencies of the first and the second VCO were set (by choosing properly the values of R_1, R_2, C_1 and R_1^*, R_2^*, C_1^*) to 55.67 kHz and 52.49 kHz, respectively, so that all circuits (and especially the op-amps) are well within their normal operating range [61].

9.5 Experimental Incomplete Synchronization

Although incomplete synchronization phenomena, such as those described in the *Synchronization* section of this contribution, have been theoretically studied and thoroughly described, there is not much experimental verification of them. The three circuits outlined above provided experimental demonstration of intermittent synchronization. In the lines that follow the *Crisis-induced*, the "*On-Off*" and the "*In-Out*", experimental intermittent synchronization are presented and verified.

9.5.1 Experimental Crisis Induced Intermittent Synchronization

Crisis-induced intermittency is a phenomenon usually encountered in the behavior of chaotic circuits and it is featured by an abrupt qualitative structure alteration of a chaotic attractor to an unstable trajectory in the corresponding phase space. This alteration is due to attractor collision between the two distinct chaotic states (in the beginning and the end of the phenomenon) [68]. It is reported that this kind of transitional situation characterizes a system's synchronization to desynchronization route, as well [61].

In this section the intermittent behavior of the phase error signal, in the experimental PLL master-slave configuration appearing in Fig. 9.18 is presented. This simple, one way connection (no feedback path or loop) between the two PLLs is widely considered as stable. However, as reported in [61] the presence of chaotic behavior at certain conditions arises. This behavior is induced by a crisis between two chaotic modes of operation. The phenomenon is observed at the limit of phase locked frequency range, obliging the PLL system to undergo a route from full output frequency synchronization to full desynchronization. The phase error signal ceases to be periodic and becomes chaotic exhibiting *Crisis-induced* intermittency.

The frequency, f_{in}, of signal u_{in} and the values of the potentiometers (P) served as control parameters. It was experimentally established, that for most of the values of the control parameters, in the frequency interval under consideration (30 kHz down to 14 kHz), both PLLs synchronized and the signals, u_1 and u_2, at the outputs of the LPFs remained periodic. However, for specific values of the control parameters accumulated at the two limits of the frequency hold-in range of the PLLs, chaos sets on.

In the experimental study presented hereby the results obtained, for ratio values of the potentiometers P_1 and P_2 equal to 0.196 and 0.174, respectively, in the frequency range $14.300\,\text{kHz} < f_{in} < 14.380\,\text{kHz}$ are reported. It should be noted that the frequency was adjusted with a precision of ± 0.1 Hz.

For 14.380 kHz both phase error signals, u_1 and u_2, were periodic and both PLLs were synchronized.

Shifting frequency from 14.380 kHz down to lower values, a bifurcation in u_2 was observed in the form of an intermittent behavior of this signal, while u_1 remained periodic. Signal u_2 consisted of long intervals of non-periodic oscillations at an upper level, occasionally interrupted by non-periodic oscillations at a lower level. The master PLL remained locked, while the slave one was intermittently locked. By further shifting down the frequency, these lower level oscillations became denser and of longer duration. In Fig. 9.19, signal u_2, in the form of time-series, in three typical cases, is presented before crisis sets on, during crisis and after the end of the crisis phenomenon.

The attractors that characterize the slave PLL were reconstructed and properly embedded in the original phase space, according to Takens theory [69]. It is apparent that the attractor in Fig. 9.20b consists of two sub-attractors, the ones presented in Fig. 9.20a, b. Note that he trajectories are orbiting for longer time intervals in the upper sub-attractor—Fig. 9.20a—occasionally escaping for shorter time intervals, to the lower one—Fig. 9.20c.

Quantitative confirmation of the crisis phenomenon just described, was accomplished by employing the Ott, Grebogi and Yorke theory about crises in dynamical systems [70]. According to this, the transition probability $P(\tau)$ of time intervals τ, for which the orbit stays in (a periodic or) one of the chaotic attractors i.e. the known laminar lengths, obeys the following law [41]:

$$P(\tau) = \frac{1}{\langle\tau\rangle} e^{\left(\frac{-\tau}{\langle\tau\rangle}\right)} \tag{9.11}$$

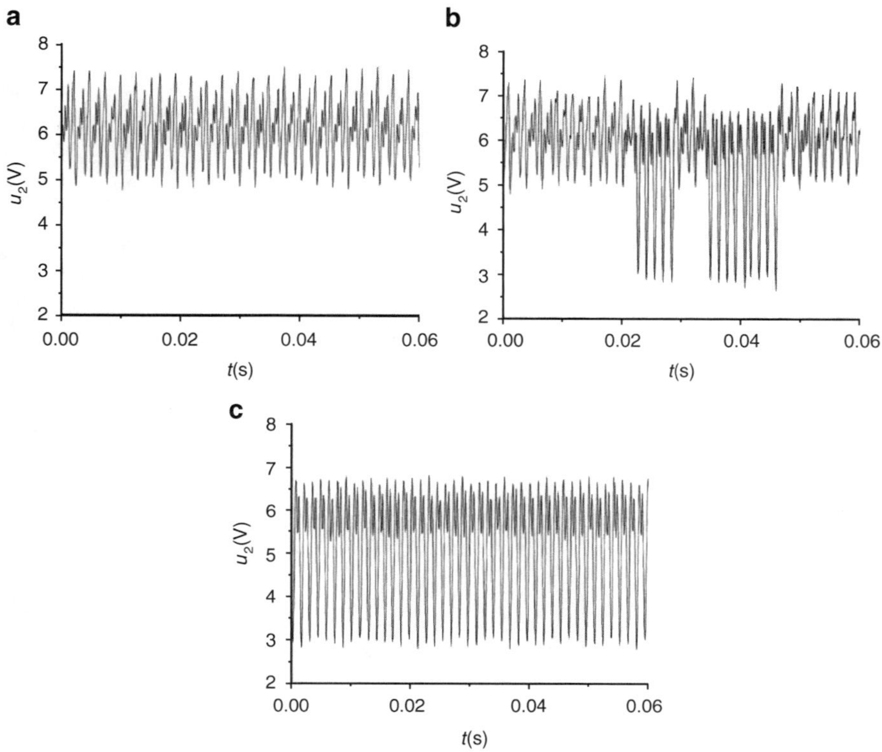

Fig. 9.19 Time-series of phase error signal u2 (in the slave PLL) for three representative cases: (**a**) $f_{\text{in}} = 14.390\,\text{kHz}$ before crisis sets on, (**b**) $f_{\text{in}} = 14.356\,\text{kHz}$ during crisis and (**c**) $f_{\text{in}} = 14.300\,\text{kHz}$ after the end of crisis

$<\tau>$ stands for the average time between successive transitions. It is important to mention that this law holds for a smooth distribution of the initial conditions, something that really applies in this case, due to the way that time-series sampling was registered [71, 72].

A typical distribution plot of probability $P(\tau)$ vs. τ appears in Fig. 9.21 for $f_{in} = 14.358\,\text{kHz}$. An exponential decay fitting of (9.11) provides an estimate for factor $1/<\tau>$ and can lead to the calculation of $<\tau>$. For this particular plot, the value obtained was $<\tau> = 3.858\,\text{ms}$.

Similar results were also obtained for every input frequency f_{in} in the range between 14.370 kHz down to 14.340 kHz, where the crisis takes place. These results were exploited in calculating the crisis critical exponent γ. The average time $<\tau>$ of the laminar lengths between two successive transitions from one chaotic to another chaotic (or periodic) periodic mode of operation is decreasing with increasing frequency f_{in}, according to [41, 70]:

$$\langle\tau\rangle \propto (f_{in} - f_{\text{crit}})^{-\gamma} = \varepsilon^{-\gamma} \qquad (9.12)$$

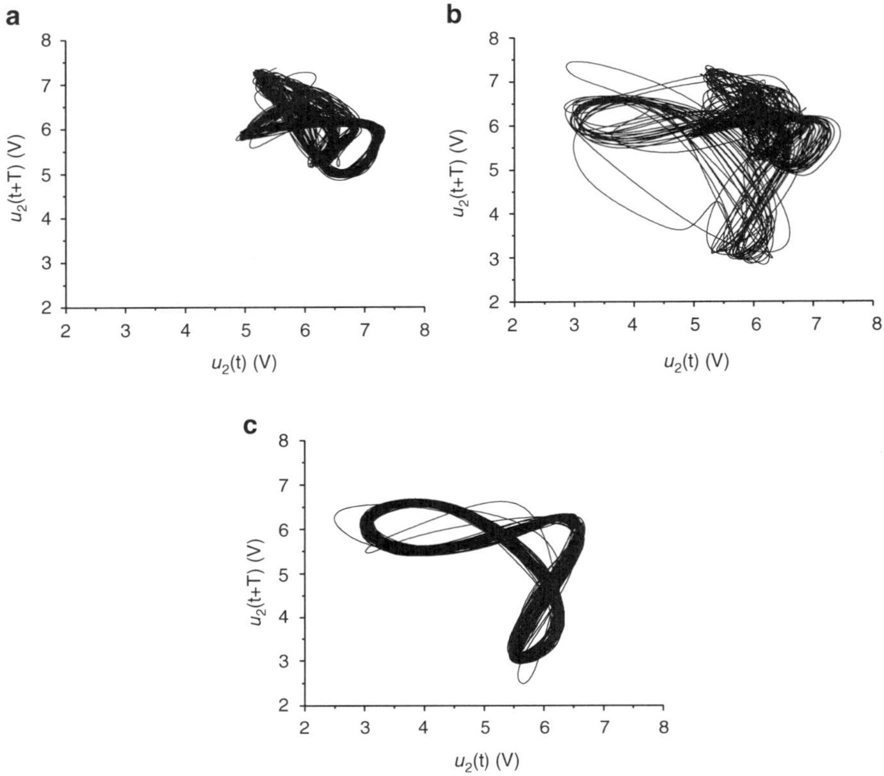

Fig. 9.20 The corresponding phase portraits for the three cases presented in Fig. 9.19

Fig. 9.21 Typical distribution plot of probability $P(\tau)$ vs. τ for (**a**) $f_{\text{in}} = 14.358\,\text{kHz}$

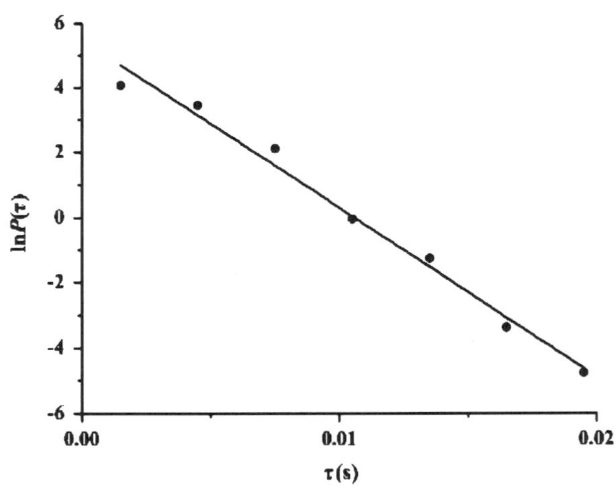

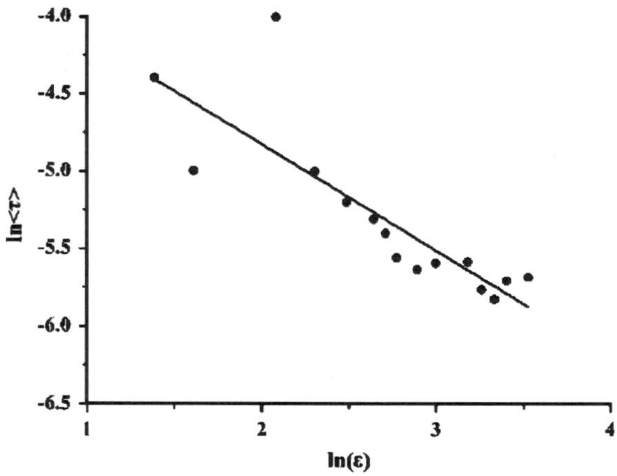

Fig. 9.22 Double logarithmic plots of $<\tau>$ vs. ε

Here, f_{crit} stands for the critical frequency at which the crisis is initiated, ε is the change of the crisis parameter $(f_{in} - f_{\text{crit}})$ and γ is the critical exponent of the crisis, characterizing its distributions. Double logarithmic plots of $<\tau>$ vs. ε, for $f_{crit} = 14.374\,\text{kHz}$ (where he crisis phenomenon experimentally started), is presented in Fig. 9.22. In this plot mean laminar lengths $<\tau>$ were estimated from the distribution plots of probability $P(\tau)$, as the one presented in the previous figure (Fig. 9.21).

Again an exponential decay fitting of (9.12), finally provided the critical exponent of the crisis γ which holds the estimated value $\gamma = 0.687$. The obtained value of γ is well above the theoretically predicted lower limit of 0.5 [73, 74], but it is not that high so that we could observe transitions of quite long duration frequently [68].

9.5.2 *Experimental "On-Off" Intermittent Synchronization*

As already mentioned "*On-Off*" intermittency is a phenomenon encountered in chaotic dynamical system-synchronization. It appears in both bidirectional and unidirectional couplings. The transition under question is characterized by specific distributions [52–54] and the desynchronization demonstrated seems to occupy almost the whole synchronization phase space, with no preference to some direction.

In this section the experimental "*On-Off*" intermittent behavior of two double scroll circuits both unidirectionally [58] and bidirectionally [57]coupled, is presented. The circuits (Fig. 9.1) demonstrating this particular route from a synchronized state of operation to a fully desynchronized one, have been already described in the previous sections.

In both cases coupling resistance R_C, in the denominator of coupling coefficient ξ in (9.8), served as the system's control parameter and its value was changed from almost zero up to $100\,\text{k}\Omega$ in the case of bidirectional coupling and up to

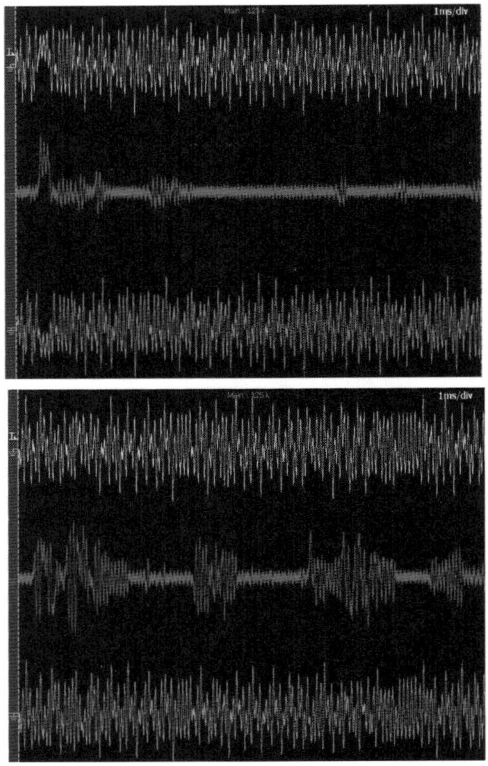

Fig. 9.23 Time-series $y_1(t)$ (*upper*) and $y_2(t)$ (*lower*) snapshots, from each sub-circuit for two characteristic cases of incomplete synchronization

$70\,\mathrm{k}\Omega$ in the case of unidirectional coupling. Although, in these regions each circuit remained chaotic exhibiting a double-scroll attractor, the coupled system underwent a transition from full to incomplete synchronization, ending to complete desynchronization.

9.5.2.1 "On-Off" Intermittency in Bidirectional Coupling

In the case of bidirectional resistive coupling the system parameter values, resulting from the circuits' component values, have already been quoted in Table 9.1 [57]. It is noted that both synchronized circuits were identical.

This resistive coupling served as the control parameter for studying the phenomenon of incomplete synchronization. Its critical value i.e. the value for which the incomplete synchronization begins, was experimentally found to be $R_C = 9.60\,\mathrm{k}\Omega$ [57].

As long as R_C was below the critical value $R_{C-\mathrm{crit}} = 9.60\,\mathrm{k}\Omega$, the two identical double-scroll circuits remained fully synchronized, as already illustrated in Figs. 9.5, 9.6 and 9.7. As expected in the case of complete synchronization, both the time-series and the corresponding circuit attractors are identical. The difference

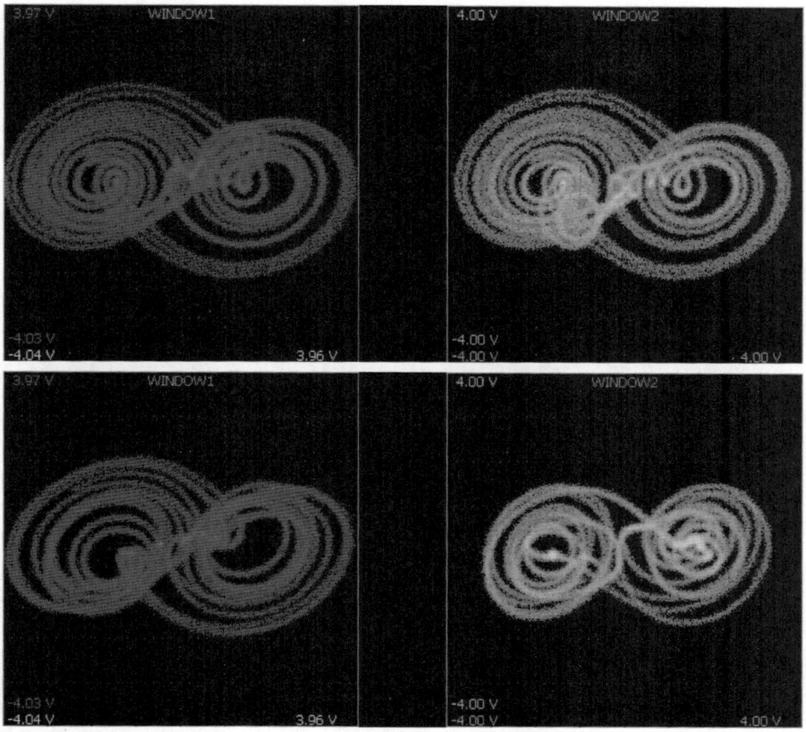

Fig. 9.24 Phase portraits of each double-scroll circuit for the two typical case of Fig. 9.23

signal is continuously almost zero and the synchronization phase portrait strictly remains along the diagonal.

For values of R_C higher than 9.60 kΩ and in the range $9.60\,\text{k}\Omega \leq R_C \leq 80.00\,\text{k}\Omega$ the two circuits were incompletely (intermittently) synchronized. In Fig. 9.23 time-series $y_1(t)$ (upper) and $y_2(t)$ (lower) of each double-scroll circuit, together with their difference signal, are presented in the case of $R_C = 30.00\,\text{k}\Omega$ and $R_C = 60.00\,\text{k}\Omega$, respectively. In Fig. 9.24 the individual phase portraits of each circuit (x_1 vs. y_1 and x_2 vs. y_2) are shown while, the corresponding synchronization phase portraits ($y_1(t)$ vs. $y_2(t)$) appear in Fig. 9.25, confirming each circuit's chaotic mode of operation. These two representative cases correspond to weak and strong incomplete (intermittent) synchronization. The difference signal is almost zero for long time spaces, bursting occasionally at significantly non-zero values. These bursts become of longer duration and appear more frequently, with increasing values of coupling parameter R_C. As expected in the case of incomplete synchronization, both the time-series and the attractors are intermittently correlated. This is confirmed in Fig. 9.24, where the corresponding phase portraits are shown. Although the attractors are quite similar a careful examination reveals that they are not identical. Moreover, the synchronization phase portraits in Fig. 9.25 consist of trajectories

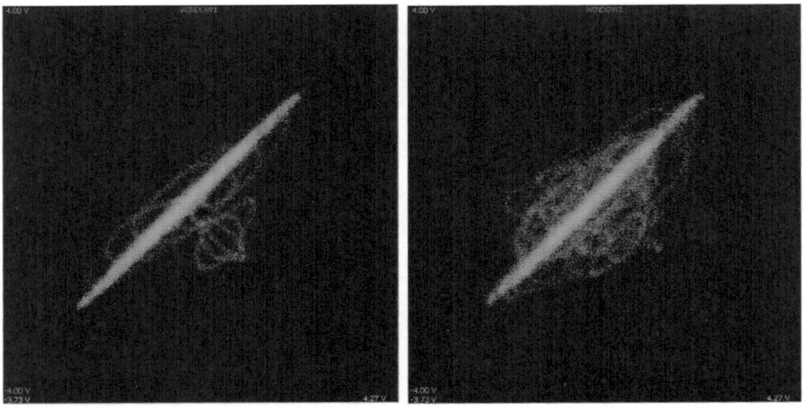

Fig. 9.25 The corresponding to Figs. 9.23 and 9.24, synchronization phase portraits for the cases of "*On-Off*" type incomplete synchronization

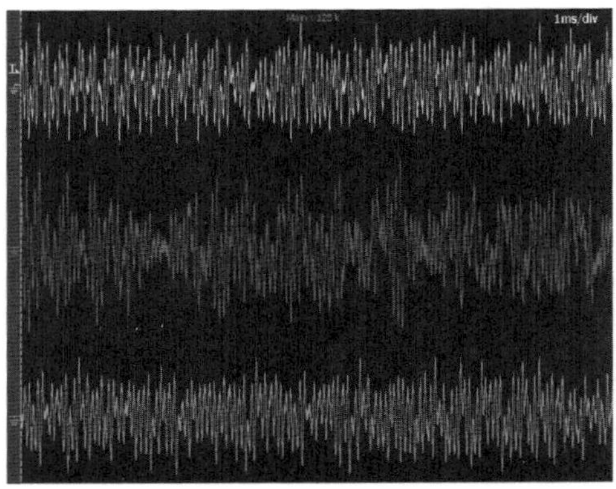

Fig. 9.26 Time-series y_1 (*upper*) and y_2 (*lower*) snapshots, in the case of full desynchronization

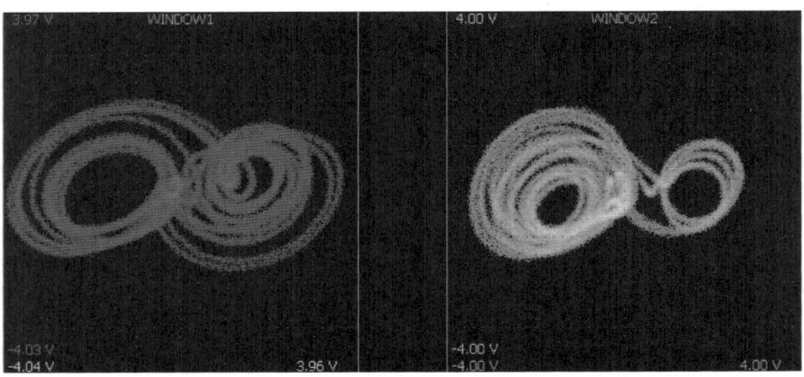

Fig. 9.27 Phase portraits of each sub-circuit for the case of full desynchronization of Fig. 9.26

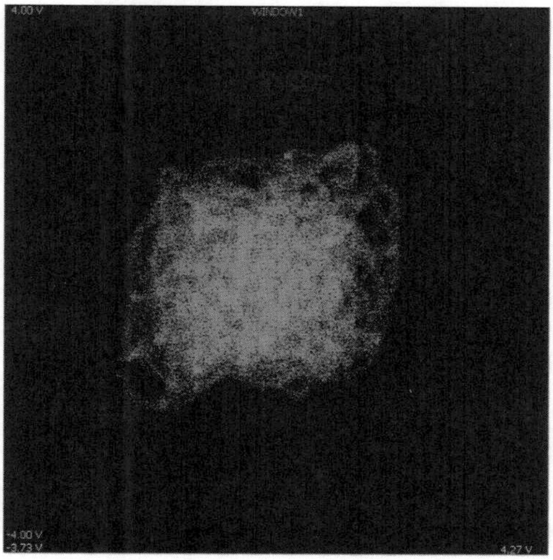

Fig. 9.28 The corresponding to Figs. 9.26 and 9.27, full desynchronization phase portrait. Intermittent bursts have no preference to direction in the phase space

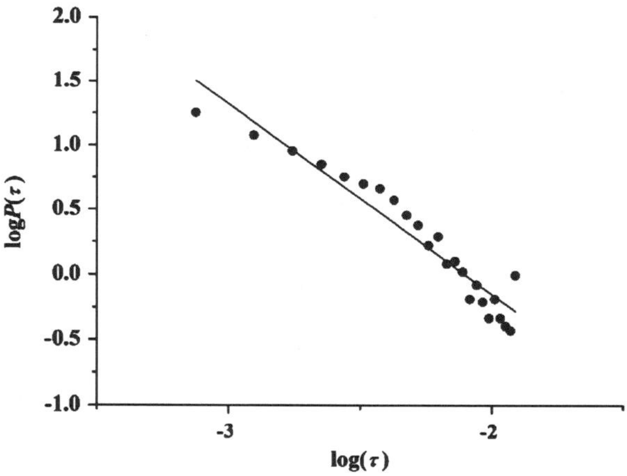

Fig. 9.29 A representative ($R_C = 20\,k\Omega$) distribution P(τ) of durations τ of laminar lengths in the case of bidirectional coupling. *Full circles* denote experimental data, while the line represents a proper fitting with a power law of the form of (9.1)

spending most of the time on the diagonal, only temporarily escaping from it. These escapes correspond to the bursts in the difference signal.

Finally, for $R_C > 90.00\,k\Omega$ the system circuits become fully desynchronized. In this case their $y(t)$ time-series' as well as their difference appear in Fig. 9.26, while in Fig. 9.27 the corresponding chaotic attractors are presented. In Fig. 9.28 synchronization phase portrait appears. In this case of full desynchronization, the time-series and the attractors are continuously uncorrelated, as confirmed by the difference signal, which possesses almost always non-zero values. Consequently, the attractors of each double-scroll circuit are completely different. Apparently, the synchronization attractor is fully blown out of the diagonal, covering most of the synchronization phase space.

The qualitative evaluation of the dynamics associated to this transition, suggested that it had the characteristics of "*On-Off*" intermittency [52–54]. In order to verify

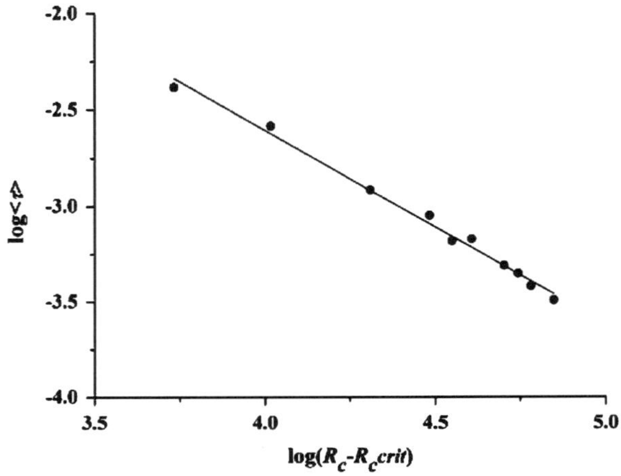

Fig. 9.30 Double-log plot of the laminar length mean duration $<\tau>$ vs. the difference $(R_C - R_{C-crit})$. *Full circles* denote experimental data, while the *line* represents a fitting of the form of (9.5)

this assumption, laminar length distributions and the scaling of the mean laminar lengths with the difference $(R_C - R_{C-crit})$, were checked, according to theoretically predicted laws described in (9.1) and (9.2) in the *Synchronization* section of this chapter.

In Fig. 9.29, a representative distribution $P(\tau)$ of laminar lengths durations τ of the difference signal, for $R_C = 20.00\,\mathrm{k}\Omega$ is shown. The proper linear fitting of the experimental data (full circles) revealed a slope β equal to $\beta = -1.52$. Similar distributions for other values of control parameter R_C provided slope values for β not substantially deviating from –1.50, supporting the classification of this intermittency as one of the "*On-Off*" type [52–54].

Finally, in Fig. 9.30 mean duration of laminar lengths $<\tau>$ vs. the difference $(R_C - R_{C-crit})$, in a double logarithmic plot, appears. Again a linear fitting of the experimental data (full circles) resulted to a straight line with a slope γ equal to $\gamma = -1.005$, further supporting the "*On-Off*" classification of the observed intermittency [52–54].

9.5.2.2 "On-Off" Intermittency in Unidirectional Coupling

In the case of unidirectional coupling the system parameter values, resulting from the circuits' component values, appear in Table 9.3 [58]. Again both synchronized double-scroll circuits were identical but this time the system behaves as a master-slave scheme.

Coupling resistance R_C served as the control parameter of the desynchronization route and its critical value was experimentally found to be $R_C = 5.75\,\mathrm{k}\Omega$ [58]; almost half of the value that it possessed in the bidirectional coupling case [57].

As long as R_C was below the critical value, the two identical double-scroll circuits remained fully synchronized, as already illustrated in Figs. 9.9, 9.10 and 9.11. Again, as expected in the case of complete synchronization, each circuit corresponding time-series and their attractors are identical, while the difference

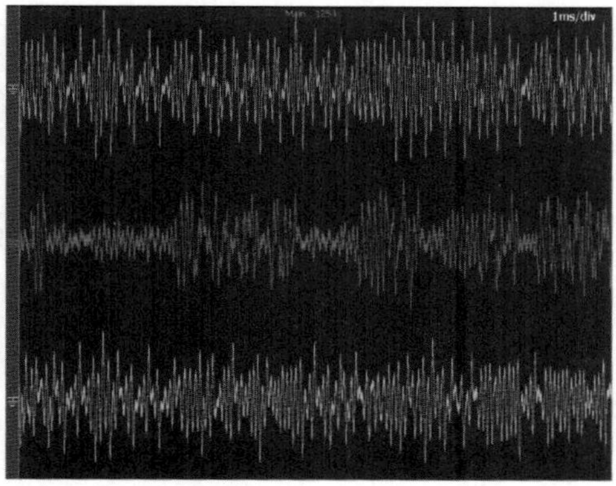

Fig. 9.31 Time-series y_1 (*upper*) and y_2 (*lower*) snapshots, in the case unidirectional incomplete synchronization, for $R_C = 30.00\,\mathrm{k}\Omega$

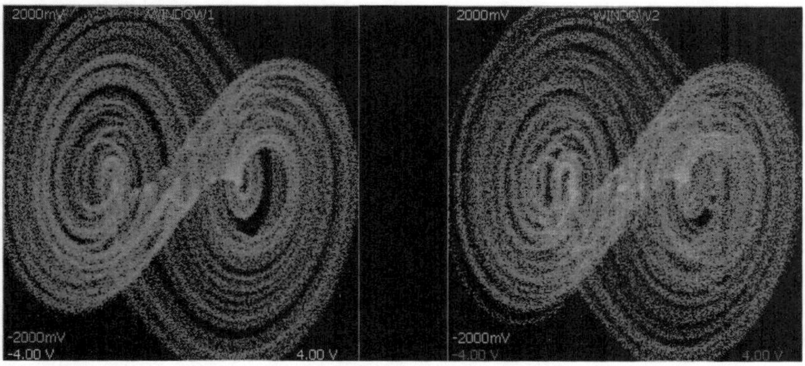

Fig. 9.32 Phase portraits of each sub-circuit for the case of incomplete synchronization of Fig. 9.31

signal is continuously zero and the synchronization phase portrait strictly remains on the diagonal.

For values of R_C higher than its critical value and in the region that ranges from $R_C = 5.75\,\mathrm{k}\Omega$ to $R_C = 60.00\,\mathrm{k}\Omega$ the two circuits were again incompletely (intermittently) synchronized. In Fig. 9.31 time-series $x_1(t)$ (upper) and $x_2(t)$ (lower) of each double-scroll circuit and their difference signal, are presented in a representative case of $R_C = 30.00\,\mathrm{k}\Omega$. In Fig. 9.32 the individual phase portrait of each circuit (x_1 vs. y_1 and x_2 vs. y_2) is illustrated. The corresponding synchronization phase portrait ($x_1(t)$ vs. $x_2(t)$) appear in Fig. 9.33, always confirming each circuit's chaotic mode of operation. Apparently, both the time-series and the attractors are intermittently correlated, something that also corroborated by the synchronization phase portrait.

For $R_C > 60.00\,\mathrm{k}\Omega$ the system exits the previously described region of intermittent synchronization and it fully desynchronizes. Although, the sub-circuits are still unidirectionally and resistively coupled, still exhibiting a double-scroll chaotic behavior, the master circuit ceases to drive the operational behavior of the slave

Fig. 9.33 The corresponding to Figs. 9.31 and 9.32, full desynchronization phase portrait. Likewise the bidirectional coupling Intermittent bursts have no preference to direction in the phase space

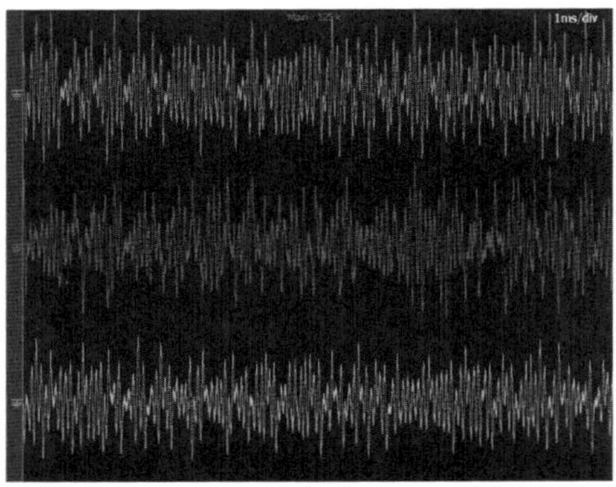

Fig. 9.34 Time-series y_1 (*upper*) and y_2 (*lower*) snapshots, in the case unidirectional full desynchronization, for $R_C = 70.00\,\mathrm{k}\Omega$

circuit. Consequently, they become fully desynchronized. In Fig. 9.34 the times-series' $x_1(t)$ (upper) and $x_2(t)$ (lower) sub-circuit with their difference signal (in the middle) are presented, while in Fig. 9.35 each sub-circuit phase portrait appear and in Fig. 9.36 the corresponding synchronization phase portrait is illustrated. Note that the control parameter was set at $R_C = 70.00\,\mathrm{k}\Omega$. It is obvious by the last three figures that the attractors are continuously uncorrelated.

In order to gain an insight of the mechanism governing the transition from full synchronization to complete desynchronization in this master–slave double-scroll system, experimental laminar length distributions and the scaling of the mean laminar lengths with the deviation of the control parameter from its critical value $(R_C - R_{C\text{-crit}})$, were again examined.

Likewise the case of bidirectional coupling [57], the qualitative evaluation of the dynamics associated to this transition, suggested again that it had the characteristics of "*On-Off*" intermittency [52–54]. In specific, proper linear fitting on the experimental data of laminar length distribution, according to the power law

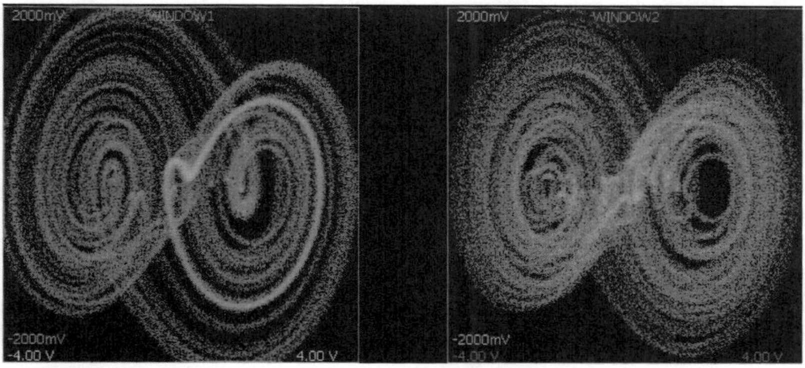

Fig. 9.35 Phase portraits of each sub-circuit in the case of full desynchronization of Fig. 9.34

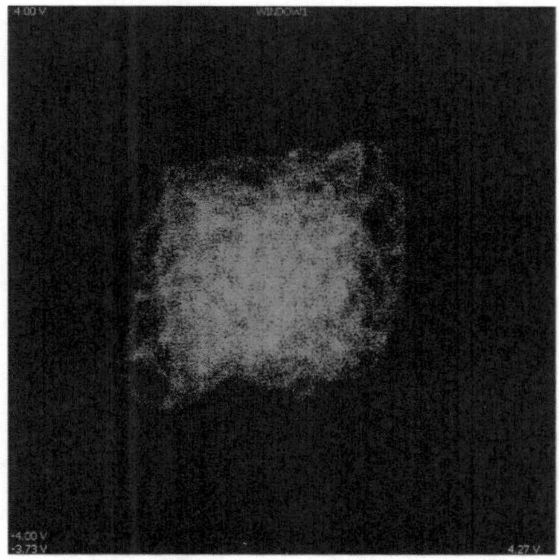

Fig. 9.36 The corresponding to Figs. 9.34 and 9.35, full desynchronization phase portrait

in (9.1) and for different values (in the region of incomplete synchronization) of the control parameter R_C, provided slope values for β not substantially deviating from –1.50, as theoretically foreseen for the type of intermittency "*On-Off*". A typical distribution $P(\tau)$ of laminar lengths durations τ of the difference signal, for $R_C = 30.00\,\text{k}\Omega$, is presented in Fig. 9.37. The value of β obtained in this case was $\beta = -1.496$.

Finally, this was further confirmed by determining the slope γ exhibited by mean duration of laminar lengths $<\tau>$ vs. the difference $(R_C - R_{C\text{-crit}})$ distribution, appearing in Fig. 9.38. A least square fitting of the experimental data (full circles) leaded to a straight line with a slope γ equal to $\gamma = -1.093$, clearly confirming the following power law of (9.2). This result is in full accordance with the classification of the observed intermittency as an "*On–Off*" type [52–54].

In this case of unidirectional coupling, the desynchronization phenomenon scaled into a narrower range (up to $60\,\text{k}\Omega$) compared to the bidirectional case (up to

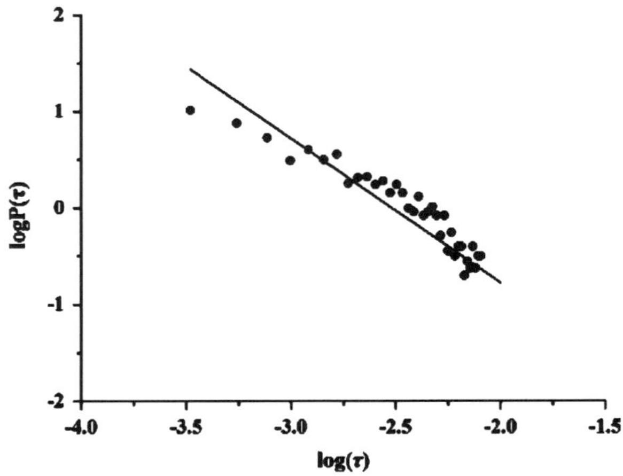

Fig. 9.37 A typical ($R_C = 30\,\mathrm{k}\Omega$) distribution P($\tau$) of durations τ of laminar lengths in the case of unidirectional coupling

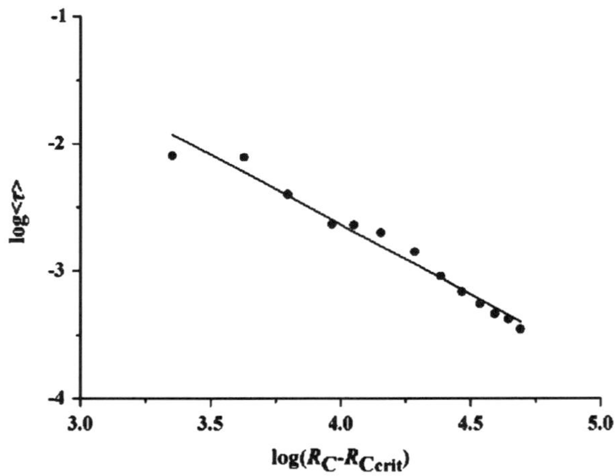

Fig. 9.38 Double-log plot of the laminar length mean duration $< \tau >$ vs. the difference ($R_C - R_{C-\mathrm{crit}}$) in the case of unidirectional coupling. Full circles denote experimental data, while the line represents a fitting of the form of (9.5)

$90\,\mathrm{K}\Omega$). This is easily explained by the fact that in unidirectional coupling the master-circuit dynamics evolve independently, while the slave-circuit dynamics evolution is influenced by the master [57, 58]. This results to earlier desynchronization. On the contrary, in bidirectional coupling both the circuits influence one another, resulting to putting back full desynchronization.

9.5.3 *In-Out Intermittent Synchronization*

As already mentioned "*In-Out*" intermittency is a phenomenon encountered in chaotic dynamical system synchronization. The transition under question is characterized by specific distributions [52, 55, 56] and desynchronization appears to demonstrate certain structures in the synchronization phase space, due to transverse instability resulting to blowing out the system attractor from a lower dimensional subspace [52].

In this section the experimental "*In-Out*" intermittent behavior in the cases of two circuit-synchronized systems i.e. a double-scroll bidirectionally [60] and a 4th order unidirectionally [59] coupled, is presented. The circuits demonstrating this particular route from a synchronized state of operation to a fully desynchronized one have been already described in the previous sections (Figs. 9.1 and 9.12). In the first case coupling resistance R_C served as the system's control parameter while in the case of the 4th order synchronized circuits, parameter mismatch was the reason of the route to complete desynchronization.

9.5.3.1 "In-Out" Intermittency in Bidirectionally Coupled Double Scroll Circuits

In the case of bidirectional resistive coupling between the two double-scroll circuits appearing in Fig. 9.1, next to the "On-Off" intermittent synchronization already described (for $9.60\,\text{k}\Omega \leq R_C \leq 80.00\,\text{k}\Omega$), the system also exhibited an "*In-Out*" intermittent behavior in a narrow range of the control parameter values i.e. in the range $8.70\,\text{k}\Omega \geq R_C \geq 7.10\,\text{k}\Omega$. In this case, the incomplete synchronization phenomenon evolution is described in details in the following lines [60].

As long as R_C possessed a value larger than, the experimentally determined, critical value $R_{C-\text{crit}} = 8.70\,\text{k}\Omega$, the two identical double-scroll circuits remained fully synchronized in the way that is illustrated in Figs. 9.5, 9.6 and 9.7. Of course, this happened for values of R_C lower than $9.60\,\text{k}\Omega$, since at this value the onset of "*On-Off*" intermittent synchronization arises [57].

For values lower than $8.70\,\text{k}\Omega$ incomplete synchronization became stronger. In Fig. 9.39 three representative snapshots of time-series, x_1 (upper) and x_2 (lower) together with their difference signal $(x_1 - x_2)$ in the middle, are presented. It is apparent from all three snapshots that the appearing bursts in the difference signal (in the middle) occur exclusively, when each double-scroll circuit trajectories reside only on one of the two scrolls in the phase space. What is noteworthy is that for each circuit, trajectories simultaneously limit themselves to one scroll of the attractor and these scrolls are opposite or complimentary to one another, as illustrated in Fig. 9.40 [60].

Three representative synchronization phase portraits (x_1 vs.x_2), depicting the transition from full synchronization to full desynchronization through a region of incompletely synchronized behavior, appear in Fig. 9.41. In the first phase portrait the synchronized case is presented for $R_C = 8.8\,\text{k}\Omega$, just above the critical value, with the trajectory resting along the diagonal, as expected. In the second, phase portrait for $R_C = 7.9\,\text{k}\Omega$ a hybrid illustration appears, consisting of both the diagonal and a wing-like structure below the diagonal. This behavior of the system trajectories depicts a special case of incomplete synchronization, with the diagonal corresponding to synchronized mode of operation and the escapes to the

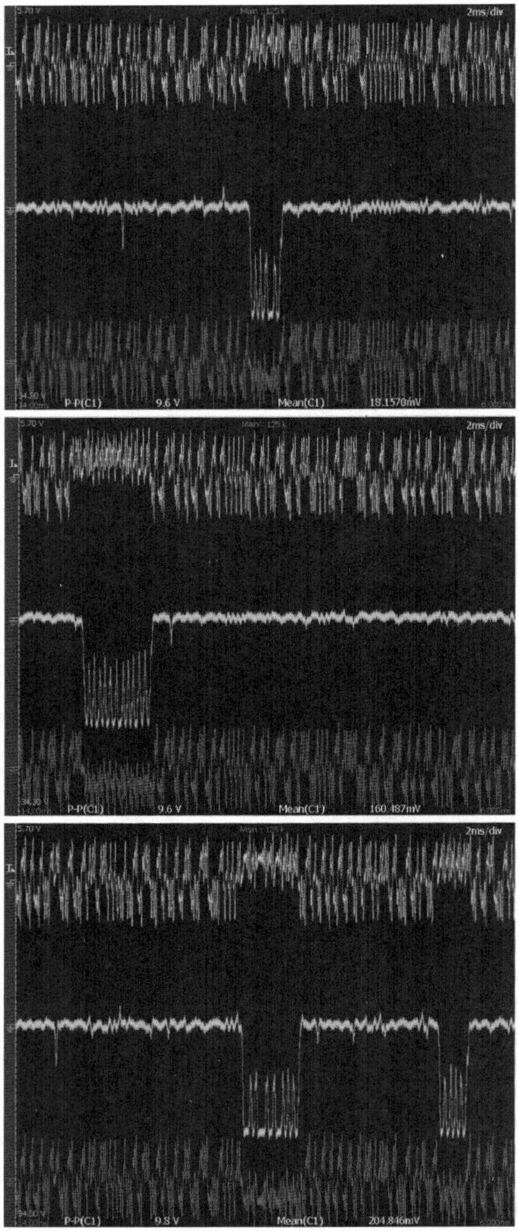

Fig. 9.39 Time-series snapshots of x_1 (*upper side*) and x_2 (*lower side*) and their difference (in the *middle*), in three typical cases of intermittent synchronization: (**a**) $R_C = 8.20\,\text{k}\Omega$, (**b**) $R_C = 8.00\,\text{k}\Omega$ and (**c**) $R_C = 7.80\,\text{k}\Omega$

wing corresponding to desynchronization bursts of the difference signal. What is remarkable in these bursts is that they are spatially limited within the appearing wing-like structure. Finally, in the third phase portrait for $R_C = 7.1\,\text{k}\Omega$ the diagonal disappears and only the wing-like structure remains. Thus, desynchronization prevails and the trajectory is always out of the diagonal. However, this kind of

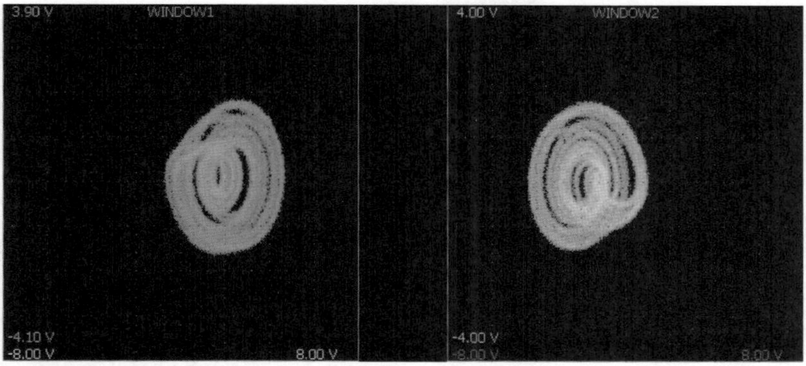

Fig. 9.40 The attractors of each double-scroll circuit when they limit themselves in one scroll, thus leading to desynchronization bursts. Note that each double-scroll circuit trajectory limits to only one scroll, being one complimentary to each other

desynchronization is spatially limited and the trajectory remains strictly within the wing-like structure, portraying an intermittent synchronization of "*In-Out*" type [52].

In order confirm the "*In-Out*" nature of this intermittent synchronization behavior, implied by the corresponding phase portraits in Fig. 9.41, experimental laminar length distributions and the scaling of the mean laminar lengths with the deviation of the control parameter from its critical value ($R_C - R_{C-crit}$), were again measured and evaluated.

The theoretical approach of this phenomenon dictates an experimental laminar length distribution (for a constant value of the control parameter) that verifies (9.3). In specific, an almost linear region and a shoulder followed by another almost linear region of different slope appear, as theoretically predicted [52, 53].

Proper linear fitting on the experimental data of the upper linear region of laminar length distribution, according to the power law in (9.3) and for different values (in the region of incomplete synchronization) of the control parameter R_C, provided slope values for β not substantially deviating from –1.50, as theoretically foreseen for intermittency of "*In-Out*" type [52]. A typical distribution $P(\tau)$ of laminar lengths durations τ of the difference signal, for $R_C = 8.20\,\text{k}\Omega$, is presented in Fig. 9.42. The value of β obtained in this case was $\beta = -1.494$.

Experimental data in the rest (lower part) region, demonstrates a shoulder and then another, rather narrow, linear part with different slope, for large values of laminar lengths as expected by theory. Taking into account the estimated value of β and the fact that synchronization bursts out of the diagonal correspond to jumps between two different manifolds (the diagonal and wing-like structure, respectively), the classification of this intermittency as one of the "*In-Out*" kind, can be safely supported [52, 55, 56].

Finally, as far as the mean duration $<\tau>$ of laminar lengths versus the difference ($R_C - R_{C-crit}$) is regarded, (9.2) holds also in the case of "*In-Out*" intermittency. In

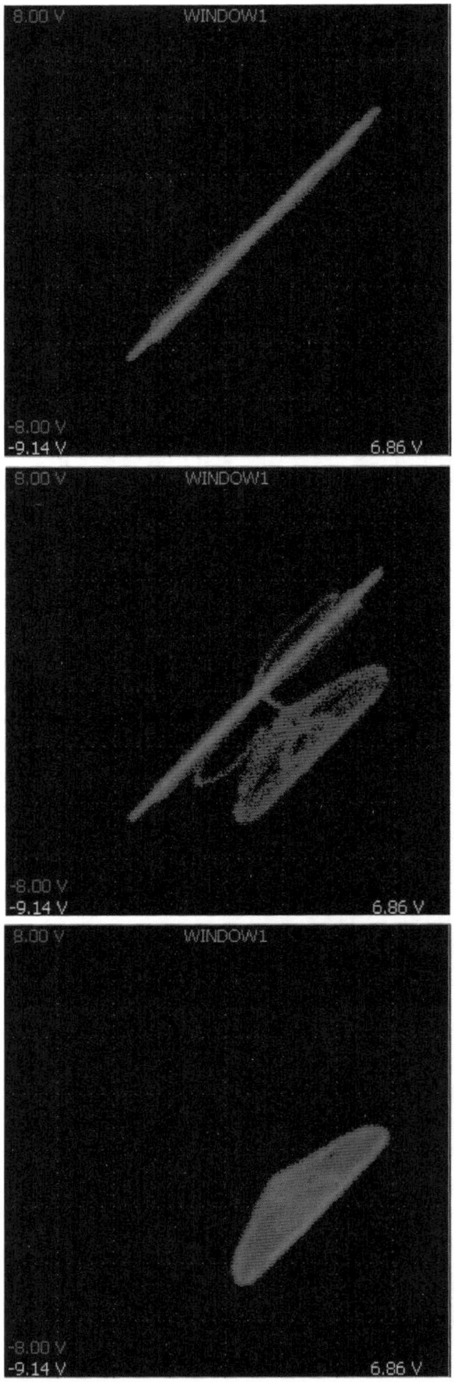

Fig. 9.41 Three synchronization phase portrait snapshots (x_1 vs. x_2) in the cases of full synchronization ($R_C = 8.8\,\mathrm{k}\Omega$), intermittent synchronization ($R_C = 7.9\,\mathrm{k}\Omega$) and desynchronization ($R_C = 7.1\,\mathrm{k}\Omega$)

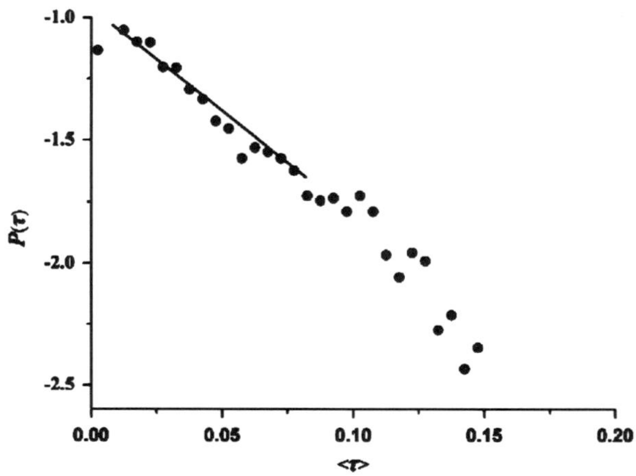

Fig. 9.42 A typical semi-logarithmic plot of P(τ) vs.τ in the case of $R_C = 8.20\,\mathrm{k}\Omega$

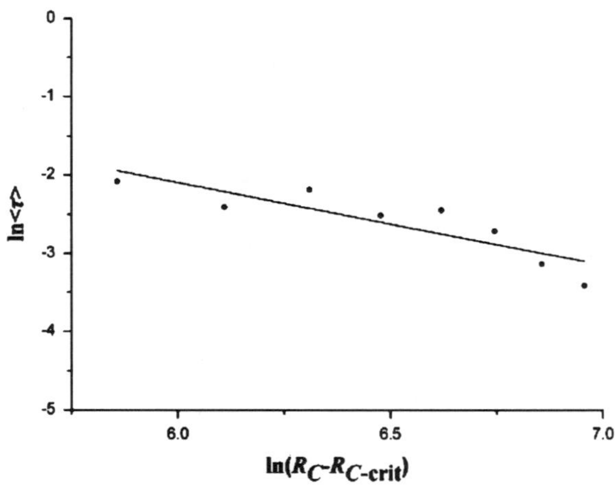

Fig. 9.43 Double logarithmic plot of $<\tau>$ vs. $(R_C - R_{C\text{-crit}})$. The line stands for the fitting of (9.2)

Fig. 9.43, a linear fitting of experimental data results for $<\tau>$ vs. $(R_C - R_{C\text{-crit}})$ in double logarithmic plots, provides with a slope $\gamma = -1.051$, as predicted in theory, further confirming the "*In-Out*" classification [52, 55].

9.5.3.2 "In-Out" Intermittency in Unidirectionally Coupled Double 4th Order Circuits

Two unidirectionally, resistively coupled 4th order circuits already, previously in this contribution, presented and illustrated in Fig. 9.12, exhibit an incomplete synchronization behavior region of the "*In-Out*" type, but this time by employing a controlled parameter mismatch as the phenomenon's control parameter [59].

In specific, capacitance C_{12} of the driven circuit served as the system's control parameter and its value was changed from 6.60 nF (identical to the corresponding capacitor of the driving circuit) down to 5.90 nF. In this region, although, both circuits remain chaotic, they undergo a transition from full to incomplete synchronization, ending to complete desynchronization [59].

In details, as long as C_{12} had the same value to that of $C_{11} = 6.60\,\text{nF}$, the two identical sub-circuits remained fully synchronized. Time-series of the master circuit (upper) u_{C11}, the slave circuit (lower) u_{C12} and their almost zero difference signal $(u_{C11} - u_{C21})$, have already been presented in Fig. 9.15. Likewise, in Fig. 9.16 the identical phase portraits, u_{C11} vs. u_{C21}, of the master circuit, and u_{C12} vs. u_{C22}, of the slave circuit, in the case of full synchronization have appeared, while in the synchronization phase portrait, with trajectories residing on the diagonal, has been presented in Fig. 9.16.

Beginning from the control parameter's critical value $C_{12\text{crit}} = 6.60\,\text{nF}$ and in the range $6.60\,\text{nF} > C_{12} > 5.90\,\text{nF}$ the two sub-circuits are incompletely synchronized. In Fig. 9.44 a representative example ($C_{12} = 6.30\,\text{nF}$) of this incomplete (intermittent) synchronization appears. In the first figure, time-series for the master u_{C11} (upper) and the slave u_{C12} (lower) circuit, as well as their difference $u_{C11} - u_{C21}$ (in the middle), are presented. The difference signal is almost zero for long time spaces, bursting occasionally at significantly non-zero values. It should be noted that these bursts become of longer duration and appeared more frequently, with decreasing values of C_{12}. In the second figure the master and the slave circuits' phase portraits (u_{C11} vs. u_{C21} and u_{C12} vs. u_{C22}) appear. As expected in the case of incomplete synchronization, both the time-series and the attractors are intermittently correlated. In the third figure, the corresponding synchronization phase portrait (u_{C11}vs. u_{C12}) is illustrated. It consists of trajectories spending most of the time on the diagonal, only temporarily escaping from it. These escapes correspond to the bursts in the difference signal.

Finally, for low enough values of $C_{12}(C_{12} \leq 5.90\,\text{nF})$, the two circuits are becoming fully desynchronized. In Fig. 9.45, their times-series and their difference, the attractors and the corresponding synchronization phase portrait appear again. In this case of full desynchronization, the time-series and the attractors are continuously uncorrelated. This is confirmed by the difference signal $(u_{C11} - u_{C21})$ in Fig. 9.45a which possesses almost always non-zero values. Consequently, the attractor on the right side (driven circuit) is blown out in comparison to the attractor on the left side (driving circuit). As it is expected in the case of full desynchronization, the synchronization phase portrait is fully blown out of the diagonal, but what is important is the fact that it still retains a certain structure clearly hinting that this intermittent behavior is of the "*In-Out*" type [52].

To confirm this suggestion the plots of the distribution of laminar lengths were checked. In Fig. 9.46 a representative distribution $P(\tau)$ of durations τ of the laminar lengths, for $C_{12} = 6.40\,\text{nF}$ is presented in double logarithmic plot. The curve of this plot consist mainly of a linear part (upper side) with a slope β equal to $\beta = -1.491$, as theoretically expected i.e. a slope with $\beta = -1.5$. All experimental distributions of laminar lengths, for other values of the control parameter C_{12}, resulted to slope values (β) very close to –1.5. On the other hand, experimental data in the rest (lower side) part of the plot in Fig. 9.46, demonstrates a shoulder

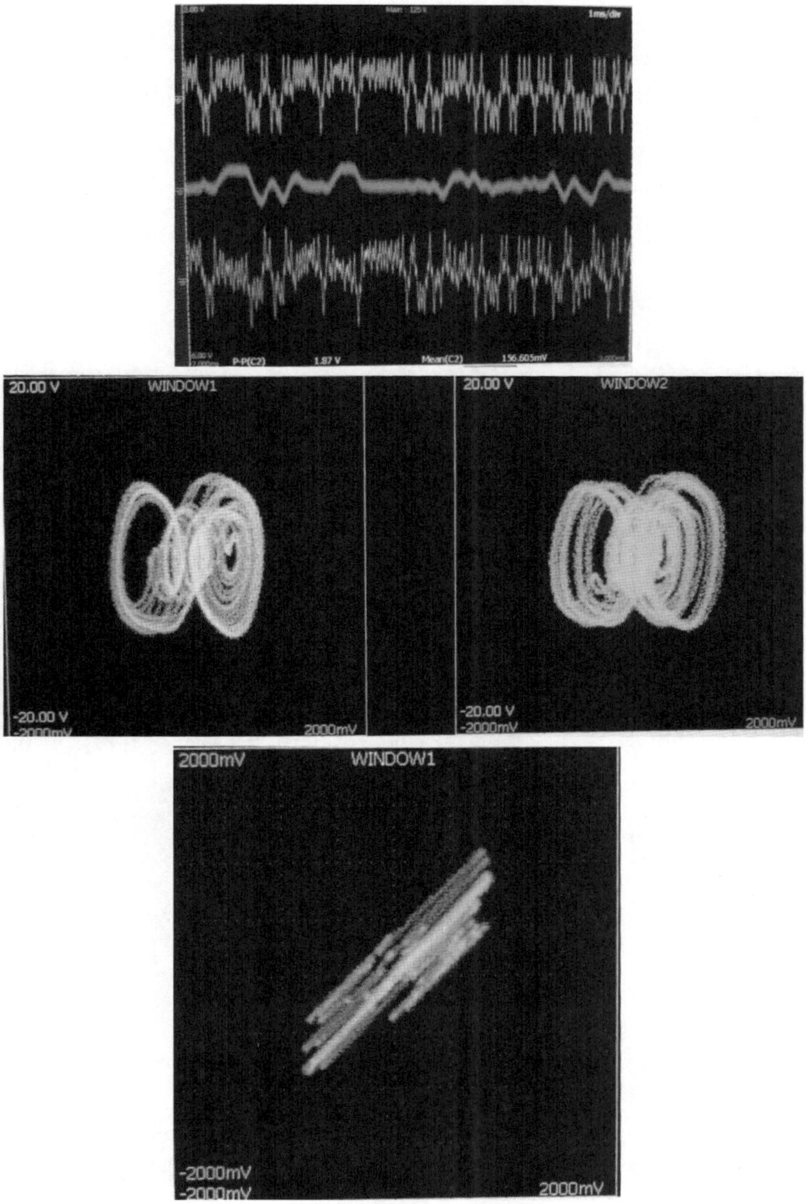

Fig. 9.44 Representative time-series, phase portraits and synchronization phase portrait in the case of "*In-Out*" intermittent synchronization, for $C_{12} \leq 6.60\,\mathrm{nF}$

and then another, almost linear part with a different slope for large values of laminar lengths, as expected by theory. Taking into account, the estimated value of β and the fact that synchronization bursts retain a certain structure corresponding to jumps

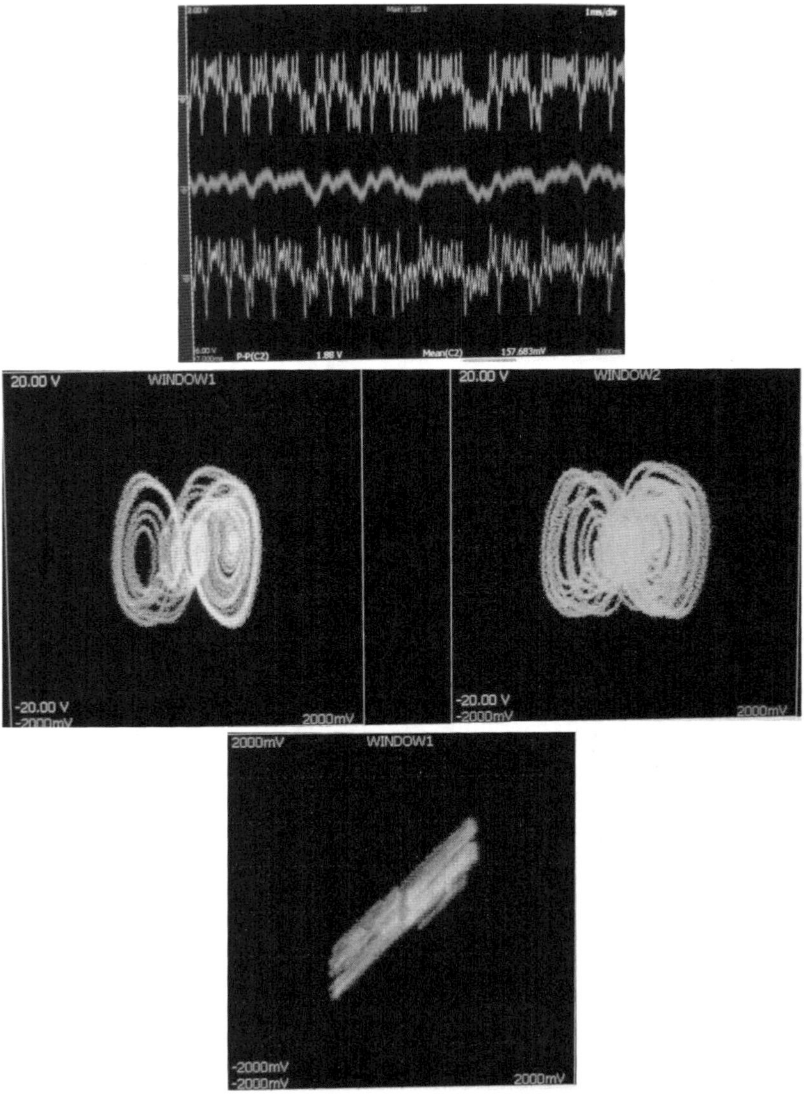

Fig. 9.45 Time-series, phase portraits and synchronization phase portrait in the case of full desynchronization, for C_{12} $\leq$5.90 nF

between two different manifolds (the diagonal and structured desynchronization phase portrait) the classification of this intermittency as one of the "*In-Out*" kind, can be safely supported, according to the related literature [52, 55, 56].

Checking the mean duration $< \tau >$ of laminar lengths versus the difference $(R_C - R_{C-\text{crit}})$ it was confirmed by the linear fitting of experimental data results (Fig. 9.47) that (9.2) holds also in this case, with a slope estimated to be $\gamma = -1.035$, further confirming the "*In-Out*" classification.

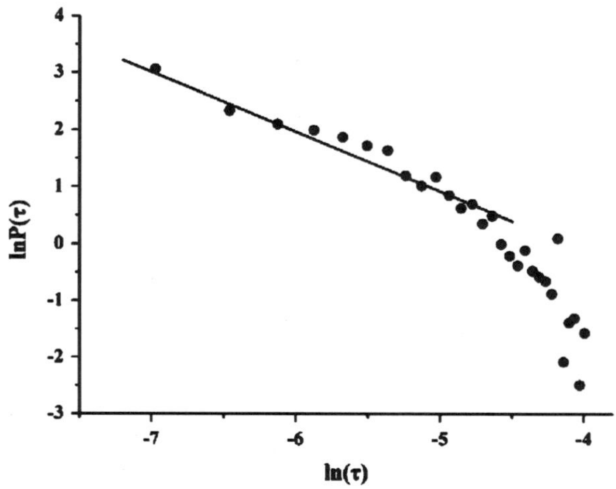

Fig. 9.46 Representative distributions $P(\tau)$ of durations τ of the laminar lengths for $C_{12} = 6.40\,\text{nF}$, in a double logarithmic plot. The line represents a linear fitting fulfilling the first term of the power law in (9.3)

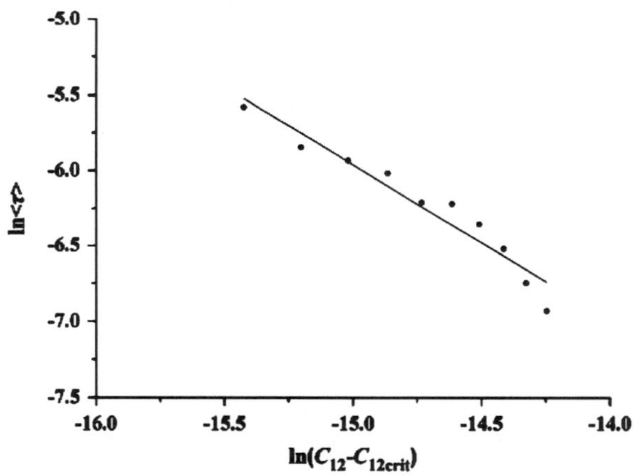

Fig. 9.47 Double logarithmic plot of the mean duration $< \tau >$ of laminar lengths vs. the difference $(C_{12} - C_{12\text{crit}})$. *Full circles* denote experimental data, while the *line* represents a fitting with a power law of the form of (9.2)

Finally, it should be mentioned that in this case of synchronization desynchronization route of 4th order coupled circuits, another feature of "*In-Out*" intermittent behavior was encountered: that of increasing burst heights, in the difference signal, with increasing $(C_{12} - C_{12\text{crit}})$ as theoretically predicted [52, 55, 56].

9.6 Discussion

In the present chapter, experimental results and their evaluation, in different cases of coupled, nonlinear electronic circuits, undergoing synchronization to desynchronization routes, have been presented.

These incomplete synchronization cases included three routes, namely the *crisis* induced intermittency, the "*On-Off*" intermittency and the "*In-Out*" intermittency.

All three cases were studied in the whole range the phenomena were developed, both qualitative and quantitative. Decisive factors for classifying the intermittency type of these transitions were: morphology of synchronization phase portraits and laminar length distribution, as well as scaling of mean duration $<\tau>$ of laminar lengths according to the deviation of the phenomenon control parameter from its critical value.

Although, intermittency is an anticipated route from a system's synchronized state to a fully desynchronized one, other routes to desynchronization could not be explicitly closed out. However, in the reported in this chapter synchronized systems and for the specific sets of parameters, routes like period doubling or quasi-periodicity were not observed.

It would be interesting to examine the potential of these systems to undergo other kinds of desynchronization routes. The authors believe that the existence of such routes could emerge in synchronization-desynchronization, if the systems are not in a chaotic state of operation but at the edge of chaos i.e. at the stage of passing from periodicity to chaos.

Finally, since desynchronization routes also appear for certain parameter mismatch, it would be interesting to consider the case of synchronization desynchronization of two otherwise identical, coupled circuits each one undergoing a different route to chaos.

References

1. Demetrakos, D.: Abridged Dictionary of the Greek Language. (in Greek) Giovanis Publications, Athens (1970)
2. Win, M.Z., Dardari, D., Molisch, A.F., Wiesbeck, W., Zhang, J.: History and applications of UWB. In: Special Issue on UWB technology and emerging applications, Proceedings of the IEEE, vol. 97, issue 2, pp. 198–204 (2009)
3. Chambers, W.G., Frey, D.: Comments on chaotic digital encoding: an approach to secure communication and reply. IEEE Trans. CAS II **46**, 1445–1448 (1993)
4. Chien, T., Liao, T.-L.: Design of secure digital communication systems using chaotic modulation, cryptography and chaotic synchronization. J. Chaos Solitons Fractals **24**, 241–255 (2005)
5. Carroll, T.L.: Noise-resistant chaotic synchronization. Phys. Rev. E **64**, 1–4 (2001)
6. Stavrinides, S.G., Anagnostopoulos, A.N., Miliou, A.N., Valaristos, A., Magafas, L., Kosmatopoulos, K., Papaioannou, S.: A digital chaotic synchronized communication system. J. Eng. Sci. Tech. Rev. **2**(1), 82–86 (2009)
7. Schwartz, W., Abel, A.: Chaos in communications—An Introduction. Dresden (2003). Available via http://rfic.ucsd.edu/chaos/ws2000/Dresden_1.pdf. Accessed 15 Sept 2009
8. Pecora, L.M., Carroll, T.L.: Synchronization in chaotic systems. Phys. Rev. Lett. **64**, 821–823 (1990)
9. Chong, C.-C., Yong, S.K., Lee, S.S.: UWB direct chaotic communication technology. IEEE Antenn. Wireless Propag. Lett. **4**, 316–319 (2005)
10. Chong, C.-C., Yong, S.K.: UWB direct chaotic communication technology for low-rate WPAN applications. IEEE Trans. Vehicular Tech. **57**(3), 1527–1536 (2008)
11. Ogorzalek, M.J.: Taming chaos—Part I: Synchronization. IEEE Trans. CAS I **40**, 693–699 (1993)

12. Frey, D. Chaotic digital encoding: An approach to secure communication. IEEE Trans. CAS II **40**, 660–666 (1993)
13. Yang, T., Chua, L.O.: Impulsive stabilization for control and synchronization of chaotic systems: Theory and application to secure communication. IEEE Trans. CAS I **44**, 976–988 (1997)
14. Yang, T., Wu C.W., Chua, L.O. Cryptography based on chaotic systems. IEEE Trans. CAS I **44**, 469–472 (1997)
15. Yang, X., Wu, T.X., Jaggard, D.L.: Synchronization recovery of chaotic wave through an imperfect channel. IEEE Antenn. Wireless Propag. Lett. **1**, 154–156 (2002)
16. Miliou, A.N., Stavrinides, S.G., Valaristos, A.P., Anagnostopoulos, A.N.: Nonlinear electronic circuit – Part II: Synchronization in a chaotic MODEM scheme. Nonlinear Anal.: TMA **71**, e21–e31 (2009)
17. Tamaševičious, A., Čenys, A., Mycolaitis, G., Namajunas, A.: Synchronizing hyperchaos in infinite-dimensional dynamical systems. J. Chaos Solitons Fractals **9**, 1403–1408 (1998)
18. Kennedy, M.P., Rovatti, R., Setti, G. (eds.): Chaotic Electronics in Telecommunications. CRC, Boca Raton (2000)
19. Dmitriev, A.S., Efremova, E.V., Kuzmin, L.V., Miliou, A.N., Panas, A.I., Starkov, S.O.: Secure transmission of analog information using chaos. In: Banerjee, S. (ed.) Chaos Synchronization and Cryptography for Secure Communications: Applications for Encryption, pp. 337–360. IGI Global, PA, USA (2010)
20. Hasler, M.: Electrical circuits with chaotic behavior. Proc. IEEE **75**(8), 1009–1021 (1987)
21. Matsumoto, T.: Chaos in electronic circuits. Proc. IEEE **75**(8), 1033–1057 (1987)
22. Ogorzalek, M.: Special issue on nonlinear dynamics in electronic systems. Int. J. Electron. **79**(6), 717–952 (1995)
23. van der Pol, B., van der Mark, J.: Frequency demultiplication. Nature **120**, 363–364 (1927)
24. Abraham, R., Ueda, Y. (ed.): The Chaos avant-garde. World Scientific, Singapore (2000)
25. Matsumoto, T., Chua, L.O., Tanaka, S.: Simplest non-chaotic non-autonomous circuit. Phys. Rev. A **30**, 1155 (1984)
26. Brorson, D.S., Dewey, D., Linsay, P.S.: Self-replicating attractor of a driven semiconductor oscillator. Phys. Rev. A, **28**, 1201 (1983)
27. Su, Z., Rollins, R.W., Hunt, E.R.: Simulation and characterization of strange attractors in driven diode resonator systems. Phys. Rev. A **40**, 2698 (1989)
28. Chua, L.O., Hasler, M., Neirynck, J., Verburgh, P.: Dynamics of a piecewise-linear resonant circuit. IEEE Trans. CAS **29**(8), 535–547 (1982)
29. Chua, L.O., Komuro, M., Matsumoto, T.: The double scroll family. IEEE CAS **33**(11), 1073–1118 (1986)
30. Chen, G., Ueta, T. (eds.): Chaos in Circuits and Systems. World Scientific, Singapore (2002)
31. Scholl, E., Schuster, H.G. (eds.): Handbook of Chaos Control, 2nd edn. Wiley-VCH, New York (2008)
32. Orgorzalek, M.J.: Chaos and Complexity in Nonlinear Electronic Circuits. World Scientific, Singapore (1997)
33. Dmitriev, A.S., Panas, A.I., Starkov, S.O., et al.: Method for transmission information using chaotic signals. Canadian Patent 2,416,702, granted and issued 2011/07/19
34. Kim, J.-H., et al (Samsung Electronics Co Ltd): Communication apparatus using chaotic signal and method thereof. US patent 8005168. Filing date: 28 Aug 2007. Issue date: 23 Aug 2011
35. Boccaletti, S., Kurths, J., Osipov, G., Valladares, D.L., Zhou, C.S.: The synchronization of chaotic systems. Phys. Rep. **366**, 1–101 (2002)
36. Huygens, C.: Horologium Oscillatorium, Paris, France, 1673. The Pendulum clock (English translation). Iowa State University Press (1986)
37. Pikovsky, A., Rosenblum, M., Kurths, J.: Synchronization: A universal concept in nonlinear sciences 1st paperback edition. In: The Cambridge Nonlinear Science Series. Cambridge University Press, London (2003)
38. Vazqueza, F., Zanette, D.H.: Epidemics and chaotic synchronization in recombining monogamous populations. Phys. D **239**, 1922–1928 (2010)

39. Volos, C.K., Kyprianidis, I.M., Stavrinides, S.G., Stouboulos, I.N., Magafas, L., Anagnostopoulos, A.N.: Nonlinear financial dynamics from an engineer's point of view. In: Proceedings of 1st International Conference on Econophysics (ICE 2011), Kavala, Greece, (2011)
40. Georganta, Z., Magafas, L., Stavrinides, S.G., Hanias, M. (eds.): Special issue on Econophysics. J. Eng. Sci. Tech. Rev. **4**(3), 207–302 (2011)
41. Shuster, H.G.: Deterministic Chaos. VCH, Weinhaim (2005)
42. Miliou, A.N, Valaristos, A.P., Stavrinides, S.G., Kyritsi, K.G., Anagnostopoulos, A.N.: Characterization of a non-autonomous second order non-linear circuit for secure data transmission. Chaos Solitons Fractals **33**(4), 1248–1255 (2007)
43. Volos, C.K., Kyprianidis, I.M., Stouboulos, I.N., Stavrinides, S.G., Anagnostopoulos, A.N., Ozer, M.: The concept of unidirectionally coupled nonlinear circuits via a memristor. Acta Physica Polonica A **121**(1), 268–270 (2012)
44. Rulkov, N.F.: Images of synchronized chaos: Experiments with circuits. Chaos **6,** 262–279 (1996)
45. Shahverdiev, E.M., Shorea, K.A.: Synchronization of chaos in unidirectionally and bidirectionally coupled multiple time delay laser diodes with electro-optical feedback. Optic. Comm. **282**(2), 310–315 (2009)
46. Liu, L., Li, R., Che, Y., Han, C.: Chaos synchronization of coupled neurons via h-infinity control with cooperative weights neural network. In: Advances in Future Computer and Control Systems. Advances in Intelligent and Soft Computing, vol. 160, pp. 369–374 (2012)
47. Rulkov, N.F., Sushchik, M.M., Tsimring, L.S., Abarbanel, H.D.I.: Generalized synchronization of chaos in directionally coupled chaotic systems. Phys. Rev. E **51**, 980–994 (1995)
48. Volos, C.K., Kyprianidis, I.M., Stavrinides, S.G., Stouboulos, I.N., Anagnostopoulos, A.N.: Inverse lag synchronization in mutually coupled nonlinear circuits. In: Proceedings of 14th WSEAS (2010) International Conference on Communications, pp. 31–37, Corfu, Greece (2010)
49. Park, J.H.: Functional projective synchronization of chaotic systems via nonlinear feedback control. Int. J. Nonlinear Sci. Numer. Simulat. **10**(10), 1301–1306 (2011)
50. Tse, C.K.: Complex Behavior of Switching Power Converters. CRC, Boca Raton (2004)
51. Li, C., Liao, X., Zhang, X.: Impulsive synchronization of chaotic systems. Chaos **15**, 023104 (2005)
52. Chen, Q., Hong, Y., Chen, G., Hill, D.J.: Intermittent Phenomena in switched systems with high coupling strengths. IEEE Trans. CAS I **53**(12), 2692–2704 (2006)
53. Toniolo, C., Provenzale, A., Spiegel, E.A.: Signature of on-off intermittency in measured signals. Phys. Rev. E **66**, 066209 1–12 (2002)
54. Yu, Y.H., Kwak, K., Lim, T.K.: On-off intermittency in an experimental synchronization process. Phys. Lett. A, **198**, 34–38 (1995)
55. Covas, E., Tavakol, R., Aswin, P., Tworkowski, A., Brooke, J.M.: In-out intermittency in partial differential equation and ordinary differential equation models. Chaos **11**(2), 404–409 (2001)
56. Aswin, P., Covas, E., Tavakol, R.: Influence of noise on scalings for in-out intermittency. Phys. Rev. E **64**, 066204 (2001)
57. Kyprianidis, I.M., Volos, C.K., Stavrinides, S.G., Stouboulos, I.N., Anagnostopoulos, A.N.: On-off intermittent synchronization between two bidirectionally coupled double scroll circuits. Comm. Nonlinear Sci. Numer. Simulat. **15**, 2192–2200 (2010)
58. Kyprianidis, I.M., Volos, C.K., Stavrinides, S.G., Stouboulos, I.N., Anagnostopoulos, A.N.: Master-Slave double-scroll circuit incomplete synchronization. J. Eng. Sci. Tech. Rev. **3**, 41–45 (2010)
59. Kyprianidis, I.M., Papadopoulou, M., Stavrinides, S.G., Stouboulos, I.N., Anagnostopoulos, A.N.: In-out intermittent loss of synchronization in two unidirectionally coupled nonlinear 4rth order autonomous circuits. Int. J. Bifurcat. Chaos **20**(10), 1–11 (2010)
60. Kyprianidis, I.M., Volos, C.K., Stavrinides, S.G., Stouboulos, I.N., Anagnostopoulos, A.N., Miliou, A.N.: Experimental in-out synchronization-intermittency between two bidirectionally coupled circuits. J. Concr. Appl. Math. **9**(3), 262–269 (2011)

61. Stavrinides, S.G., Miliou, A.N., Anagnostopoulos, A.N., Konstantakos, V., Laopoulos, Th., Magafas, L.: Desynchronization crisis induced intermittency in a master-slave PLL configuration. Chaos Solitons Fractals **42**(1), 33–39 (2009)
62. Elwakil, A.S., Salama, K.N., Kennedy, M.P.: A system for chaos generation and its implementation in monolithic form. In: Proceedings of IEEE ISCAS'00, Geneva, Switzerland, vol. 5, pp. 217–220 (2000)
63. Elwakil, A.S., Salama, K.N., Kennedy, M.P.: An equation for generating chaos and its monolithic implementation. Int. J. Bifurcat. Chaos **6**, 2885–2896 (2002)
64. Lü, J., Chen, G., Yu, X., Leung, H.: Design and analysis of multi-scroll chaotic attractors from saturated function series. IEEE Trans. CAS I **51**, 2476–2490 (2001)
65. Chua, L.O., Lin, G.N.: Canonical realization of Chua's circuit family. IEEE Trans. CAS I **37**, 885–902 (1990)
66. Chua, L.O., Lin, G.N.: Intermittency in a piecewise-linear circuit. IEEE Trans. CAS I **38**, 510–520 (1991)
67. Ling, B.W.-K., Iu, H.H.-C., Lam, H.-K. (eds.): Control of Chaos in Nonlinear Circuits and Systems. Nonlinear Science Series A. World Scientific, Singapore (2009)
68. Ott, E., Chaos in Dynamical Systems, 2nd edn. Cambridge University Press, London (2002)
69. Takens, F.: Dynamical Systems and Turbulence. Lecture Notes in Mathematics, p. 898. Springer, New York (1981)
70. Grebogi, C., Ott, E., Yorke, J.A.: Crises, sudden changes in chaotic attractors and transient chaos. Phys. D **7**, 181–200 (1983)
71. Stavrinides, S.G., Laopoulos, Th., Anagnostopoulos, A.N.: An automated acquisition setup for the analysis of chaotic systems. In: Proceedings of IEEE IDAACS'05, pp. 628–632, Sofia, Bulgaria (2005)
72. Stavrinides, S.G., Laopoulos, Th., Anagnostopoulos, A.N.: An automated acquisition setup for the evaluation of intermittency statistics. In: Proceedings of IEEE IDAACS'09, pp. 111–116, Cosenza (Rende), Italy (2009)
73. Grebogi, C., Ott, E., Romeiras, F., Yorke, J.A.: Critical exponents for crisis induced intermittency. Phys. Rev. A **36**, 5365 (1987)
74. Grebogi, C., Ott, E., Yorke, J. A.: Critical exponents of chaotic transients in nonlinear dynamical systems. Phys. Rev. Lett. **57**, 1284 (1986)

Chapter 10
Projective Synchronization of Delayed Chaotic Systems

S. Jeeva Sathya Theesar, P. Balasubramaniam, and Santo Banerjee

Abstract This chapter is focused to study various analytical conditions for occurrence of projective synchronization of time varying delayed chaotic systems. Synchronization criteria which depend on the time delay are presented via linear matrix inequalities and these conditions have been extended to chaotic neural networks with time-varying delay. Non-differentiable time varying delays can be present in the system and the more relaxed condition gives the upper delay bounds for ensuring the projective synchronization. To achieve the desired proportional relation between the drive and response system, an active nonlinear control technique using an observer technique is applied and the control is practically realizable. Various numerical results on Ikeda system, Lu attractor and modified delayed-Chua system are exhibited to show the effectiveness of the proposed results.

10.1 Introduction

Projective synchronization (PS) is a dynamical phenomena where state of the drive system $x(t)$ and that of the response system $y(t)$ synchronizes with respect to the scaling factor α such that there is a proportional relation $y(t) = \alpha x(t)$. This special

S.J.S. Theesar
Institute of Mathematical Sciences, University of Malaya, 50603 Kuala Lumpur, Malaysia
e-mail: sjstheesar@gmail.com

P. Balasubramaniam
Department of Mathematics, Gandhigram Rural University, Gandhigram 624302, Tamilnadu, India
e-mail: balugru@gmail.com

S. Banerjee (✉)
Institute for Mathematical Research, University Putra Malaysia, Malaysia

Department of Complexity and Network Dynamics, International Science Association (ISCASS), Ankara, Turkey
e-mail: santoban@gmail.com

S. Banerjee and L. Rondoni (eds.), *Applications of Chaos and Nonlinear Dynamics in Science and Engineering 3*, Understanding Complex Systems,
DOI 10.1007/978-3-642-34017-8_10,

type of synchronization behaviour was first observed by Gonzales-Miranda [11], by studying the invariance properties under amplification and displacement of the chaotic attractors. Mainieri and Rehacek [24] first named this phenomena, where two identical chaotic systems could be synchronized up to a scaling factor, as projective synchronization. The study on PS showed useful insight to various real time applications from secure communication [2, 6, 22] to Cellular Neural Networks (CNN) [12] due to its proportionality relation. In secure communication problem, this characteristic used for M-nary digital communication for fast transfer of data [6]. Thus researcher dwelled into the study of various types of projective synchronization in finite dimensional systems [19, 31], infinite dimensional systems [5, 9, 29], multiple-delayed systems [15, 16], neural networks [21, 25] and complex networks [10], etc.

In existing results [22, 29, 30], the performance of the projective synchronization was controlled by the scaling factor. In the view point of control theory, however the scaling scaling factor, the projective synchronization must occur and the nonlinear observer based control technique [12] came into play. An observer is a dynamic system designed to be driven by the output of another dynamic system (plant) and having the property that the state of the observer converges to the state of the plant [12]. On the other hand, lag between the states of a chaotic system due to signal transmission can not be neglected and further delay can detonate the successful secure communication. In fact, even for finite dimensional system, the delayed feedback control technique [14, 26] was introduced for practical realization of infinite dimensional structure and further time delay in chaotic system helped to have better secure communication system. Very recently, the study of PS using observer technique [9] was proposed in time delay system considering the above facts. The theoretical analysis for projective synchronization of time delay system lacks pertinence as the obtained condition is applicable only for lower order delay system and differentiable time varying delay with a strong restriction. Also, the result in [9] does not even depend on the time delay and can not abe applied to time varying delays. Also it is worth noting that if delay time is known priori, the time-delayed system becomes quite simple and the message encrypted by the chaotic dynamic can easily be extracted by the common attack methods. Thus more general case of time varying delay such as non differentiable delays and modulated delays [18] must be considered.

Motivated by the above discussion and considering the practical applicability of general case of time delays, this study overcomes the above mentioned drawbacks. Firstly, PS problem is generalized to system with time varying delay. The delay can be non-differentiable but bounded by a constant. The active nonlinear observer is used for control and the linear error state feedback gain is obtained from sufficient linear matrix inequality [4] conditions. The absolute stability of error dynamics ensures the synchronization between the drive and response systems. By making use of Krasovskii-Lyapunov approach [20], the sufficient conditions have been obtained.

Can we have a general condition so that given two time-varying delayed system achieve PS with given scaling factor α? Answer to this question is presented in this first part of the present study. In the next part, we intended to look for generalized projective synchronization (GPS) and GPS of interacting network of

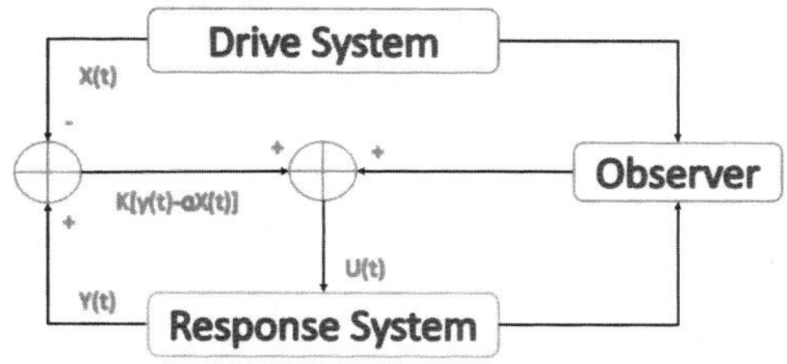

Fig. 10.1 Schema of projective synchronization with observer control

chaotic systems. In the present chapter, applicability of the sufficient conditions to chaotic neural networks [1,23,27] and multiple delayed systems are presented. The conditions are well tested on Ikeda systems [2, 17], Lu chaotic attractor [23], and newly introduced delayed-Chua chaotic system. Numerical results clearly shows the effectiveness of the proposed method and the sufficient conditions.

This chapter is organized as follows. In the next section, the definition of PS between two general delayed chaotic systems is introduced and the main problem is formulated. In Sect. 10.3, the sufficient conditions for achieving PS are derived for two identical delayed chaotic systems and its various extensions. Section 10.4 takes the PS of multiple delayed chaotic systems and Sect. 10.5 is devoted to some applications of the proposed scheme. In Sect. 10.6, numerical examples to verify the scheme are presented. Conclusions are given in the final section.

10.2 Projective Synchronization Formulation

Consider the coupled chaotic systems with time varying delay

$$\dot{x}(t) = f(x, x_{\tau(t)}) \tag{10.1}$$

$$\dot{y}(t) = f(y, y_{\tau(t)}) + u(x, y) \tag{10.2}$$

where $x, y \in \Re^n$, $f : \Re^n \to \Re^n$ is a nonlinear vector field, $u(x, y)$ is the active nonlinear control term and $x_{\tau(t)} = x(t-\tau(t))$. The main objective of this study is to achieve projective synchronization between (10.1) and (10.2). Let the error between two dynamics be $e = y(t) - \alpha x(t)$. If $e \to 0$ as $t \to \infty$, projective synchronization between the coupled system is achieved with respect to the nonlinear control $u(x, y)$ to be designed. PS synchronization scheme with nonlinear observer based control is presented in Fig. 10.1.

Definition 1. A projective observer is a dynamical system in which the state of the master is proportional by a constant scaling factor α with the state of the slave's. Let the output of the system (1) by $z = s(x, x_{\tau(t)})$. The dynamical system

$$\dot{y}(t) = f(y, y_{\tau(t)}) + g(z - s(y, y_{\tau(t)})) \tag{10.3}$$

is then said to be nonlinear projective observer system (1) if $y \rightarrow \alpha x(t)$, where $g : \Re^n \rightarrow \Re^n$ is a suitable chosen nonlinear function. Thus synchronization manifold of the system (10.1) and (10.2) is $y = \alpha x(t)$.

Model (10.1) and (10.2) can be written in the following form:

$$\dot{x}(t) = Ax + \mu_1 B_1\phi_1(x) + \mu_1 B_2\phi_2\left(x_{\tau(t)}\right) \tag{10.4}$$

$$\dot{y}(t) = Ay + \mu_2 B_1\phi_1\left(\frac{y}{\alpha}\right) + \mu_2 B_2\phi_2\left(\frac{y_{\tau(t)}}{\alpha}\right) + g(z - s(y, y_{\tau(t)})) \tag{10.5}$$

where $x(t),\ y(t) \in \Re^n$, $A \in \Re^{n\times n}$, $B_j \in \Re^{n\times n}$, $\phi_i(x) = \left[\phi_{j1}, \phi_{j2}, \cdots, \phi_{jn}\right]^T : \Re^n \rightarrow \Re^n$ is a nonlinear function of x, satisfying the generalized sector bounding condition

$$\left[\phi_{ji}(\xi) - l_{ji}^{+}\xi\right]\left[\phi_{ji}(\xi) - l_{ji}^{-}\xi\right] \leqslant 0, \quad \forall\xi, \tag{10.6}$$

for all $i = 1, 2, \ldots, n$, and $j = 1, 2$, where $\phi_{ji}(\cdot)$ belongs to the sector $\left[l_{ji}^{-}, l_{ji}^{+}\right]$, characterizing the nonlinear behavior of the system which forms large class of chaotic and hyperchaotic systems. Here $\tau(t)$ is the time varying delay such that $0 \leqslant \tau(t) \leqslant \tau_M < \infty$ and τ_M is a scalar.

Given a system, let $s(x, x_{\tau(t)}) = B_1\phi_1(x) + B_2\phi_2(x_{\tau(t)}) + Cx$ be the nonlinear observer of synchronizing signal, and without loss of generality, choose $g(z_1 - s(y, y_{\tau(t)}))$ such that

$$u(x, y) = B_1\left[\phi_1(y) - \phi_1(\alpha x)\right] + B_2\left[\phi_2(y_{\tau(t)}) - \phi_2(x_{\tau(t)})\right] - KC\left[y - \alpha x\right] \tag{10.7}$$

where $K \in \Re^{n\times n}$ is the coupling strength to be determined later. The synchronization error between two dynamics be $e = y - \alpha x$. Thus the error dynamic can be obtained as

$$\begin{aligned}\dot{e} &= \dot{y} - \alpha\dot{x}\\ &= (A - KC)e + B_1\tilde{\phi}_1(e) + B_2\tilde{\phi}_2(e_{\tau(t)}),\end{aligned} \tag{10.8}$$

where $\tilde{\phi}_1(e) = \phi_1(y) - \phi_1(\alpha x)$, $\tilde{\phi}_2(e_{\tau(t)}) = \phi_2(y_{\tau(t)}) - \phi_2(\alpha x_{\tau(t)})$, and $\mu_2 = \alpha\mu_1$. From (10.6), $\tilde{\phi}_j(\xi)$ satisfies

$$\left[\tilde{\phi}_{ji}(\xi) - l_{ji}^{+}\xi\right]\left[\tilde{\phi}_{ji}(\xi) - l_{ji}^{-}\xi\right] \leqslant 0, \quad \forall\xi. \tag{10.9}$$

The initial condition corresponding to the error dynamical system (10.8) is given as $e(\theta) = \varphi(\theta)$, $\theta \in [-\tau_M, 0]$, $\varphi \in \mathbb{M}$, where $\mathbb{M}$ is the Banach space of absolutely continuous functions from $[-\tau_M, 0]$ to $\Re^n$ with square-integrable derivative and with the norm

$$\|\varphi\|_{\mathbb{M}} = \left[\|\varphi(0)\|^2 + \int_{-\tau_M}^{0} \|\varphi(s)\|^2 ds + \int_{-\tau_M}^{0} \|\dot{\varphi}(\theta)\|^2 d\theta\right]^{1/2},$$

where the vector norm $\|\cdot\|$ represents the Euclidean norm. Few useful results used in obtaining the condition are stated below.

Lemma 1. *[4] (Schur complement) For a given matrix* $S = \begin{bmatrix} S_{11} & S_{12} \\ S_{12}^T & S_{22} \end{bmatrix} > 0$, *where* $S_{11} = S_{11}^T$, $S_{12} = S_{12}^T$, *is equivalent to any one of the following conditions:*

$$(1) \quad S_{22} > 0, \; S_{11} - S_{12}S_{22}^{-1}S_{12}^T > 0$$
$$(2) \quad S_{11} > 0, \; S_{11} - S_{12}^T S_{11}^{-1} S_{12} > 0$$

Lemma 2. *[13] For any constant matrix* $X \in \mathbb{R}^{n\times n}$, $X = X^T > 0$, *there exists positive scalar* τ_M *such that* $0 \leqslant \tau(t) \leqslant \tau_M$, *and a vector-valued function* $\dot{x}$: $[-\tau_M, 0] \rightarrow \mathbb{R}^n$, *the integration* $-\tau_M \int_{t-\tau_M}^{t} \dot{x}^T(s)X\dot{x}(s)ds$ *is well defined,*

$$-\tau_M \int_{t-\tau_M}^{t} \dot{x}^T(s)X\dot{x}(s)ds \leq \begin{bmatrix} x(t) \\ x(t-\tau_M) \end{bmatrix}^T \begin{bmatrix} -X & X \\ \star & -X \end{bmatrix} \begin{bmatrix} x(t) \\ x(t-\tau_M) \end{bmatrix}.$$

In this section we formulated the PS problem between (10.4) and (10.5) with nonlinear observer controller (10.7). The asymptotical stability of closed-loop error dynamics (10.8) ensures the PS and the design of the control gain K is presented in the following section as sufficient conditions.

10.3 Analytical Conditions

In this section, we will derive the projective synchronization criteria based on the stability theory of time varying delayed systems. For this objective, consider the Krasovskii-Lyapunov functional as

$$V(t, e_t, \dot{e}_t) = e^T(t)Pe(t) + \int_{-\tau_M}^{0} e^T(t+\xi)\, Z_1\, e(t+\xi)d\xi$$
$$+\tau_M \int_{-\tau_M}^{0} \int_{t+\theta}^{t} \dot{e}^T(\xi)\, Z_2\, \dot{e}(\xi)d\xi d\theta \qquad (10.10)$$

where $e_t = e(t+s)$, $\forall s \in [-\tau_M, 0]$, and P, Z_1, Z_2 are positive definite matrices of appropriate dimensions. It is clear that $V(t, e_t, \dot{e}_t)$ is a quadratic functional depending on derivatives. The asymptotic stability of the error system described by (10.8) with (10.9) is assured if there exist $\varepsilon_i > 0$ $(i = 1, 2, 3)$ such that

$$\begin{aligned} \varepsilon_1 \|e(t)\|^2 \leqslant V(t, e_t, \dot{e}_t) &\leqslant \varepsilon_2 \|e_t\|_{\mathbb{M}}^2, \\ \dot{V}(t, e_t, \dot{e}_t) \qquad &\leqslant \qquad -\varepsilon_3 \|e(t)\|^2. \end{aligned}$$

We now state and establish the following theorem.

Theorem 1. *For the scalar $\tau_M > 0$, the system described by (10.8) and (10.9) is globally asymptotically stable with prescribed nonlinear observer control (10.7), if there exist symmetric positive definite matrices P, Z_1, Z_2, diagonal matrices $U > 0$, $V > 0$, and any matrix G with compatible dimensions such that the following inequality holds:*

$$\Xi = \begin{bmatrix} \eta_{11} & Z_2 & 0 & F_{12}U + GB_1 & GB_2 & \eta_{16} \\ \star & \eta_{22} & Z_2 & 0 & F_{22}V & 0 \\ \star & \star & -Z_2 - Z_1 & 0 & 0 & 0 \\ \star & \star & \star & -U & 0 & (GB_1)^T \\ \star & \star & \star & \star & -V & (GB_2)^T \\ \star & \star & \star & \star & \star & \eta_{66} \end{bmatrix} < 0 \qquad (10.11)$$

where
$\eta_{11} = Z_1 - Z_2 + 2(GA - LC) - F_{11}U$;
$\eta_{22} = -2Z_2 - F_{21}V$;
$\eta_{16} = -G + P + (GA - LC)^T$;
$\eta_{66} = -2G + \tau_M^2 Z_2$,
and the control gain is given by $K = G^{-1}L$.

Proof. Consider the Krasovskii-Lyapunov functional given in (10.10). Taking the time derivative of (10.10) along the trajectories of (10.8) yields

$$\begin{aligned} \dot{V}(t, e_t, \dot{e}_t) = {} & 2e^T(t)P\dot{e}(t) + e^T(t)Z_1 e(t) - e^T(t - \tau_M)Z_1 e(t - \tau_M) \\ & + \tau_M^2 \dot{e}^T(t)Z_2\dot{e}(t) - \tau_M \int_{t-\tau_M}^{t} \dot{e}^T(\theta)\, Z_2\, \dot{e}(\theta)d\theta \end{aligned} \qquad (10.12)$$

In view of Lemma 2, the integral term in the above equation (10.12) can be written as

$$-\tau_M \int_{t-\tau_M}^{t} \dot{e}^T(\theta) Z_2 \dot{e}(\theta) d\theta \leqslant \tilde{\xi}^T(t) \Omega \tilde{\xi}(t), \qquad (10.13)$$

where $\tilde{\xi}(t) = \begin{bmatrix} e \\ e_{\tau(t)} \\ e_{\tau_M} \end{bmatrix}$, and $\Omega = \begin{bmatrix} -Z_2 & Z_2 & 0 \\ \star & -2Z_2 & Z_2 \\ \star & \star & -Z_2 \end{bmatrix}$. From the sector bounding conditions, for all $i = 1, 2, \ldots, n$

$$\begin{bmatrix} e \\ \tilde{\phi}_1(e) \end{bmatrix}^T \begin{bmatrix} l_{1i}^- l_{1i}^+ \hat{e}_i \hat{e}_i^T & -\frac{l_{1i}^- + l_{1i}^+}{2} \hat{e}_i \hat{e}_i^T \\ \star & \hat{e}_i \hat{e}_i^T \end{bmatrix} \begin{bmatrix} e \\ \tilde{\phi}_1(e) \end{bmatrix} \leqslant 0 \tag{10.14}$$

$$\begin{bmatrix} e_{\tau(t)} \\ \tilde{\phi}_2(e_{\tau(t)}) \end{bmatrix}^T \begin{bmatrix} l_{2i}^- l_{2i}^+ \hat{e}_i \hat{e}_i^T & -\frac{l_{2i}^- + l_{2i}^+}{2} \hat{e}_i \hat{e}_i^T \\ \star & \hat{e}_i \hat{e}_i^T \end{bmatrix} \begin{bmatrix} e_{\tau(t)} \\ \tilde{\phi}_2(e_{\tau(t)}) \end{bmatrix} \leqslant 0 \tag{10.15}$$

where $\hat{e}_i$ denotes the unit column vector having 1 on its ith row and zeros elsewhere. Thus for any appropriately dimensioned diagonal matrices $U > 0$, $V > 0$ it can be obtained that

$$0 \leqslant \begin{bmatrix} e \\ \tilde{\phi}_1(e) \end{bmatrix}^T \begin{bmatrix} -F_{11}U & F_{12}U \\ \star & -U \end{bmatrix} \begin{bmatrix} e \\ \tilde{\phi}_1(e) \end{bmatrix} \tag{10.16}$$

$$0 \leqslant \begin{bmatrix} e_{\tau(t)} \\ \tilde{\phi}_2(e_{\tau(t)}) \end{bmatrix}^T \begin{bmatrix} -F_{21}V & F_{22}V \\ \star & -V \end{bmatrix} \begin{bmatrix} e_{\tau(t)} \\ \tilde{\phi}_2(e_{\tau(t)}) \end{bmatrix}, \tag{10.17}$$

where $F_{j1} = diag\left\{ l_{j1}^- l_{j1}^+, \, l_{j2}^- l_{j2}^+, \cdots l_{jn}^- l_{jn}^+, \right\}$, and

$F_{j2} = diag\left\{ \frac{l_{j1}^- + l_{j1}^+}{2}, \frac{l_{j2}^- + l_{j2}^+}{2}, \ldots \frac{l_{jn}^- + l_{jn}^+}{2} \right\}$. On the other hand one can have the following equation for appropriately dimensioned matrix G,

$$0 = \left[2e^T G + 2\dot{e}^T G \right] \left[-\dot{e} + (A - KC)e + B_1 \tilde{\phi}_1(e) + B_2 \tilde{\phi}_2(e_{\tau(t)}) \right]. \tag{10.18}$$

Thus, (10.12) can be written as

$$\begin{aligned} \dot{V}(e_t) &\leqslant 2e^T(t)P\dot{e}(t) + e^T(t)Z_1 e(t) - e^T(t - \tau_M)Z_1 e(t - \tau_M) + \tau_M^2 \dot{e}^T(t) Z_2 \dot{e}(t) \\ &\quad + \xi^T(t)\Omega\xi(t) + \begin{bmatrix} e \\ \tilde{\phi}_1(e) \end{bmatrix}^T \begin{bmatrix} -F_{11}U & F_{12}U \\ \star & -U \end{bmatrix} \begin{bmatrix} e \\ \tilde{\phi}_1(e) \end{bmatrix} \\ &\quad + \begin{bmatrix} e_{\tau(t)} \\ \tilde{\phi}_2(e_{\tau(t)}) \end{bmatrix}^T \begin{bmatrix} -F_{21}V & F_{22}V \\ \star & -V \end{bmatrix} \begin{bmatrix} e_{\tau(t)} \\ \tilde{\phi}_2(e_{\tau(t)}) \end{bmatrix} - 2e^T G\dot{e} + 2e^T G(A - KC)e \\ &\quad + 2e^T G B_1 \tilde{\phi}_1(e) + 2e^T G B_2 \tilde{\phi}_2(e_{\tau(t)}) - 2\dot{e}^T G\dot{e} + 2\dot{e}^T G(A - KC)e \\ &\quad + 2\dot{e}^T G B_1 \tilde{\phi}_1(e) + 2\dot{e}^T G B_2 \tilde{\phi}_2(e_{\tau(t)}). \end{aligned}$$

Letting $L = GK$ for $j = 1, 2, 3$, and $\zeta(t) = \left[\xi(t) \; \tilde{\phi}_1(e) \; \tilde{\phi}_2(e_{\tau(t)}) \; \dot{e}(t) \right]$, we have

$$\dot{V}(t, e_t, \dot{e}_t) \leq \zeta^T(t)\Xi\zeta(t) < 0, \tag{10.19}$$

for all $\zeta(t) \neq 0$ with $\Xi < 0$. Thus there exists a positive constant $\varepsilon > 0$ such that $\dot{V}(t, e_t, \dot{e}_t) \leqslant -\varepsilon \|e(t)\|^2$. This completes the proof.

Very recently, projective synchronization conditions for modulated time-varying delays have been discussed in [18]. In fact, the condition cannot be applied to higher order time varying delayed chaotic system where as Theorem 1 is more general case in terms of time delay. Also PS sufficient condition presented in [9] can not be applied to synchronize (10.4) and (10.5). Moreover, in secure communication problem, time delay plays a vital role in synchronization for encrypted transmitted signal and receiving signal for decryption. The synchronization scheme presented in this chapter can be applied directly as general sufficient condition in order to have a successful communication. It must be noted that the above theorem provides sufficient conditions for PS of time-varying delayed chaotic systems for bounded delays such that $0 \leqslant \tau(t) \leqslant \tau_M$. Thus the condition can be used for chaotic system with non-differentiable time varying delays and modulated delays such as $\tau(t) = \tau_0 + a_0 e^{|\sin(\omega_0 t)|}$ and $\tau(t) = \tau_0 + a_0|\sin(\omega_0 t)|$ and the criteria yields the upper-bound of time varying delay. If the derivative of time varying delay exists such that $\dot{\tau}(t) \leqslant \delta$, then we deduce the criteria as below.

Theorem 2. *For the scalars $\tau_M > 0$ and $\delta > 0$, the system described by (10.8) and (10.9) is globally asymptotically stable with prescribed nonlinear observer control (10.7), if there exist symmetric positive definite matrices P, Z_1, Z_2, Z_3 diagonal matrices $U > 0$, $V > 0$, and any matrix G with compatible dimensions such that the following inequality holds:*

$$\Xi = \begin{bmatrix} \eta_{11} & Z_2 & 0 & F_{12}U + GB_1 & GB_2 & \eta_{16} \\ \star & \eta_{22} & Z_2 & 0 & F_{22}V & 0 \\ \star & \star & -Z_2 - Z_1 & 0 & 0 & 0 \\ \star & \star & \star & -U & 0 & (GB_1)^T \\ \star & \star & \star & \star & -V & (GB_2)^T \\ \star & \star & \star & \star & \star & \eta_{66} \end{bmatrix} < 0 \qquad (10.20)$$

where

$\eta_{11} = Z_1 - Z_2 + Z_3 + 2(GA - LC) - F_{11}U;$

$\eta_{22} = -2Z_2 - F_{21}V - (1-\delta)Z_3;$

$\eta_{16} = -G + P + (GA - LC)^T;$

$\eta_{66} = -2G + \tau_M^2 Z_2,$

and the control gain is given by $K = G^{-1}L$.

Proof. Adding the functional $\int_{t-\tau(t)}^{t} e^T(\xi)\, Z_3\, e(\xi) d\xi$ to right hand side of (10.10) and following the same procedure, we get (10.20), Hence the detailed derivation is omitted.

The above theorem is presented for PS of chaotic systems with differentiable time-varying delays and modulated delays such as $\tau(t) = \tau_0 + a_0 e^{\sin(\omega_0 t)}$ and $\tau(t) = \tau_0 + a_0 \sin(\omega_0 t)$. To the best of author's knowledge, the above sufficient conditions are very less conservative than the existing results [18] in the literature.

For constant delay case, we have $\dot{\tau}(t) = \delta = 0$, and the criteria for projective synchronization is then deduced to (10.20) with $\eta_{22} = -2Z_2 - F_{21}V - Z_3$. In another way, we can get the condition for constant delay system given below without using $\tau(t)$ relation in the inequality (10.13).

Corollary 1. *For the constant delay* $\tau > 0$*, the system described by (10.8) and (10.9) with constant delay is globally asymptotically stable with prescribed nonlinear observer control (10.7), if there exist symmetric positive definite matrices* P, Z_1, Z_2*, diagonal matrices* $U > 0$, $V > 0$*, and any matrix* G *with compatible dimensions such that the following inequality holds:*

$$\Xi = \begin{bmatrix} \eta_{11} & Z_2 & F_{12}U + GB_1 & GB_2 & \eta_{15} \\ \star & -Z_2 - Z_1 & 0 & F_{22}V & 0 \\ \star & \star & -U & 0 & (GB_1)^T \\ \star & \star & \star & -V & (GB_2)^T \\ \star & \star & \star & \star & \eta_{55} \end{bmatrix} < 0 \qquad (10.21)$$

where

$\eta_{11} = Z_1 - Z_2 + 2(GA - LC) - F_{11}U;$
$\eta_{15} = -G + P + (GA - LC)^T;$
$\eta_{55} = -2G + \tau^2 Z_2,$
and the control gain is given by $K = G^{-1}L$.

In [9], PS for constant time-delay has been discussed for lower order chaotic systems. The condition does not depend on time-delay bearing nonlinear term is not suitable and failure one in case of multi-dimensional system. On the other hand, the above Theorem 2 ensures the PS of systems with constant time delay and can yield the applicable delay bound to achieve the synchronization. The above criteria improve the drawbacks of the above existing results.

Corollary 2. *In case, there exist only delayed nonlinear terms in the systems for models such as Ikeda system [17], then the system and the conditions can be deduced as follows. For the scalar* $\tau_M > 0$*, the system described by (10.8) with* $B_1 = 0$*, and* $B_2 = B$ *is globally asymptotically stable, if there exist symmetric positive definite matrices* P, Z_1, Z_2*, diagonal matrix* $U > 0$*, and any matrix* G *with compatible dimensions such that the following inequality holds:*

$$\Xi = \begin{bmatrix} \eta_{11} & Z_2 & 0 & 0 & \eta_{16} \\ \star & \eta_{22} & Z_2 & F_2U + GB & 0 \\ \star & \star & -Z_2 - Z_1 & 0 & 0 \\ \star & \star & \star & -U & (GB)^T \\ \star & \star & \star & \star & \eta_{66} \end{bmatrix} < 0 \qquad (10.22)$$

where

$\eta_{11} = Z_1 - Z_2 + 2(GA - LC);$
$\eta_{22} = -2Z_2 - F_1U;$
$\eta_{16} = -G + P + (GA - LC)^T;$
$\eta_{66} = -2G + \tau_M^2 Z_2,$
and the linear error state feedback control gain is given by $K = G^{-1}L$.

In this section, a more generalized sufficient condition for PS of delayed chaotic systems has been investigated. In secure communication problem, having single delay may be vulnerable to break the corresponding communication by means of conventional reconstruction techniques. To overcome this drawback, multi-delayed driving signal must be used and the new synchronization scheme must be obtained. This shows insight to more practical significance of the study of PS in chaotic system with multiple delays. In the following section, we present the PS sufficient conditions for multiple delayed chaotic systems.

10.4 Systems with Multiple Time-Delays

Time-delay chaotic systems arise in many applications from diverse areas such as economy, biology, population dynamics, traffic flow and communication systems. Synchronization of time-invariant time-delay systems itself is a notoriously complex task. In addition, consideration of multiple delays totally hampers the existing analysis which are limited to certain cases. There is still no comprehensive treatment for PS of most general multiple time-delay systems without strict limitations. All the existing techniques are case-specific and derived only for lower order time-delay systems. The main objective of this section is to develop a synchronization analysis scheme for the most general multiple time-delay systems, relaxing all the mentioned limitations and better than existing results.

If the system contains multiple time delays, then we can model the projective synchronization problem as follows:

$$\dot{x}(t) = Ax + \mu_1 B_1 \phi_1(x) + \mu_1 B_2 \sum_{k=1}^{N} \phi_2^{(k)}\left(x_{\tau_k(t)}\right) \tag{10.23}$$

$$\dot{y}(t) = Ay + \mu_2 B_1 \phi_1\left(\frac{y}{\alpha}\right) + \mu_2 B_2 \sum_{k=1}^{N} \phi_2^{(k)}\left(\frac{y_{\tau_k(t)}}{\alpha}\right) + g(z - s(y, y_{\tau_1}, \ldots, y_{\tau_N})) \tag{10.24}$$

such that $\phi_2^{(k)}(x) = \left[\phi_{21}^{(k)}, \phi_{j2}^{(k)}, \cdots, \phi_{jn}^{(k)}\right]^T : \Re^n \to \Re^n$ is a nonlinear function of x, satisfying the generalized sector bounding condition

$$\left[\phi_{2i}^{(k)}(\xi) - l_{2i}^{(k)+}\xi\right]\left[\phi_{2i}^{(k)}(\xi) - l_{2i}^{(k)-}\xi\right] \leqslant 0, \quad \forall \xi, \tag{10.25}$$

for all $k = 1, 2, \ldots, N$, $i = 1, 2, \ldots, n$, and $j = 1, 2$, where $\phi_{2i}^{(k)}(\cdot)$ belongs to the sector $\left[l_{2i}^{(k)-}, l_{2i}^{(k)+}\right]$. Here $\tau_k(t)$ is the time varying delay such that $0 \leqslant \tau_k(t) \leqslant \tau_k < \infty$ and $\tau_M = \max\left\{\tau_1, \tau_2, \cdots \tau_N\right\}$. Then the nonlinear observer of synchronizing signal is $s(x, x_{\tau_1(t)}, \ldots, x_{\tau_N(t)}) = B_1\phi_1(x) + B_2 \sum_{k=1}^{N} \phi_2^{(k)}(x_{\tau(t)}) + Cx$, and the controller can be chosen such that

$$u(x, y) = B_1 [\phi_1(y) - \phi_1(\alpha x)] + B_2 \sum_{k=1}^{N} \left[\phi_2^{(k)}(y_{\tau_k}) - \phi_2^{(k)}(x_{\tau_k})\right] - KC [y - \alpha x] \tag{10.26}$$

where $K \in \Re^{n\times n}$ is the coupling strength. Then we have

$$\dot{e} = (A - KC)e + B_1\tilde{\phi}_1(e) + B_2 \sum_{k=1}^{N} \tilde{\phi}_2^{(k)}(e_{\tau_k(t)}), \tag{10.27}$$

where $\tilde{\phi}_1(e) = \phi_1(y) - \phi_1(\alpha x)$, $\tilde{\phi}_2^{(k)}(e_{\tau_k(t)}) = \phi_2^{(k)}(y_{\tau_k(t)}) - \phi_2^{(k)}(\alpha x_{\tau_k(t)})$, and $\mu_2 = \alpha\mu_1$. From (10.25), $\tilde{\phi}_j^{(k)}(\xi)$ satisfies

$$\left[\tilde{\phi}_{ji}^{(k)}(\xi) - l_{ji}^{(k)+}\xi\right]\left[\tilde{\phi}_{ji}^{(k)}(\xi) - l_{ji}^{(k)-}\xi\right] \leqslant 0, \quad \forall\xi. \tag{10.28}$$

Then the required K-L functional can be given as

$$\begin{aligned} V(t, e_t, \dot{e}_t) = e^T(t)Pe(t) + \sum_{k=1}^{N} \Bigg\{ \int_{-\tau_k}^{0} e^T(t+\xi)\, Z_{1k}\, e(t+\xi)d\xi \\ +\tau_k \int_{-\tau_k}^{0} \int_{t+\theta}^{t} \dot{e}^T(\xi)\, Z_{2k}\, \dot{e}(\xi)d\xi d\theta \Bigg\}, \end{aligned} \tag{10.29}$$

and the inequality (10.17) can be written as

$$0 \leqslant \begin{bmatrix} e_{\tau_k(t)} \\ \tilde{\phi}_{2k}(e_{\tau_k(t)}) \end{bmatrix}^T \begin{bmatrix} -F_{21}^{(k)}V_k & F_{22}^{(k)}V_k \\ \star & -V_k \end{bmatrix} \begin{bmatrix} e_{\tau(t)} \\ \tilde{\phi}_2(e_{\tau(t)}) \end{bmatrix}, \tag{10.30}$$

where

$$\begin{aligned} F_{21}^{(k)} &= diag\left\{ l_{j1}^{(k)-} l_{j1}^{(k)+}, l_{j2}^{(k)-} l_{j2}^{(k)+}, \cdots l_{jn}^{(k)-} l_{jn}^{(k)+}, \right\}, \text{ and} \\ F_{j2}^{(k)} &= diag\left\{ \frac{l_{j1}^{(k)-} + l_{j1}^{(k)+}}{2}, \frac{l_{j2}^{(k)-} + l_{j2}^{(k)+}}{2}, \cdots \frac{l_{jn}^{(k)-} + l_{jn}^{(k)+}}{2} \right\}. \end{aligned}$$

The sufficient condition ensuring the PS of the multiple delayed chaotic systems is presented in the following theorem.

Theorem 3. *For the scalars $\tau_k > 0$, the system described by (10.27) with (10.28) is globally asymptotically stable with prescribed nonlinear observer control (10.26), if there exist symmetric positive definite matrices P, Z_{1k}, Z_{2k}, diagonal matrices*

$U > 0$, $V_k > 0$, for $k = 1, 2, \dots, N$ and any matrix G with compatible dimensions such that the following inequality holds:

$$\Xi < 0 \tag{10.31}$$

where

$$\Xi = \begin{bmatrix} \eta_{11} & Z_{21} & 0 & \cdots & 0 & 0 & F_{12}U+GB_1 & GB_2 & \cdots & GB_2 & \eta_{13N+3} \\ \star & \eta_{22} & Z_{21} & \cdots & Z_{2N-1} & Z_{2N} & 0 & F_{22}^{(1)}V_1 & \cdots & 0 & 0 \\ \star & \star & \eta_{33} & \cdots & 0 & 0 & 0 & 0 & \cdots & 0 & 0 \\ \vdots & \vdots & \vdots & \ddots & \vdots & \vdots & \vdots & \vdots & \dots & \vdots & \vdots \\ \star & \star & \star & \cdots & -2Z_{2N} & 0 & 0 & 0 & \dots & F_{22}^{(N)}V_N & 0 \\ \star & \star & \star & \cdots & \star & -Z_{1N}-Z_{2N} & 0 & 0 & \dots & 0 & 0 \\ \star & \star & \star & \dots & \star & \star & -U & 0 & \dots & 0 & (GB_1)^T \\ \star & \star & \star & \dots & \star & \star & \star & -V_1 & \dots & 0 & (GB_2)^T \\ \vdots & \vdots & \vdots & \vdots & \vdots & \vdots & \vdots & \vdots & \ddots & \vdots & \vdots \\ \star & \star & \star & \dots & \star & \star & \star & \star & \dots & V_N & (GB_2)^T \\ \star & \star & \star & \dots & \star & \star & \star & \star & \dots & \star & -2G+\tau^2 Z_2 \end{bmatrix}$$

$\eta_{11} = \sum_{k=1}^{N} Z_{1k} - \sum_{k=1}^{N} Z_{2k} + 2(GA - LC) - F_{11}U$;
$\eta_{22} = -2Z_{21} - F_{21}V$;
$\eta_{33} = -Z_{11} - Z_{21}$;
$\eta_{13N+3} = -G + P + (GA - LC)^T$;
and the control gain is given by $K = G^{-1}L$.

Though the concept of PS for single delayed chaotic system has been well established in the literature, PS in multiple time delayed system has been unexplored area. Theorem 3 presents more generalized conditions for PS of multiple delayed systems.

10.5 Some Applications

In the above two sections, we have presented the conditions for the delayed systems with sector bounded nonlinearities. Moreover, special cases of PS are complete synchronization (CS) in which the scaling factor $\alpha = 1$ and Anti-synchronization (AS) for $\alpha = -1$. The above obtained conditions can be applied to study CS and AS outmost. In order to have a successful secure communication, the present study of obtaining more general sufficient condition on proposed PS scheme with non-differentiable time delay plays a vital role in synchronization for encrypted transmitted signal and receiving signal for decryption. Further, these conditions can be applied to the neural networks with time varying delay.

As a special case of complex network, delayed neural networks (DNN) are also exposed complex dynamical properties even chaotic attractor [23]. Since then the study of synchronization in DNNs has been paid much attentions ([27] and [1], the reference cited therein). On the other hand, PS problem is first extended to CNN [7] and presented in [12] for finite dimensional chaotic systems. But for DNNs, the study in [25] presents the adaptive synchronization scheme. It must be noted that still there is a room for better condition for PS of DNNs with non-differentiable time varying delays. In this chapter, we extend the above synchronization schema presented in Sect. 10.2 and sufficient criteria presented in Sect. 10.3 to DNNs with more general time varying delay. Based on [27], unidirectional-coupled neural networks model can be given as

$$\dot{x}_i(t) = -b_i x_i(t) + \sum_{j=1}^{n} c_{ij} f_j(x_j(t)) + \sum_{j=1}^{n} d_{ij} f_j(x_j(t-\tau(t))) + I_i \quad i = 1, 2, \cdots, n, \tag{10.32}$$

where $n \geq 2$ denotes the number of neurons in the networks, x_i and y_i are the state variables associated with ith neuron of master and slave systems, $b_i x_i(t)$ is an appropriately behaved function imposing bounds on the solution of the master neural networks and c_{ij} and d_{ij} indicate the interconnection strength among the neurons without and with time-varying delays, respectively. The neuron activation function f_j describes the neuron response to each other, I_i denotes the constant external input. Equation (10.32) can then be transformed to the following compact form [1]:

$$\dot{x}(t) = -Bx(t) + Cf\big(x(t)\big) + Df\big(x(t-\tau(t))\big), \tag{10.33}$$

where $x(\cdot) = [x_1(\cdot), x_2(\cdot), \cdots, x_n(\cdot)]^T$, $f(x(\cdot)) = [f_1(x_1(\cdot), f_2(x_2(\cdot)), \cdots, f_n(x_n(\cdot))]^T$, $C = (c_{ij})_{(n\times n)}$, and $D = (d_{ij})_{(n\times n)}$, $B = diag\{b_1, b_2, \cdots b_n\}$. Thus the above model (10.33) is equivalent to (10.4) if $A = -B$, $B_1 = C$, $B_2 = D$, and $f(x(\cdot)) = \phi_1(x(t)) = \phi_2(x_{\tau(t)}(t))$. The condition can be easily deduced which is left to the reader.

10.6 Numerical Examples

In order to illustrate the effectiveness of the proposed scheme and obtained sufficient conditions in the previous sections, Ikeda system [2, 17], Lu attractor [23] and a new modified delayed-Chua system were used as examples.

10.6.1 Ikeda System

Consider unidirectional coupled Ikeda system in order to verify the above analytical conditions numerically:

$$\dot{x} = -ax + \mu_1 m_1 \sin(x_{\tau(t)}) \tag{10.34}$$

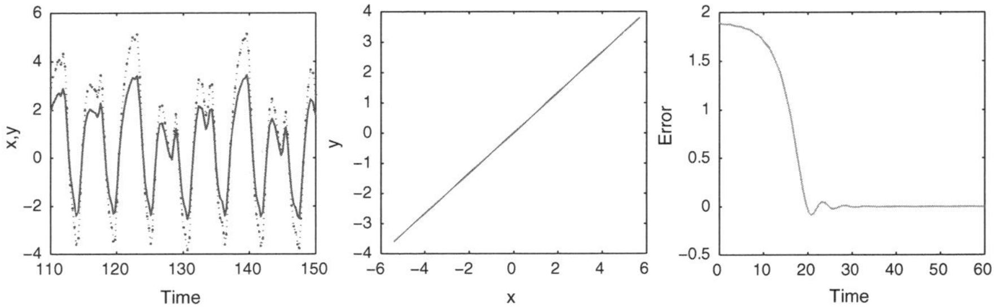

Fig. 10.2 Projective synchronization between Ikeda system (10.34) is depicted. Here $\alpha = 1.5$, $\tau(t) = 0.3 + 0.05|sin(t)|$ and $K = 0.966$. Synchronization manifold $y = 1.5x$ is plotted and projective synchronization achieved by $e \rightarrow 0$

Physically, x is the phase lag of the electric field across the resonator, a is the realization co efficient for the dynamical variable, and m_1 is the laser intensity injected into the system. The Ikeda model was introduced to describe the dynamics of an optical bistable resonator and is well known for delay-induced chaotic behavior.

System (10.28) is chaotic for the set of parameter values $a = 1$, $m_1 = 4$. The synchronization manifold along with error trajectories are presented in Fig. 10.2. Solving LMI presented in Theorem 1 for (10.28), we get $P = 4.4742 \times 10^8$, $Z_1 = 3.6391 \times 10^8$, $Z_2 = 5.4493 \times 10^{-7}$, the coupling strength $K = 0.966$, and the upper bound of the time varying delay as $\tau = 0.3535$. $\tau(t) = 0.3 + 0.05|sin(t)|$ is the most suitable time varying delay. For constant delay case, $\tau = 5$, the control gain achieved is $K = 0.9169$, for $\tau = 2$, the control gain achieved is $K = 0.9496$ and $K = 0.9415$ is obtained for $\tau = 1$. The error between the drive and response system is controlled and asymptotically reaches zero to ensure the synchronization with the given scaling factor $\alpha = 1.5$. It is noted that the coupling strength is very small which is practically applicable.

10.6.2 Lu Attractor

It has been shown that DNN are also exposed complex dynamical properties even chaotic attractor [23]. In order to test the obtained sufficient conditions on DNN, consider the following two dimensional chaotic system known as Lu attractor [23] with time varying delay:

$$\dot{x}(t) = -Ax(t) + B_1\phi_1(x(t)) + B_2\phi_2(x(t-\tau(t))), \tag{10.35}$$

where $A = diag\{1, 1\}$, $B_1 = \begin{bmatrix} 2.0 & -0.1 \\ -5.0 & 4.5 \end{bmatrix}$, $B_2 = \begin{bmatrix} -1.5 & -0.1 \\ -0.2 & -4.0 \end{bmatrix}$, $\phi_i(x_i(\cdot)) = tanh(x_i(\cdot))$. The chaotic attractor formed by (10.35) along with PS manifold for

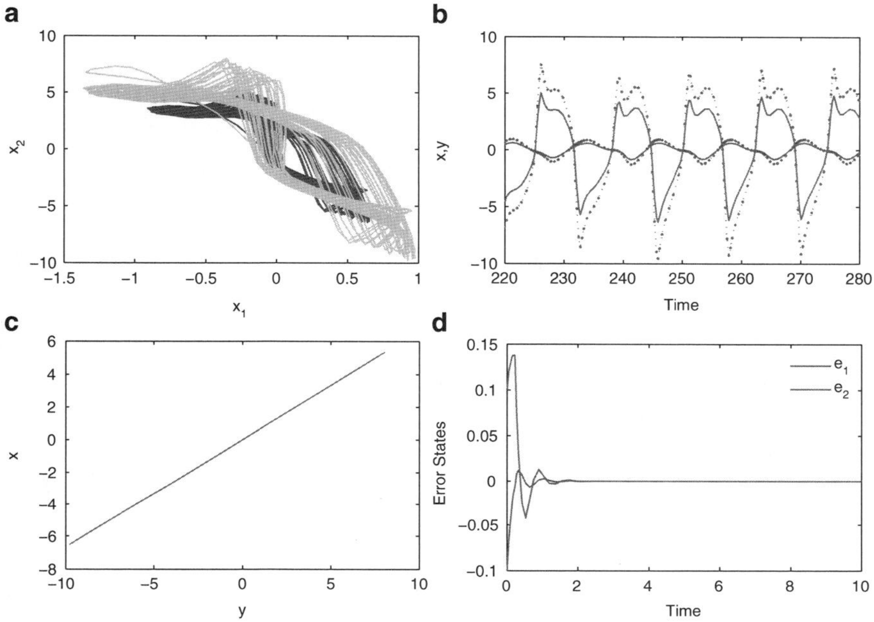

Fig. 10.3 Projective synchronization between identical Lu attractor given (10.35) along with $\alpha = 1.5$, and $\tau(t) = 1 + 0.1|\sin t|$, synchronization manifold $y = 1.5x$ is plotted and projective synchronization achieved by $e \to 0$

$\alpha = 1.5$ is presented in Fig. 10.3. Solving the inequality (10.11) presented in the Theorem 1 by using Matlab LMI toolbox, we get $P = 10^4 \begin{bmatrix} 3.6925 & -0.2585 \\ -0.2585 & 3.7763 \end{bmatrix}$, $Z_1 = 10^4 \begin{bmatrix} 3.1563 & 0.0458 \\ 0.0458 & 3.2009 \end{bmatrix}$, $Z_2 = 10^4 \begin{bmatrix} 1.3865 & -0.0553 \\ -0.0553 & 0.0022 \end{bmatrix}$, the control gain $K = \begin{bmatrix} 0.9286 & 0.0028 \\ -0.0603 & 1.0024 \end{bmatrix}$ and the upper bound of the time delay $\tau_M = 0.3528$. Figure 10.3 shows the PS between identical Lu attractors given by (10.35) along with $\alpha = 1.5$, and $\tau(t) = 1 + 0.1|\sin t|$. It is observed that synchronization manifold $y = 1.5x$ showing the occurrence of PS and assured by the closed-loop error trajectories reaching zero asymptotically.

10.6.3 Delayed-Chua System

Chua circuit system [3, 8, 28] is ubiquitous real-world example of a chaotic system which is autonomous in time. In fact, there have been vast study of synchronization of Chua system under different control techniques [14, 26] and references cited there in. The attractor depends on the nonlinear resistor of Chua circuit, and the delay

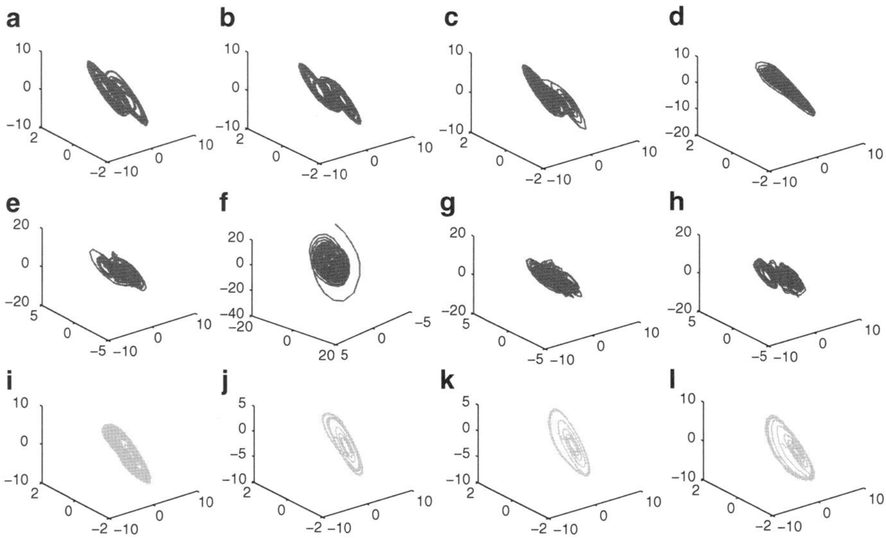

Fig. 10.4 Chua attractor formed by (10.36) with various time delays. (**a**) $0.6|\sin t|$, (**b**) $0.1 + 0.1e^{\sin t}$, (**c**) $0.2 + 0.1e^{\sin t}$, (**d**) $0.2 + 0.2e^{\sin t}$, (**e**) $0.2 + 0.5005 \sin t$, (**f**) $0.2 \sin t$, (**g**) $0.53 \sin t$, (**h**) $0.6041e^{\sin t}$, (**i**) $0.3 + 0.2e^{\sin t}$, (**j**) 0.4, (**k**) 0.7, (**l**) 0.8

inside the nonlinear resistor may be taken into account. Thus we are modifying the Chua circuit model with time varying delay present in the nonlinear term in order to adequate the present study. Consider the following modified time delayed Chua's system:

$$\begin{aligned} \dot{x}_1(t) &= a(x_2(t) - \phi(x_1(t - \tau(t)))) \\ \dot{x}_2(t) &= x_1(t) - x_2(t) + x_3(t) \\ \dot{x}_3(t) &= -bx_2(t) \end{aligned} \tag{10.36}$$

with modified nonlinear characteristic of Chua's diode with time delay

$$\phi(x(t)) = m_1 x_1(t) + \frac{1}{2}(m_0 - m_1)(|x_1(t - \tau(t)) + c| - |x_1(t - \tau(t)) - c|),$$

The parameters are used from Chua model as $a = 9, b = 14.28, c = 1, m_0 = -(1/7)$, $m_1 = 2/7$. With these parameters, chaotic attractors formed by (10.36) are given in Figs. 10.4–10.6, for showing phase space attractors with different time delays. It is interesting to see that the the double scroll attractor merges to single attractor form. The gallery of attractors presented in Figs. 10.4–10.6 are simulated by solving (10.36) using Runge-Kutta method for delay differential equations. It is seen from

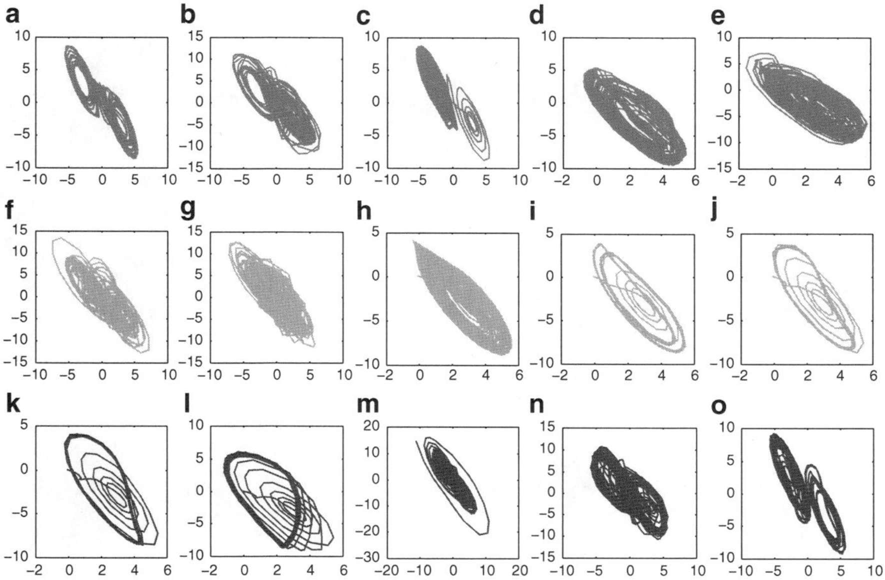

Fig. 10.5 2-D chaotic phase space (x_1 *vs* x_3) portrait of delayed Chua system given in (10.36) with different time delays. (**a**) 0.1, (**b**) $0.6041e^{\sin t}$, (**c**) $0.2 + 0.1e^{\sin t}$, (**d**) $0.3 + 0.2e^{\sin t}$, (**e**) $0.2 + 0.2e^{\sin t}$, (**f**) $0.2 + 0.5005e^{\sin t}$, (**g**) $0.53 \sin t$, (**h**) 0.3, (**i**) 0.4, (**j**) 0.6, (**k**) 0.7, (**l**) 0.8, (**m**) $0.2 \sin t$, (**n**) $0.2 + 0.8|\sin t|$, (**o**) $0.6|\sin t|$

the simulations that the modified model may show light on the practical significance on communication system. Now for the present study, the system (10.36) is represented in the form (10.5) with $A = \begin{bmatrix} -am_1 & a & 0 \\ 1 & -1 & 1 \\ 0 & -b & 0 \end{bmatrix}$, $B = \begin{bmatrix} -a(m_0 - m_1) \\ 0 \\ 0 \end{bmatrix}$, $C = D = \begin{pmatrix} 1 & 0 & 0 \end{pmatrix}$, and $\phi(\xi) = (|\xi+1| - |\xi-1|)$ belonging to the sector $\begin{bmatrix} 0 & 1 \end{bmatrix}$. By using the LMI presented in Theorem 1, the condition is solved in Matlab LMI toolbox and we obtain the solution as $P = 10^8 \begin{bmatrix} 2.3805 & -0.1486 & -0.0857 \\ -0.1486 & 1.7790 & 0.5546 \\ -0.0857 & 0.5546 & 2.2136 \end{bmatrix}$, $Z_1 =$

$$10^8 \begin{bmatrix} 1.8369 & -0.0598 & 0.0043 \\ -0.0598 & 2.2432 & -0.0214 \\ 0.0043 & -0.0214 & 1.8738 \end{bmatrix}, \quad Z_1 = 10^8 \begin{bmatrix} 0.0000 & -0.0000 & -0.0000 \\ -0.0000 & 2.8060 & 0.0878 \\ -0.0000 & 0.0878 & 1.9425 \end{bmatrix}.$$

Also the control gain $K = G^{-1}L = \begin{bmatrix} 1.0000 & 0.4370 & 0.0031 \\ -0.0000 & 0.3846 & -2.1089 \\ 0.0000 & 0.5000 & 1.5734 \end{bmatrix}$ is obtained for

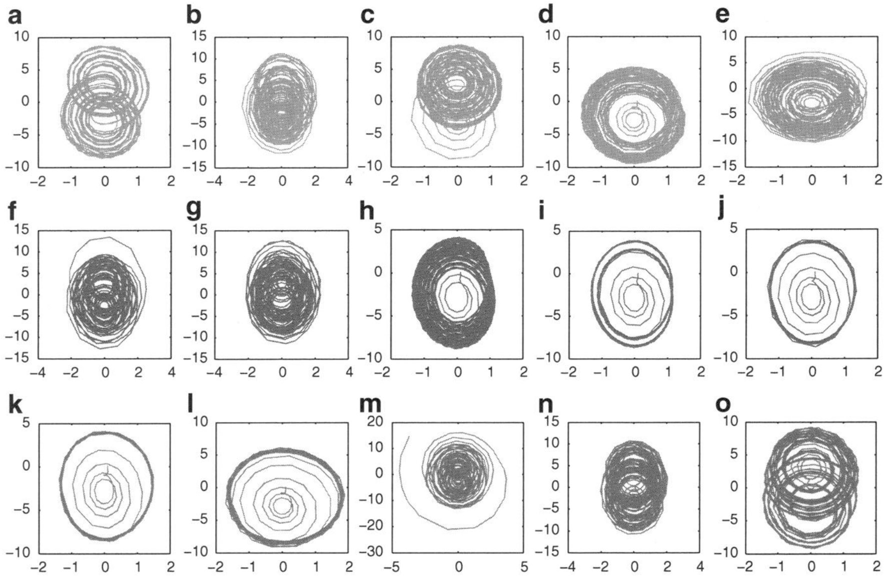

Fig. 10.6 2-D chaotic phase space (x_2 *vs* x_3) portrait of delayed Chua system given in (10.36) with different time delays. (**a**) 0.1, (**b**) $0.6041e^{\sin t}$, (**c**) $0.2 + 0.1e^{\sin t}$, (**d**) $0.3 + 0.2e^{\sin t}$, (**e**) $0.2 + 0.2e^{\sin t}$, (**f**) $0.2 + 0.5005e^{\sin t}$, (**g**) $0.53 \sin t$, (**h**) 0.3, (**i**) 0.4, (**j**) 0.6, (**k**) 0.7, (**l**) 0.8, (**m**) $0.2 \sin t$, (**n**) $0.2 + 0.8|\sin t|$, (**o**) $0.6|\sin t|$

the upper delay bound $\tau_{M=0.3666}$ of the time delay. Figure 10.7 shows that PS between identical modified Chua system with time-varying delay given (10.36) occurs for $\alpha = 1.5$, and $\tau(t) = 0.25 + 0.1e^{\sin t}$ along with its synchronization manifold $y = 1.5x$. The closed-loop error is plotted with nonlinear observer control and PS is achieved as the error reaching zero asymptotically.

10.7 Concluding Remarks

Summarizing the above, the more general analytical conditions for projective synchronization of delayed chaotic systems is studied in this chapter. With suitable nonlinear observer based control, the proposed scheme improves the existing results. Our work overcomes the drawbacks as well as new complex nature of the time varying delay has been taken into account. The conditions have further extended and applied to neural networks and chaotic systems with multiple time delays. In the upcoming part of the present study, we intended to provide analytical conditions for generalized projective synchronization of time varying delays and network of interacting chaotic systems with functional coupling. The practical importance of the obtained conditions by simulating the numerical results on Ikeda system, Lu

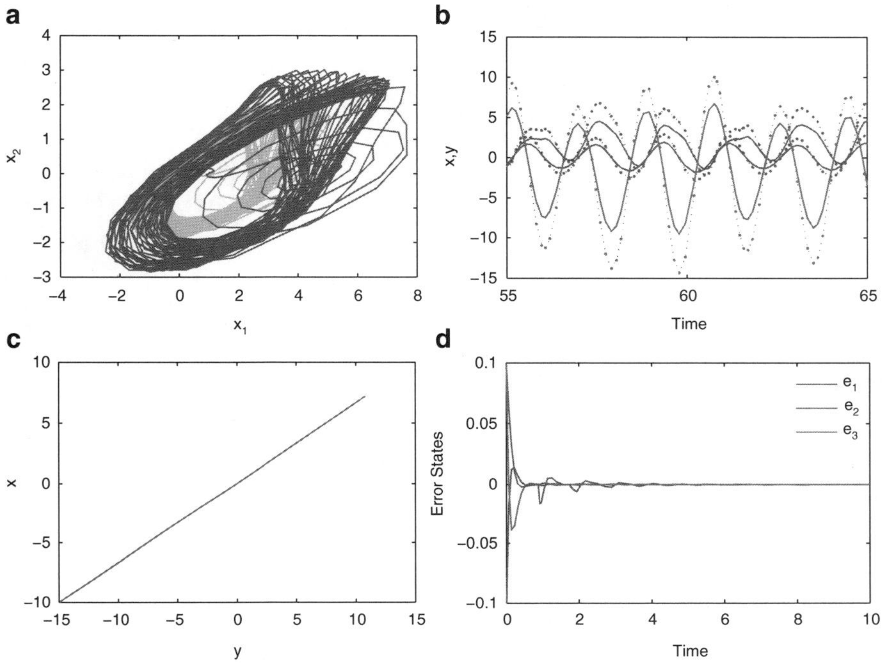

Fig. 10.7 Projective synchronization between identical modified Chua system with time-varying delay given (10.36) along with $\alpha = 1.5$, and $\tau(t) = 0.25 + 0.1e^{\sin t}$, synchronization manifold $y = 1.5x$ is plotted and projective synchronization achieved by $e \rightarrow 0$

attractor and delayed-Chua system were effectively exhibited and will show a new light on the significance of usage of time varying delay in secure communication problems.

References

1. Balasubramaniam, P., Chandran, R., Theesar, S.J.S.: Synchronization of chaotic nonlinear continuous neural networks with time-varying delay. Cognit. Neurodynam. **5**(4), 361–371 (2011)
2. Banerjee, S., Ghosh, D., Roy, A., Roychowdhury, A.: Synchronization between two different time-delayed systems and image encryption. Euro Phys. Lett. **81**, 1–6 (2006)
3. Bilotta, E., Pantano, P.: A Gallery of Chua Attractors. World Scientific, Singapore (2008)
4. Boyd, S., Ghaoui, E.L., Feron, E., Balakrishnan, V.: Linear Matrix Inequalities in System and Control Theory. SIAM, Philadelphia (1994)
5. Cao, T., Ho, D., Yang, Y.: Projective synchronization of calss of delayed chaotic system via impulsive contol. Phys. Lett. A **373**, 3128–3133 (2009)
6. Chee, C., Xu, D.: Secure digital communication using controlled projective synchronization of chaos. Chaos Solitons and Fractals **23**, 1063–1070 (2005)

7. Chua, L., Roska, T.: The CNN paradigm. IEEE Trans. Circ. Syst. I: Fund. Theor. Appl. **40**, 147–156 (1993)
8. Chua, L.O.: A zoo of strange attractors from the canonical chua circuits. In: Proceedings of the 35th Midwest Symposium on Circuits and Systems, IEEE, pp. 916–926 (1992)
9. Feng, C.: Projective synchronization between two different time-delayed chaotic systems using active control approach. Nonlinear Dynam. **62**, 453–459 (2010)
10. Feng, C., Xu, X., Wang, S., Wang, Y.: Projective-anticipating, projective, and projective-lag synchronization of time-delayed chaotic systems on random networks. Chaos **18**, 023117 (2008)
11. Gonzalez-Miranda, J.: Amplification and displacement of chaotic attractors by means of unidirectional chaotic driving. Phys. Rev. E **57**, 7321–7324 (1998)
12. Grassi, G., Miller, D.: Projective synchronization via linear observer: Application to time-delay, continous-time and discrete-time systems. Int. J. Bifurcat. Chaos **17**(4), 1337–1342 (2007)
13. Han, Q.: Absolute stability of time-delay systems with sector-bounded nonlinearity. Automatica **41**, 2171–2176 (2005)
14. He, Y., Wen, G., Wang, Q.: Delay-dependent synchronization criterion for lur'e systems with delay feedback control. Int. J. Bifurcat. Chaos **16**(10), 3087–3087 (2006)
15. Hoang, T., Nakagawa, M.: Anticipating and projective-anticipating synchronization of coupled multidelay feedback systems. Phys. Lett. A **365**, 407–411 (2007)
16. Hoang, T., Nakagawa, M.: A secure communication system using projective-lag and/or projective-anticipating synchronizations of coupled multi-delay feedback systems. Chaos Solitons Fractals **38**, 1423–1438 (2008)
17. Ikeda, K., Kondo, K., Akimoto, O.: Successive higher-harmonic bifurcations in systems with delayed feedback. Phys. Rev. Lett. **49**, 1467–1470 (1982)
18. Banerjee, S., Jeeva, S.T.S., Kurths, J.: Generalized variable projective synchronization of time delayed systems. Chaos **23**, 013118 (2013)
19. Jia, Q.: Projective synchronization of new hyperchaotic lorenz system. Phys. Lett. A **370**, 40–455 (2007)
20. Krasovskii, N.N.: Stability of Motion. Stanford University Press, Standford (1963)
21. Li, C., Yang, S.: Synchronization in linearly coupled dynamical networks with distributed time delays. Int. J. Bifurcat. Chaos **18**, 2039–2047 (2008)
22. Li, Z., Xu, D.: A secure communication scheme using projective chaos synchronization. Chaos Solitons Fractals **22**(2), 477–481 (2004)
23. Lu, H.: Chaotic attractos in delayed neural networks. Phys. Lett. A **298**, 110–116 (2002)
24. Mainieri, R., Rehacek, J.: Projective synchronization in three-dimensional chaotic systems. Phys. Rev. Lett. **82**, 3042–3045 (1999)
25. Rao, P., Wu, Z., Liu, M.: Adaptive projective synchronization of dynamical networks with distributed time delays. Nonlinear Dynam. **67**, 1729–1736 (2012)
26. Souza, F., Palhares, R., Mendes, E.: Robust h_∞ control for master-slave synchronization of $lur'e$ systems with time-delay feedback control. Int. J. Bifurcat. Chaos **18**(4), 1161–1173 (2008)
27. Theesar, S., Balasubramaniam, P., Chandran, R.: Delay-dependent exponential synchronization criteria for chaotic neural networks with time-varying delays. Braz. J. Phys. **42**, 207–218 (2012)
28. Torres, L., Aguire, L.: Inductorless chua's circuit. Electron. Lett. **36**(3), 1915–1916 (2000)
29. Wen, G., Xu, D.: Nonlinear observer control of full-state projective synchronization in chaotic contontinous-time delay systems. Chaos Solitons and Fractals **26**, 71–77 (2005)
30. Xu, D.: Control of projective synchronizationin chaotic systems. Phys. Rev. E **63**, 27201–27204 (2001)
31. Xu, D., Zhigang, L.: Controlled projective synchronization in nonparitally-linear chaotic systems. Int. J. Bifurcat. Chaos **12**(6), 1395–1402 (2002)

Printed by Printforce, the Netherlands